Environmental Chemistry for a Sustainable World

Volume 65

Other Publications by the Editors

Books

Environmental Chemistry
http://www.springer.com/978-3-540-22860-8

Organic Contaminants in Riverine and Groundwater Systems
http://www.springer.com/978-3-540-31169-0

Sustainable Agriculture
Volume 1: http://www.springer.com/978-90-481-2665-1
Volume 2: http://www.springer.com/978-94-007-0393-3

Book series

Environmental Chemistry for a Sustainable World
http://www.springer.com/series/11480

Sustainable Agriculture Reviews
http://www.springer.com/series/8380

Journals

Environmental Chemistry Letters
http://www.springer.com/10311

More information about this series at http://www.springer.com/series/11480

Nadia Morin-Crini • Eric Lichtfouse
Grégorio Crini
Editors

Emerging Contaminants Vol. 1

Occurrence and Impact

Springer

Editors
Nadia Morin-Crini
Chrono-environnement, UMR 6249
Université Bourgogne Franche-Comté
Besançon, France

Eric Lichtfouse
Aix Marseille University, CNRS, IRD, INRA, Coll France, CEREGE
Aix en Provence, France

Grégorio Crini
Chrono-Environnement, UMR 6249
Université Bourgogne Franche-Comté
Besançon, France

ISSN 2213-7114 ISSN 2213-7122 (electronic)
Environmental Chemistry for a Sustainable World
ISBN 978-3-030-69078-6 ISBN 978-3-030-69079-3 (eBook)
https://doi.org/10.1007/978-3-030-69079-3

This Springer imprint is published by the registered company Springer Nature Switzerland AG
The registered company address is: Gewerbestrasse 11, 6330 Cham, Switzerland

Preface

Pollution is nothing but the resources we are not harvesting. We allow them to disperse because we've been ignorant of their value.
Buckminster Fuller

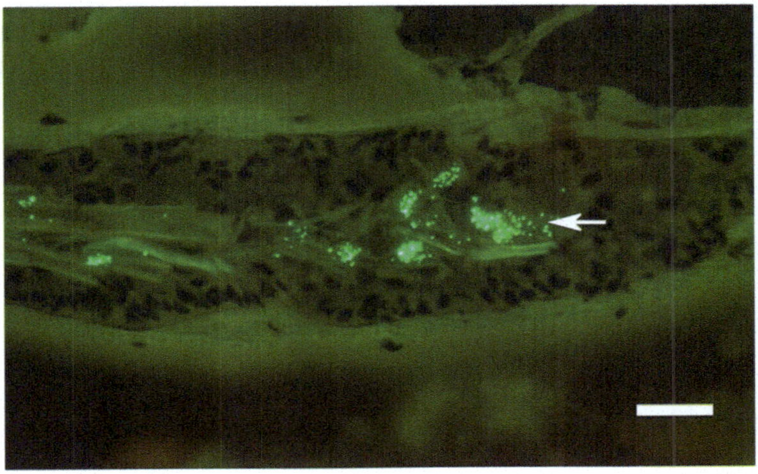

Figure. Fate of polyethylene beads and fragments after artificial ingestion by Antarctic krill, showing fragmentation and accumulation of beads in the midgut. White bar scale 50 μm. Reprinted from Dawson et al., Turning microplastics into nanoplastics through digestive fragmentation by Antarctic krill. Nature Communications 9, 1001 (2018). https://doi.org/10.1038/s41467-018-03465-9. Creative Commons CC BY

This book, entitled *Occurrence and Impact of Emerging Contaminants*, is the first volume on emerging contaminants published in the series Environmental Chemistry for a Sustainable World and written by 64 international contributors from 24 countries. In this volume, the chapters focus on the presence of emerging contaminants in the water compartment and their potential health effects.

The first chapter by Nadia Morin-Crini et al. summarizes the current state of knowledge on emerging contaminants, with focus on contaminants in aquatic systems. The second chapter by Anna Jurado et al. reviews the occurrence, fate, and associated risks of organic micropollutants included from the Watch List of the European groundwater bodies. Antimicrobial resistance is discussed in the next two chapters where Steffi Rocchi et al. disclose the emergence of pathogenic fungi resistant to triazole antifungal drugs in Chap. 3, and Charles P. Henriot et al. reveal the fate of antibiotic-resistant bacteria in the environment. In the last chapter, Ewa Konduracka and Paweł Rostoff discuss the relationship between outdoor air pollution and cardiovascular disorders in Chap. 5. The second volume focuses on methods for remediation of emerging contaminants.

The editors extend their thanks to all the authors who contributed to this book for their efforts in producing timely and high-quality chapters. The creation of this book would not have been possible without the assistance of several colleagues and friends deserving acknowledgment. They have helped by choosing contributors and reviewing chapters and in many other ways. Finally, we would like to thank the staff of Springer Nature for their highly professional editing of the publication.

Besançon, France Nadia Morin-Crini

Aix en Provence, France Eric Lichtfouse

Besançon, France Grégorio Crini

Contents

Contributors

Vysetti Balaram CSIR – National Geophysical Research Institute (NGRI), Hyderabad, Telangana, India

Xavier Bertrand UMR 6249 Chrono-environnement, CNRS, Université Bourgogne Franche-Comté, Besançon, France
Hygiène Hospitalière, Centre Hospitalier Universitaire de Besançon, Besançon, France

Gudrun Bornette UMR 6249 Chrono-environnement, CNRS, Université Bourgogne Franche-Comté, Besançon, France

Eric Carmona Department of Effect-Directed Analysis, Helmholtz-Centre for Environmental Research (UFZ), Leipzig, Germany

Grégorio Crini Laboratoire Chrono-environnement, UMR 6249, UFR Sciences et Techniques, Université Bourgogne Franche-Comté, Besançon, France

Christophe Dagot UMR Inserm 1092, Agents anti-microbiens, CBRS – Université de Limoges, Limoges, France

Liliana Giraldo Facultad de Ciencias, Departamento de Química, Universidad Nacional de Colombia, Bogotá, Colombia

Chloé Godeau Laboratoire Chrono-environnement, UMR 6249, UFR Sciences et Techniques, Université Bourgogne Franche-Comté, Besançon, France

Charles P. Henriot UMR 6249 Chrono-environnement, CNRS, Université Bourgogne Franche-Comté, Besançon, France

Didier Hocquet Hygiène Hospitalière, Centre Hospitalier Régional Universitaire de Besançon, Besançon, France
UMR 6249 Chrono-environnement, CNRS, Université Bourgogne Franche-Comté, Besançon, France

Anna Jurado Institute for Groundwater Management, Technische Universität Dresden, Dresden, Germany
GHS, Institute of Environmental Assessment and Water Research (IDAEA), Severo Ochoa Excellence Center of the Spanish Council for Scientific Research (CSIC), Barcelona, Spain

Ewa Konduracka Department of Coronary Disease and Heart Failure, Jagiellonian University Medical College, John Paul II Hospital, Kraków, Poland

Cui Li Research Center for Eco-Environmental Sciences, Chinese Academy of Sciences, Beijing, China

Eric Lichtfouse Aix Marseille University, CNRS, IRD, INRA, Coll France, CEREGE, Aix en Provence, France

Guorui Liu Research Center for Eco-Environmental Sciences, Chinese Academy of Sciences, Beijing, China

Zhijiang Lu Institute of Soil and Water Resources and Environmental Science, College of Environmental and Resource Sciences, Zhejiang University, Hangzhou, China

Daniel Martak UMR 6249 Chrono-environnement, CNRS, Université Bourgogne Franche-Comté, Besançon, France
Hygiène Hospitalière, Centre Hospitalier Universitaire de Besançon, Besançon, France

Juan Carlos Moreno-Piraján Facultad de Ciencias, Departamento de Química, Universidad de los Andes (Colombia), Bogotá, Colombia

Nadia Morin-Crini Laboratoire Chrono-environnement, UMR 6249, UFR Sciences et Techniques, Université Bourgogne Franche-Comté, Besançon, France

Abhishek Pandey School of Studies in Pharmaceutical Sciences, Jiwaji University, Gwalior, Madhya Pradesh, India

Fabienne Petit Normandie Université, UniRouen, UniCaen, CNRS UMR M2C, Rouen, France
Sorbonne Universités, UPMC, CNRS, EPHE, UMR 7619 METIS, Paris, France

Lorenzo A. Picos-Corrales Facultad de Ciencias Químico Biológicas, Universidad Autónoma de Sinaloa, Ciudad Universitaria, Culiacán, Sinaloa, Mexico

Ana Rita Lado Ribeiro Laboratory of Separation and Reaction Engineering – Laboratory of Catalysis and Materials (LSRE-LCM), Faculdade de Engenharia, Universidade do Porto, Porto, Portugal

Steffi Rocchi Service de Parasitologie-Mycologie, CHRU Jean Minjoz, Besançon, France
Laboratoire Chrono-environnement, UMR 6249, UFR Santé, Université Bourgogne Franche-Comté, Besançon, France

Paweł Rostoff Department of Coronary Disease and Heart Failure, Jagiellonian University Medical College, John Paul II Hospital, Kraków, Poland

M. Silvia Díaz-Cruz Department of Environmental Chemistry, Institute of Environmental Assessment and Water Research (IDAEA), Severo Ochoa Excellence Center of the Spanish Council for Scientific Research (CSIC), Barcelona, Spain

Eveline Snelders Laboratory of Genetics, Wageningen University, Wageningen, The Netherlands

Friederike Stock German Federal Institute of Hydrology, Koblenz, Germany

Margarida Ribau Teixeira CENSE, Center for Environmental and Sustainability Research, and University of Algarve, Faculty of Sciences and Technology, Faro, Portugal

Edward Topp Agriculture and Agri-Food Canada, University of Western Ontario, London, ON, Canada

Giangiacomo Torri Istituto di Chimica e Biochimica G. Ronzoni, Milan, Italy

Marc Walther Chair of Forest Biometrics and Forest Systems Analysis, Faculty of Environmental Sciences, Department of Forest Sciences, Technische Universität Dresden, Dresden, Germany

About the Editors

Nadia Morin-Crini is an analytical chemist at the University of Bourgogne Franche-Comté, Besançon, France. Her research interests include the extraction and analysis of chemicals from various solid and liquid environmental matrices. She is responsible for a technological platform thatstudies old and present environments. She has published more than 130 papers in peer-reviewed scientific journals.

Eric Lichtfouse is a geochemist and professor of scientific writing at Aix-Marseille University, France. He has invented carbon-13 dating and published the book *Scientific Writing for Impact Factor Journals*. He is chief editor of the journal *Environmental Chemistry Letters* and the book series Sustainable Agriculture Reviews and Environmental Chemistry for a Sustainable World.

Grégorio Crini is an environmental polymer scientist at the University of Bourgogne Franche-Comté, Besançon, France. His research activities focus on the design of new polymer networks and the environmental aspects of polysaccharide chemistry. He has published more than 210 papers in international journals and is a highly cited researcher with an h-index of 40 and more than 13,000 citations.

Abbreviations

ACE	acetamiprid
ADI	acceptable daily intake
AMX	amoxicillin
ARB	antibiotic-resistant bacteria
ARG	antibiotic resistance gene
AZY	azithromycin
BOD	biological oxygen demand
BPA	bisphenol A
BW	body weight
C_{max}	maximum concentration
CFU	colony-forming unit
CI	confidence interval
CIMT	carotid intima-media thickness
CIP	ciprofloxacin
CLA	clarithromycin
CLO	clothianidin
COD	chemical oxygen demand
D_{ow}	pH-dependent *n*-octanol-water distribution ratio
DDT	dichlorodiphenyltrichloroethane
DT_{50}	soil half-life
DWELs	drinking water equivalent levels
DWI	drinking water intake
E1	estrone
E2	17-beta-estradiol
EC	European Commission
EC_{50}	half-maximal effective concentration
EE2	17-alpha-ethinylestradiol
ERY	erythromycin
ESBL	extended-spectrum β-lactamase
FD	frequency of detection
GC-MS	gas chromatography coupled with mass spectrometry

GUS	groundwater ubiquity score
HPLC	high-performance liquid chromatography
HPLC-UV	high-performance liquid chromatography with UV detection
HQ	hazard quotient
HR	hazard ratio
IMD	imidacloprid
IQR	interquartile range
ISPRA	Italian National Institute for Environmental Protection and Research
K_{oc}	partition coefficient between soil organic carbon and water
K_{ow}	n-octanol-water partition coefficient
LC-MS/MS	liquid chromatography coupled with (tandem) mass spectrometry
LC_{50}	lethal concentration that kills 50% of the test animals during an observation period
LOD	limit of detection
LOQ	limit of quantification
LS	light source
MEC	measured environmental concentrations
MET	methiocarb
MFM	metaflumizone
n.d.	not detected
Non-STEMI	myocardial infarction without ST-segment elevation
OR	odds ratio
pK_a	the negative base-10 logarithm of the acid dissociation constant (K_a) of a solution
PM	particulate matter
PM_{10}	coarse particulate matter
$PM_{2.5}$	fine particulate matter
$PM_{0.1}$	ultrafine particulate matter
PNEC	predicted no-effect concentration
RQ	risk quotient
RR	removal rate
RR	relative risk
SCE	saturated calomel electrode
S_w	water solubility
SHE	standard hydrogen electrode
SS	stainless steel
STEMI	ST-segment elevation myocardial infarction
THI	thiacloprid
TMX	thiamethoxam
TOC	total organic carbon
VBNC	viable but non-culturable
WWTP	wastewater treatment plant

Chapter 1
Emerging Contaminants: Analysis, Aquatic Compartments and Water Pollution

Nadia Morin-Crini, Eric Lichtfouse, Guorui Liu, Vysetti Balaram, Ana Rita Lado Ribeiro, Zhijiang Lu, Friederike Stock, Eric Carmona, Margarida Ribau Teixeira, Lorenzo A. Picos-Corrales, Juan Carlos Moreno-Piraján, Liliana Giraldo, Cui Li, Abhishek Pandey, Didier Hocquet, Giangiacomo Torri, and Grégorio Crini

Abstract Over the past two decades, the presence of so-called emerging contaminants in various environmental compartments around the world, such as water, sediments, soils and atmosphere, and in a wide variety of consumer products has become a major concern for society, public health authorities, industry (namely food industry and water sector) and the agricultural sector. Some of these substances are endocrine disruptors and others are proven carcinogens and mutagens. In particular,

N. Morin-Crini (✉) · G. Crini
Laboratoire Chrono-environnement, UMR 6249, UFR Sciences et Techniques, Université Bourgogne Franche-Comté, Besançon, France
e-mail: nadia.crini@univ-fcomte.fr; gregorio.crini@univ-fcomte.fr

E. Lichtfouse
Aix Marseille University, CNRS, IRD, INRA, Coll France, CEREGE, Aix-en-Provence, France

G. Liu · C. Li
Research Center for Eco-Environmental Sciences, Chinese Academy of Sciences, Beijing, China
e-mail: grliu@rcees.ac.cn; cuili_st@rcees.ac.cn

V. Balaram
CSIR – National Geophysical Research Institute (NGRI), Hyderabad, Telangana, India

A. R. L. Ribeiro
Laboratory of Separation and Reaction Engineering – Laboratory of Catalysis and Materials (LSRE-LCM), Faculdade de Engenharia, Universidade do Porto, Porto, Portugal
e-mail: ritalado@fe.up.pt

Z. Lu
Institute of Soil and Water Resources and Environmental Science, College of Environmental and Resource Sciences, Zhejiang University, Hangzhou, China
e-mail: zhijianglu@zju.edu.cn

N. Morin-Crini et al. (eds.), *Emerging Contaminants Vol. 1*, Environmental Chemistry for a Sustainable World 65,
https://doi.org/10.1007/978-3-030-69079-3_1

there is a growing concern about the presence of emerging contaminants in water resources.

The list of substances and products is particularly long: pesticides, pharmaceuticals, drugs of abuse (including licit and illicit drugs), cosmetics, personal care products, surfactants, cleaning products, industrial formulations and chemicals, food additives, food packaging, metalloids, rare earth elements, nanomaterials, microplastics and pathogens. Most of these substances are found not only in domestic wastewater and industrial discharges, but also in surface and ground water, and consequently in drinking water and food sources. Therefore, emerging contaminants entering the aquatic compartment may cause known, unknown or suspected

F. Stock
German Federal Institute of Hydrology, Koblenz, Germany
e-mail: Stock@bafg.de

E. Carmona
Department of Effect-Directed Analysis, Helmholtz-Centre for Environmental Research (UFZ), Leipzig, Germany
e-mail: eric-carmona.martinez@ufz.de

M. R. Teixeira
CENSE, Center for Environmental and Sustainability Research, and University of Algarve, Faculty of Sciences and Technology, Faro, Portugal
e-mail: mribau@ualg.pt

L. A. Picos-Corrales
Facultad de Ciencias Químico Biológicas, Universidad Autónoma de Sinaloa, Ciudad Universitaria, Culiacán, Sinaloa, Mexico
e-mail: lorenzo.picos.c@uas.edu.mx

J. C. Moreno-Piraján
Facultad de Ciencias, Departamento de Química, Universidad de los Andes (Colombia), Bogotá, Colombia
e-mail: jumoreno@uniandes.edu.co

L. Giraldo
Facultad de Ciencias, Departamento de Química, Universidad Nacional de Colombia, Bogotá, Colombia
e-mail: lgiraldogu@unal.edu.co

A. Pandey
School of Studies in Pharmaceutical Sciences, Jiwaji University, Gwalior, Madhya Pradesh, India

D. Hocquet
Laboratoire Chrono-environnement, UMR 6249, UFR Sciences et Techniques, Université Bourgogne Franche-Comté, Besançon, France

Hygiène Hospitalière, Centre Hospitalier Régional Universitaire de Besançon, Besançon, France
e-mail: dhocquet@chu-besancon.fr

G. Torri
Istituto di Chimica e Biochimica G. Ronzoni, Milan, Italy
e-mail: torri@ronzoni.it

adverse effects on the ecosystems and/or human health. Many molecules have been classified in recent years as priority or priority hazardous substances, and thus, the industrial world and the municipal wastewater treatment plants are particularly concerned by the release of these substances by virtue of increasingly strict regulations. However, existing conventional municipal wastewater treatment plants and drinking water treatment plants were not designed for eliminating these new contaminants. In addition, most emerging contaminants are not yet routinely monitored either in industrial discharges or in the environment. From an academic point of view, emerging substances are also of great interest to the scientific community and receive special attention. The areas of research concern not only their effects on human health and their impact on environment, but also their sources, analysis and fate in the environment, as well as their remediation. Indeed, their behavior in the environment and their effects on all living organisms remain largely unknown. Many substances are difficult to remove by conventional water treatment processes. Thus, there is an intense research activity on all these topics.

The objective of this chapter is to present a recent state of knowledge on emerging substances and their presence in the aquatic environment. After general considerations on emerging contaminants, the first part is focused on chromatographic methods coupled to mass spectrometry for their analysis and detection, and on detection of microplastics in water and sediment. The presence of alkylphenols, rare earth elements and nanoparticles in the aquatic environment is then discussed. Finally, examples of contamination described from around the world, from China to Portugal, Mexico, Colombia and Brazil, are presented.

Keywords Emerging contaminants · Substances of global concern · Pharmaceuticals · Personal care products · Pesticides · Alkylphenols · Persistent organic pollutants · Microplastics · Rare earth elements · Nanoparticles · Biological agents · Analysis · Water pollution · Aquatic compartments

1.1 Introduction

Although water quality problems are largely associated with developing countries, they also persist in developed countries and include, for example, the loss of pristine water bodies, impacts related to changes in hydromorphology, the spread of invasive species, the loss of biodiversity, and the increasing occurrence of chemical substances (Morin-Crini and Crini 2017). The water quality problems are intimately linked to human activities, whether domestic, agricultural or industrial. Indeed, population growth, agricultural intensification, rapid urbanization, increased industrial production and climate change are beginning to impact on Nature's ability to provide safe and clean water for our essential uses and functions. One of the most important global environmental problems is water pollution by chemicals (Aristizabal-Ciro et al. 2017; Hernández-Padilla et al. 2017; Morin-Crini and Crini

2017; Botero-Coy et al. 2018; Crini and Lichtfouse 2018; Dong et al. 2018; Díaz-Casallas et al. 2019; Gallego-Schmid and Tarpani 2019; Kumar et al. 2019; Reichert et al. 2019; Deviller et al. 2020). We are all concerned about this problem and the anthropogenic sources are well-know, namely domestic discharges, hospital effluents, industrial wastewaters, runoff from agriculture, livestock and aquaculture, and landfill leachates. It is recognized that, due to insufficient chemical treatment abatement, the discharge of effluents from municipal wastewater treatment plants is an important route for the appearance of chemical substances such as pharmaceuticals in the aquatic environment (Hernández-Padilla et al. 2017; Botero-Coy et al. 2018; Crini and Lichtfouse 2018; Gallego-Schmid and Tarpani 2019; Inyinbor et al. 2019; Mohapatra and Kirpalani 2019; Patel et al. 2019).

Many chemical substances such as metals, dyes, phenols, pesticides, pharmaceuticals and other so-called emerging contaminants are indeed part of our daily lives as they are used in a wide variety of products and applications. For example, they are ubiquitously in every household: in food, energy drinks, medicines, household and cleaning products, personal care products, and cosmetics, paints, electronics and computers, furniture and even our clothes (Puzyn and Mostrag-Szlichtyng 2012; Mortimer 2013; Schrenk and Chopra 2013; Duedahl-Olesen 2013; Juliano and Magrini 2017). The vast majority of chemicals, called contaminants or pollutants depending on their toxicity and impact (Chapman 2007), end up in our discharges of wastewater treatment plants and ultimately in the aquatic compartments of the environment such as rivers, lakes, coastal and marine ecosystems, seawater and ice caps (Crini and Badot 2007, 2010; Kosma et al. 2014; Gee et al. 2015; Archer et al. 2017; Ebele et al. 2017; Peña-Guzmán et al. 2019; Xu et al. 2019). As a result, chemicals can be found not only in surface water and freshwater, but also in groundwater, which is the main source of drinking water in many countries (Dévier et al. 2013; Gee et al. 2015). The presence of chemicals in general, and particularly emerging substances, in water is a cause of concern and debate, as the risk they pose to human health and the environment is not yet fully understood. Biological agents are also widely found in the environment. They include bacteria, viruses, fungi (yeasts, molds) and parasites. The majority of these agents are harmless, but some may have potential to cause ill health. The bibliography on emerging chemical and biological contaminants is particularly abundant and many topics are of interest to the scientific community (Table 1.1). In particular, the presence, occurrence, transport of emerging contaminants in the water resources and their impact on water quality are documented in thousands of publications reported worldwide during the last two decades, demonstrating an increasing concern about them.

Due to their widespread use in our daily applications, chemicals are ubiquitous not only in water and sediments, but also in soils and the atmosphere, at concentrations between few ng/L and several µg/L (Salthammer 2014, 2020; Patel et al. 2019). In recent decades, thanks to significant advances in the analysis and detection of trace contaminants, emerging contaminants have also been detected in plants (due to soil and water pollution), food products, living organisms and protected species (*e.g.*, birds, polar bears), even in areas where there is no human activity (Datta et al. 2018; Routti et al. 2019). This demonstrates their uncontrolled

Table 1.1 Selected comprehensive reviews and papers on emerging contaminants published in the last five years

General topic	References
Alkylphenols	Careghini et al. 2015; Crini et al. 2016; Tuan Omar et al. 2016; Chokwe et al. 2017; Deshayes et al. 2017; Priac et al. 2017; Acir and Guenther 2018; Salgueiro-Gonzalez et al. 2017; Zaytseva and Medvedeva 2019; Olaniyan et al. 2020; Vargas-Berrones et al. 2020a, b
Aquatic toxicity	Harmon 2015; Priac et al. 2017; Acir and Guenther 2018; Koba et al. 2018; Lin et al. 2018a; Mearns et al. 2018; Mezzelani et al. 2018; Salgueiro-Gonzalez et al. 2018; de Marchi et al. 2019; Fekadu et al. 2019; Li et al. 2019b; Patel et al. 2019; Olaniyan et al. 2020
Aquifer	Moreau et al. 2019; Burri et al. 2019; Kurwadkar 2019; Daesslé et al. 2020; Sackaria and Elango 2020
Analysis	Lu and Gan 2014a; Gago-Ferrero et al. 2015; Löder and Gerdts 2015; Bijlsma et al. 2016; Jardak et al. 2016; Tuan Omar et al. 2016; Priac et al. 2017; Salgueiro-Gonzalez et al. 2017, 2018; Acir and Guenther 2018; Botero-Coy et al. 2018; Lorenzo et al. 2018; Patel et al. 2019; Sousa et al. 2019; de Oliveira et al. 2020; Kovalakova et al. 2020
Antibiotics	Botero-Coy et al. 2018; Dong et al. 2018; Gogoi et al. 2018; Mezzelani et al. 2018; Fekadu et al. 2019; Jurado et al. 2019; Kumar et al. 2019; Ng et al. 2019; Reichert et al. 2019; García et al. 2020; Khan et al. 2020; Patel et al. 2020; Wang and Zhuan 2020
Antibiotic resistance	Zad et al. 2018; Zhao et al. 2018; Kumar et al. 2019; Liu et al. 2019; Ng et al. 2019; Reichert et al. 2019; García et al. 2020
Behavior	Jardak et al. 2016; Ebele et al. 2017; Mearns et al. 2018; Li et al. 2019a; Houde et al. 2019; de Marchi et al. 2019; Patel et al. 2019; Starling et al. 2019; Yu et al. 2019
Bioaccumulation	Tuan Omar et al. 2016; Ebele et al. 2017; Damkjaer et al. 2018; Herrera-Melian et al. 2018; Jiang et al. 2018; Mauricio et al. 2018; Mearns et al. 2018; Mezzelani et al. 2018; Salgueiro-Gonzalez et al. 2018; Wang et al. 2018b; Houde et al. 2019; Patel et al. 2019; Xu et al. 2020
Biodegradation	Lu and Gan 2014a, b; Ebele et al. 2017; Couto et al. 2019; Cao et al. 2020; de Oliveira et al. 2020
Biomagnification	Mearns et al. 2018, 2019
Bisphenol A	Machado et al. 2016; Gogoi et al. 2018; Zhao et al. 2018; Xiao et al. 2020
Caffeine	Machado et al. 2016; Sodré et al. 2018; Montagner et al. 2019; Anastopoulos et al. 2020; Rigueto et al. 2020
Cosmetics	Juliano and Magrini 2017; Li et al. 2018a; Caldas et al. 2019
Disinfection by-products	Benitez et al. 2015; Nika et al. 2016; Zad et al. 2018; Gao et al. 2019; Ng et al. 2019; Khan et al. 2020
Drinking water	Gee et al. 2015; Hamza et al. 2016; Machado et al. 2016; Gwenzi et al. 2018; Sodré et al. 2018; Caldas et al. 2019; Gao et al. 2019; Patel et al. 2019; Starling et al. 2019; Iroegbu et al. 2020; Sousa et al. 2020
Ecotoxicology	Petrie et al. 2015; Cavalheiro et al. 2017; Kumar et al. 2019; Li et al. 2019b; Patel et al. 2019; Starling et al. 2019; Xu et al. 2019

(continued)

Table 1.1 (continued)

General topic	References
Ecological risk assessment	Gwenzi et al. 2018; Matich et al. 2019; Mezzelani et al. 2018; Pedrazzani et al. 2019; Xu et al. 2019
Endocrine disrupting compounds	Różalska et al. 2015; Ma et al. 2016; Murdoch and Sanin 2016; Tuan Omar et al. 2016; Archer et al. 2017; Acir and Guenther 2018; Dong et al. 2018; Gogoi et al. 2018; Liu et al. 2019; Ng et al. 2019; Olaniyan et al. 2020; Patel et al. 2020
Environmental fate	Cavalheiro et al. 2017; Monneret 2017; Priac et al. 2017; Archundia et al. 2018; Koumaki et al. 2018; Reichert et al. 2019; Yu et al. 2019
Exposure pathways	Chokwe et al. 2017; Koba et al. 2018; Mearns et al. 2018; Li et al. 2019b; Routti et al. 2019; Wu et al. 2020b; Xiao et al. 2020
Fate	Eggen and Vogelsang 2015; Katsigiannis et al. 2015; Noutsopoulos et al. 2015; Carr et al. 2016; Archer et al. 2017; Cavalheiro et al. 2017; Ebele et al. 2017; Datta et al. 2018; Gogoi et al. 2018; Zad et al. 2018; Zhao et al. 2018; Couto et al. 2019; de Marchi et al. 2019; Jurado et al. 2019; Mohapatra and Kirpalani 2019; Ng et al. 2019; Routti et al. 2019; Starling et al. 2019
Flame retardants	Hamza et al. 2016; Lin et al. 2018a; Zhao et al. 2018; Llompart et al. 2019; Li et al. 2019a; Peña-Guzmán et al. 2019; Tongue et al. 2019; Wang et al. 2020; Wu et al. 2020b; Zhou et al. 2020; Zuiderveen et al. 2020
Food additives	Careghini et al. 2015; Moreau et al. 2019; Peña-Guzmán et al. 2019
Hormones	Barbosa et al. 2016; Gogoi et al. 2018; Mezzelani et al. 2018; Zad et al. 2018; Fekadu et al. 2019; Montagner et al. 2019; Ng et al. 2019; Peña-Guzmán et al. 2019; Sousa et al. 2019
Illicit drugs	Bijlsma et al. 2016; Archer et al. 2017; Ng et al. 2019; Peña-Guzmán et al. 2019
Indoor pollutants	Salthammer 2014, 2020; Cretescu et al. 2019
Industrial wastewater	Petrie et al. 2015; Jardak et al. 2016; Archer et al. 2017; Chokwe et al. 2017; Priac et al. 2017; Botero-Coy et al. 2018; Dong et al. 2018; Montagner et al. 2019; Moreau et al. 2019; Ng et al. 2019; Patel et al. 2019; Peña-Guzmán et al. 2019; Tolboom et al. 2019; Zhang et al. 2019; Daesslé et al. 2020
Life cycle assessment	Gallego-Schmid and Tarpani 2019
Marine organisms	Lin et al. 2018a; Mearns et al. 2018, 2019
Metabolites	Petrie et al. 2015; Acir and Guenther 2018; Koba et al. 2018; Patel et al. 2019
Metabolomics	Matich et al. 2019
Microplastics	Löder and Gerdts 2015; Oberbeckmann et al. 2015; Thompson 2015; Carr et al. 2016; Avio et al. 2017; Juliano and Magrini 2017; Barboza et al. 2018; Datta et al. 2018; Lambert and Wagner 2018; Li et al. 2018b; Lin et al. 2018a; Wagner and Lambert 2018; Zad et al. 2018; da Costa et al. 2019; Guo and Wang 2019; Mishra et al. 2019; Ng et al. 2019; Prata et al. 2019; Stock et al. 2019; Amato-Lourenço et al. 2020; Guo et al. 2020; Iroegbu et al. 2020; Kumar et al. 2020; Möller et al. 2020; Tiwari et al. 2020; Wong et al. 2020; Xu et al. 2020

(continued)

Table 1.1 (continued)

General topic	References
Modeling	Wang et al. 2016c; Petrie et al. 2015; Peng et al. 2018; Li et al. 2019b; Moreau et al. 2019; Noutsopoulos et al. 2019; Sousa et al. 2019; Daesslé et al. 2020; Gibson 2020
Monitoring	Priac et al. 2017; Mearns et al. 2018; Sousa et al. 2018, 2019; Montagner et al. 2019; Moreau et al. 2019; Peña-Guzmán et al. 2019; Gibson 2020; Patel et al. 2020
Municipal wastewater	Bijlsma et al. 2016; Archer et al. 2017; Cavalheiro et al. 2017; Deshayes et al. 2017; Chokwe et al. 2017; Ebele et al. 2017; Hernández-Padilla et al. 2017; Priac et al. 2017; Dong et al. 2018; Gogoi et al. 2018; Gallego-Schmid and Tarpani 2019; Montagner et al. 2019; Peña-Guzmán et al. 2019; Starling et al. 2019
Nanoparticles/ nanomaterials	Hamza et al. 2016; Mearns et al. 2018; Zhao et al. 2018; de Marchi et al. 2019; Matich et al. 2019; Sousa and Teixeira 2020
Occurrence	Kosma et al. 2014; Annamalai and Namasivayam 2015; Careghini et al. 2015; Eggen and Vogelsang 2015; Hao et al. 2015; Petrie et al. 2015; Barbosa et al. 2016; Jardak et al. 2016; Hamza et al. 2016; Liu et al. 2016; Xu et al. 2016; Cavalheiro et al. 2017; Chokwe et al. 2017; Ebele et al. 2017; Priac et al. 2017; Acir and Guenther 2018; Amin et al. 2018; Botero-Coy et al. 2018; Gogoi et al. 2018; Herrera-Melian et al. 2018; Mezzelani et al. 2018; Zhao et al. 2018; Couto et al. 2019; Fekadu et al. 2019; Gao et al. 2019; Jurado et al. 2019; Li et al. 2019b; Patel et al. 2019; Ng et al. 2019; Starling et al. 2019; Xu et al. 2019; Yu et al. 2019; Xu et al. 2020
Pathogens	Ng et al. 2019
Parabens	Juliano and Magrini 2017; Mearns et al. 2018; Caldas et al. 2019; Salthammer 2020
Pesticides	Barbosa et al. 2015; Hamza et al. 2016; Machado et al. 2016; Dong et al. 2018; Mearns et al. 2018; Sodré et al. 2018; Caldas et al. 2019; Houde et al. 2019; Montagner et al. 2019; Moreau et al. 2019; Ng et al. 2019; Xu et al. 2019; Patel et al. 2020
Persistence	Ebele et al. 2017; Priac et al. 2017; Mezzelani et al. 2018; Houde et al. 2019; Li et al. 2019a; Gibson 2020
Persistent organic pollutants	Eggen and Vogelsang 2015; Harmon 2015; Kallenborn et al. 2015; Zeng 2015; Hamza et al. 2016
Personal care products	Eggen and Vogelsang 2015; Archer et al. 2017; Aristizabal-Ciro et al. 2017; Ebele et al. 2017; Juliano and Magrini 2017; Dong et al. 2018; Gogoi et al. 2018; Li et al. 2018a; Caldas et al. 2019; Liu et al. 2019; Moreau et al. 2019; Ng et al. 2019; Peña-Guzmán et al. 2019; Xu et al. 2019; Patel et al. 2020
Pharmaceuticals	Eggen and Vogelsang 2015; Tuan Omar et al. 2016; Roberts et al. 2016; Archer et al. 2017; Aristizabal-Ciro et al. 2017; Ebele et al. 2017; Botero-Coy et al. 2018; Dong et al. 2018; Gogoi et al. 2018; Koba et al. 2018; Mezzelani et al. 2018; Couto et al. 2019; Fekadu et al. 2019; Li et al. 2019b; Liu et al. 2019; Moreau et al. 2019; Ng et al. 2019; Peña-Guzmán et al. 2019; Xu et al. 2019; Yu et al. 2019; Patel et al. 2019, 2020

(continued)

Table 1.1 (continued)

General topic	References
Plasticizers	Llompart et al. 2019; Patel et al. 2020; Salthammer 2020; Wang et al. 2020
Perfluoroalkyl substances/ Polyfluoroalkyl compounds	DeWitt 2015; Xiao 2017; Dong et al. 2018; Lorenzo et al. 2018; Brusseau 2019; Klemes et al. 2019; Ng et al. 2019; Pan et al. 2020; Salthammer 2020
Radionuclides	Mearns et al. 2018; Zhang et al. 2019
Rare earth elements	Gwenzi et al. 2018; Mubashir 2018; Balaram 2019; Salthammer 2020
Reclaimed water	Gilabert-Alarcón et al. 2018; Sodré et al. 2018
Regulation	Barbosa et al. 2016; Acir and Guenther 2018; Dulio et al. 2018; Mearns et al. 2018; Jurado et al. 2019; Lapworth et al. 2019; Patel et al. 2019; Sousa et al. 2019
Remediation	Barbosa et al. 2016; Hamza et al. 2016; Jardak et al. 2016; Priac et al. 2017; Crini and Lichtfouse 2018; Ma et al. 2018; Couto et al. 2019; Crini et al. 2019; Kumar et al. 2019; Llompart et al. 2019; Liu et al. 2019; Patel et al. 2019, 2020; Serna-Galvis et al. 2019; Tolboom et al. 2019; Zaytseva and Medvedeva 2019; Zhang et al. 2019; Dhangar and Kumar 2020; Godage and Gionfriddo 2020; Khan et al. 2020; Mautner 2020; Pesqueira et al. 2020; Rasheed et al. 2020; Vara and Karnena 2020; Wang and Zhuan 2020; Yadav et al. 2020; Zaied et al. 2020
Risk assessment	Thomaidi et al. 2015; Osorio et al. 2016; Archer et al. 2017; Tahar et al. 2017; Gwenzi et al. 2018; Mearns et al. 2018; Mezzelani et al. 2018; Sodré et al. 2018; Jurado et al. 2019; Montagner et al. 2019; Patel et al. 2019; Pedrazzani et al. 2019; Xu et al. 2019; Wang et al. 2020
Sediments	Careghini et al. 2015; Petrie et al. 2015; Yang et al. 2015; Bradley et al. 2016; Jin and Zhu 2016; Ebele et al. 2017; Koumaki et al. 2017; Koba et al. 2018; Mearns et al. 2018; Salgueiro-Gonzalez et al. 2018; Li et al. 2019a; Stock et al. 2019
Soils	Careghini et al. 2015; Guo et al. 2020
Sources	Thompson 2015; Jardak et al. 2016; Chokwe et al. 2017; Deshayes et al. 2017; Sodré et al. 2018; Gogoi et al. 2018; Mezzelani et al. 2018; Patel et al. 2019
Surfactants	Hamza et al. 2016; Jardak et al. 2016; Xiao 2017; Llompart et al. 2019; Patel et al. 2020
Toxicity	Lu and Gan 2014a; Harmon 2015; Chokwe et al. 2017; Ebele et al. 2017; Priac et al. 2017; Acir and Guenther 2018; Guo et al. 2020; Olaniyan et al. 2020; Xiao et al. 2020
Tracers	Mearns et al. 2018; Moreau et al. 2019
Transport	Kallenborn et al. 2015; Ebele et al. 2017; Datta et al. 2018; Koba et al. 2018; Mearns et al. 2018; Zhao et al. 2018; Li et al. 2019b; Patel et al. 2019; Yu et al. 2019
Triclosan	Kosma et al. 2014; Juliano and Magrini 2017; Machado et al. 2016; Montagner et al. 2019
UV filters	Juliano and Magrini 2017; Mearns et al. 2018; Jurado et al. 2019; Peña-Guzmán et al. 2019; Sousa et al. 2019

(continued)

Table 1.1 (continued)

General topic	References
Wastewater treatment plants	Kosma et al. 2014; Hamza et al. 2016; Carr et al. 2016; Bijlsma et al. 2016; Archer et al. 2017; Hernández-Padilla et al. 2017; Botero-Coy et al. 2018; Gallego-Schmid and Tarpani 2019; Inyinbor et al. 2019; Mohapatra and Kirpalani 2019; Patel et al. 2019
Water compartment	Careghini et al. 2015; Koumaki et al. 2015; Machado et al. 2016; Archer et al. 2017; Aristizabal-Ciro et al. 2017; Ebele et al. 2017; Dong et al. 2018; Datta et al. 2018; Burri et al. 2019; Fekadu et al. 2019; Kurwadkar 2019; Li et al. 2019a; Montagner et al. 2019; Moreau et al. 2019; Peña-Guzmán et al. 2019; Starling et al. 2019; Stock et al. 2019; Xu et al. 2019; Daesslé et al. 2020; Kroon et al. 2020; Sackaria and Elango 2020; Wong et al. 2020
Water quality	Aristizabal-Ciro et al. 2017; Hernández-Padilla et al. 2017; Botero-Coy et al. 2018; Dong et al. 2018; Díaz-Casallas et al. 2019; Gallego-Schmid and Tarpani 2019; Kumar et al. 2019; Reichert et al. 2019; Deviller et al. 2020
Water management	Díaz-Casallas et al. 2019; Gilabert-Alarcón et al. 2018
Water reuse	Dulio et al. 2018; Ma et al. 2018; Sodré et al. 2018; Inyinbor et al. 2019; Deviller et al. 2020

movement in the global environment. Indeed, the issue of contaminants/pollutants and their impact on the environment (especially on water quality), society and human health is common to the entire World, making it a global concern at the present time. However, concentration levels are extremely variable from one continent to another, from one country to another and even from one region to another. Recently, Fekadu et al. (2019), comparing levels of contamination of pharmaceuticals such as carbamazepine and sulfamethoxazole in surface waters between Africa and Europe, observed that the maximum concentrations reported in Africa were 20,000 times higher than in Europe. The authors indicated that the variation in the consumption of pharmaceuticals, the partial removal of substances in wastewater treatment plants and the direct discharge of wastewater from livestock farms are among the reasons that may explain the observed differences (Fekadu et al. 2019). Other studies showed a correlation between the presence of emerging substances in the aquatic environment and discharges from wastewater treatment plants (Stackelberg et al. 2007; Kasprzyk-Hordern et al. 2008; Watkinson et al. 2009; Ort et al. 2010; Basile et al. 2011; Gonzalez et al. 2012; Machado et al. 2016; Priac et al. 2017; Gilabert-Alarcón et al. 2018; Mezzelani et al. 2018).

The vast majority of the anthropogenic substances are currently not regulated and some have been regulated only recently, namely certain pesticides and alkylphenols known as endocrine disruptors or endocrine-disrupting chemicals (Daughton 2004; Lu and Gan 2014a; Gee et al. 2015; Petrie et al. 2015; Barbosa et al. 2016; Morin-Crini and Crini 2017; Jurado et al. 2019; Sousa et al. 2019). For example, paraquat, an herbicide used worldwide since the 1960s that caused severe and fatal poisonings, was banned in the European Union in 2003 and withdrawn from the market in 2007. Another example is the organochlorine insecticide chlordecone, introduced in 1958 and banned in 1978 in the US and in the early 1993 in the French West Indies (Rana et al. 2016). In 2000, a first list of 33 priority substances (several pesticides, the phthalate di(2-ethylhexyl)-phthalate, octylphenol, nonylphenol, solvents, among others) was established as a control measure for the next 15 years based on their significant risk to or *via* the aquatic environment (European Water Framework Directive 2000/60/EC; Decision 2455/2001/EC). In 2008, the Directive 2008/105/EC layed down the environmental quality standards (EQS) for the priority substances and certain other pollutants. Moreover, other groups of substances were also targeted to be reviewed for possible future prioritization (*e.g.*, bisphenol A, musk xylene, perfluorooctane sulphonic acid, dioxins, polychlorinated biphenyls), and in fact some of them were later included in the list of priority substances by Directive 2013/39/EU, which enlisted 45 priority subtances or group of substances. Additionally, this Directive preconized the launch of the first Watch List and recommended to include diclofenac, 17-β-estradiol and 17-α-ethinylestradiol. The non-steroidal anti-inflammatory drug diclofenac is often detected in European municipal wastewater treatment plants influents and effluents, surface water and groundwater. The first complete European watch list of substances of emerging concern (European Decision 2015/495/EU of 20 March 2015) was launched in 2015 and included three hormones (17-α-ethinylestradiol, 17-β-estradiol and estrone), four pharmaceuticals (diclofenac, azithromycin, clarithromycin and erythromycin), eight pesticides (methiocarb, oxadiazon, imidacloprid, thiacloprid, thiamethoxam, clothianidin, acetamiprid and triallate), a UV filter (2-ethylhexyl-4-methoxycinnamate) and an antioxidant (2,6-di-*tert*-butyl-4-methylphenol) commonly used as food additive (Morin-Crini and Crini 2017). This list is dynamic and the latest update was published in 2018 (*e.g.*, with the European Decision 2018/840/EU), by removing tri-allate, oxadiazon, 2,6-di-*tert*-butyl-4-methylphenol and diclofenac (due to the sufficient high-quality monitoring data already available, according to the Decision), as well as 2-ethylhexyl-4-methoxycinnamate (since data is needed for sediment rather than water), and by adding 3 candidates: the insecticide metaflumizone, and the antibiotics amoxicillin and ciprofloxacin. Recently, the substances propiconazole and 4-nonylphenol were banned in Europe in November 2018 and July 2019, respectively. Indeed, from a regulatory point of view, the situation is constantly evolving (Barbosa et al. 2016; Jurado et al. 2019; Sousa et al. 2019). The chronicle of European science, policy and legislation to develop a groundwater monitoring list can be found in the review by Lapworth et al. (2019). Water contamination by pesticides, pharmaceuticals and industrial chemicals has become a global environmental and health crisis (Morin-Crini and Crini 2017; Crini and Lichtfouse 2018; Patel et al. 2019; Tolboom et al. 2019). There is also a mobilization of the global scientific community on all subjects concerning emerging substances, from their analysis to their removal, their behavior in the environment and their impact on health (Table 1.1).

The objective of this chapter is to present a recent state of knowledge on emerging substances, considering several aspects presented in three parts. After general considerations on emerging contaminants, the first part is devoted to analysis and detection of emerging substances with the presentation of chromatographic methods coupled to mass spectrometry for their analysis and detection. The detection of microplastics in water and sediments is also discussed. The second part of this chapter presents the occurrence, fate and toxicity of alkylphenols, rare earth elements and nanoparticles in the aquatic environment. The third part presents selected examples of contamination described from around the world, from China to Portugal, and Latin America.

1.2 General Considerations on Emerging Contaminants

1.2.1 Contamination and Pollution

Contamination/Pollution arises from all sectors of human activity, and is due to natural (petroleum, minerals, etc.) and anthropogenic causes (*e.g.*, sewage treatment sludge or persistent organic pollutants produced by waste incineration) and especially to synthetic substances produced by chemical industries (*e.g.*, dyes, fertilizers, pesticides, and so on). The terms contamination/pollution and contaminant/pollutant are often used in relation to subjects like environment, food and medicine (Crini and Badot 2007, 2010; Morin-Crini and Crini 2017). Both contaminant and pollutant refer to undesirable or unwanted substances. Pollutant refers to a harmful substance but contaminant is not necessarily harmful since contamination refers simply to the presence of a chemical substance where it should not be. This means that all pollutants are contaminants, but not all contaminants are pollutants. In this chapter, both these terms were used.

1.2.2 What Is an Emerging Contaminant?

A chemical pollutant is a substance toxic for flora, fauna and for humans, and present at concentrations such that, in Nature, has repercussions for the environment and/or on health in general (Crini and Badot 2007, 2010; Sauvé and Desrosiers 2014; Harmon 2015). Examples of pollutants/micropollutants known to the public are various, including nitrates, phosphates, pesticides such as glyphosate or atrazine, detergents (cleaning products), metals, dyes, pharmaceuticals such as diclofenac or ibuprofen, licit drugs (caffeine 1,3,7-trimethyl-3-7-dihydro-1H-purine-2,6-dione), illicit drugs (cocaine, amphetamines), synthetic (17-α-ethinylestradiol) and natural (17-β-estradiol) hormones, cosmetics and personal care products (parabens, triclosan, sunscreens and UV filters, domestic biocides, disinfectants), food derivatives and additives (nitrites, caffeine, artificial sweeteners), food packaging and containers (phthalates, other plasticizers), industrial substances (bisphenol A, surfactants, chlorinated solvents such as trichloroethylene, disinfection products, biocides, lubricants, silicones, substances used in nonstick materials, ingredients used in paints), microbial pathogens, indoor pollutants, etc. Other substances, that are less known by the public, are considered substances of concern: estrogens (estriol), pesticides less known by the public but widely used such as methiocarb, triallate and oxadiazon, neonicotinoids (insecticides), 2,6-di-*tert*-butyl-4-methylphenol (anti-oxidant), UV filter (2-ethylhe xyl-4-methoxycinnamate), macrolide antibiotics used in veterinary medicine and aquaculture, volatile organic compounds, polycyclic aromatic hydrocarbons, polychlorinated biphenyls, bromine-containing flame-retardants, perfluorinated compounds and perfluorinated alkyl substances, polybrominated diphenyl ethers, alkylphenols and alkylphenol polyethoxylates, rare earth elements, radionuclides, metalloids such as selenium, and many more (Kolpin et al. 2002; Daughton 2004; Tijani et al. 2016). The public is also increasingly aware of so-called emerging substances: antibiotics, analgesics, hormones and other drugs (anti-inflammatory, anti-diabetic, antiepileptic), nanoparticles, bioterrorism and sabotage agents, and pathogens (cyanotoxins or algal toxins, mycotoxins, antibiotic-resistant organisms, etc.). The list of emerging substances and products is therefore particularly long (Tijani et al. 2016) and is also of great interest to the scientific community (Table 1.1).

The term "emerging contaminants/pollutants" refers primarily to those for which there are currently no regulations requiring monitoring or public reporting of their presence in our water supply or wastewaters. The US Environmental Protection Agency (EPA) defines an emerging pollutant as "a chemical or material which because of a recent source that it originates, or because of a new pathway that has developed, and for which a lack of published health standards exist poses a perceived, potential, or real threat to the human health or the environment" (source: US EPA 2012, Washington, DC). Their occurrence results from point (urban and industry) or diffuse (agriculture) contamination/pollution sources. Various other terms and definitions can be found in the literature. Indeed, there is no generally accepted definition, each definition and interpretation depending on the respective perspective and/or objective (Morin-Crini and Crini 2017; Salthammer 2020). In Europe, the term contaminant of emerging concern is a common term for chemicals that are currently not regulated (not submitted to a routine monitoring and/or emission control regime) but may be under scrutiny for future regulation (Dulio et al. 2018).

In the bibliography, there are several terms to denote emerging contaminants almost interchangeably such as emerging organic contaminants, emerging substances, substances of emerging concern, contaminants of emerging concern, emerging pollutants, emerging micropollutants and other similar terminologies (trace organic contaminants, micropollutants, trace organic compounds, trace pollutants, constituents of emerging concern, xenobiotic organic compounds, etc.). However, it is important to point out that all emerging pollutants are emerging contaminants, but not all emerging contaminants are emerging pollutants. This is often confuse in the literature. There are also terms that refer to subsets of emerging substances. All of these terms refer to many different types of chemicals, such as pharmaceuticals and drugs (medicines, non-prescription drugs, endocrine disrupting compounds), personal care products and cosmetics (soaps, fragrances, bisphenols, parabens, triclosan, endocrine disrupting compounds), household cleaning products, industrial compounds (surfactants, flame/fire retardants, monomers and polymers, solvents), pesticides (new formulations), and other agricultural products (fertilizers, lawn care products) and biocides (Tijani et al. 2016; Morin-Crini and Crini 2017; Patel et al. 2019). Endocrine disruptors can be found in the pharmaceutical and/or cosmetic category. They also include a range of other compounds, including pesticides and phytocompounds, as well as industrial chemicals. Many researchers consider endocrine disruptors to be a narrower subclass of emerging substances. This term refers to specific substances that have been shown to have an effect on the endocrine system of aquatic organisms or that have been designed to act on the human hormone system. Other substances, products and new technologies of concern include UV filters, corrosion inhibitors (benzothiazoles), 3D-printing (thermoplastics, additives), E-cigarettes and E-shishas, nanomaterials and nanoparticles (new products of new anthropogenic processes, nanosprays), microplastics and nanoplastics (considered as truly "new" to the environment). Biological agents and other pathogens (pathogenic bacteria, multi-drug resistant microbes) can also be defined as emerging (Morin-Crini and Crini 2017; Crini and Lichtfouse 2018; García et al. 2020). Indeed, the literature also includes discussion of antibiotic-resistant microorganisms and antibiotic-resistance genes under the broad category of emerging contaminants. More recently, nanoparticles have also been included in the class of emerging substances (Mearns et al. 2018; Zhao et al. 2018; de Marchi et al. 2019; Matich et al. 2019).

Luo et al. (2014) proposed to classify emerging contaminants in five categories, namely personal care products (fragrances, disinfectants, UV filters and insect repellents), pharmaceuticals (non-steroidal anti-inflammatory drugs, antibiotics, lipid regulator, anticonvulsants, β-blockers), steroid hormones, industrial chemicals (plasticizers, fire retardants, bisphenol A, phthalates) and surfactants (cleaning products, alkylphenols), excluding pesticides. According to Norman network (2016), over 1000 emerging chemicals are recognized and classified in four main groups, *i.e.*, drugs, personal care products, endocrine disruptors and pesticides (Norman network 2016; Dulio et al. 2018). Another simplified classification for pharmaceuticals and personal care products proposed by Ebele et al. (2017) considers only one main family of endocrine disruptors, which is subdivided into three categories: (1) personal care products including cosmetics, disinfectants, conservation agents, fragrances and UV screens; (2) nonsteroidal drugs, including antibiotics, analgesics, and other pharmaceuticals (human and veterinary drugs); and (3) steroids (estrogens, etc.). Patel et al. (2020) recently proposed four groups of emerging contaminants (excluding pesticides), *i.e.*, pharmaceuticals, endocrine disruptors (including cosmetics, hormones, xenobiotics, plastics), personal care products (including cosmetics, UV filters, disinfectants, insect repellents such as triclosan, synthetic hormones, lipid regulators and steroids), and perfluorinated compounds. Other researchers and organizations have classified emerging substances into three broad categories: pharmaceuticals, personal care products and endocrine disrupting compounds, excluding pesticides. However, these classifications can be confusing because some personal care products are also endocrine-disrupting chemicals while other drugs are not (Darbre 2015; Gee et al. 2015). In reality, there is not yet an

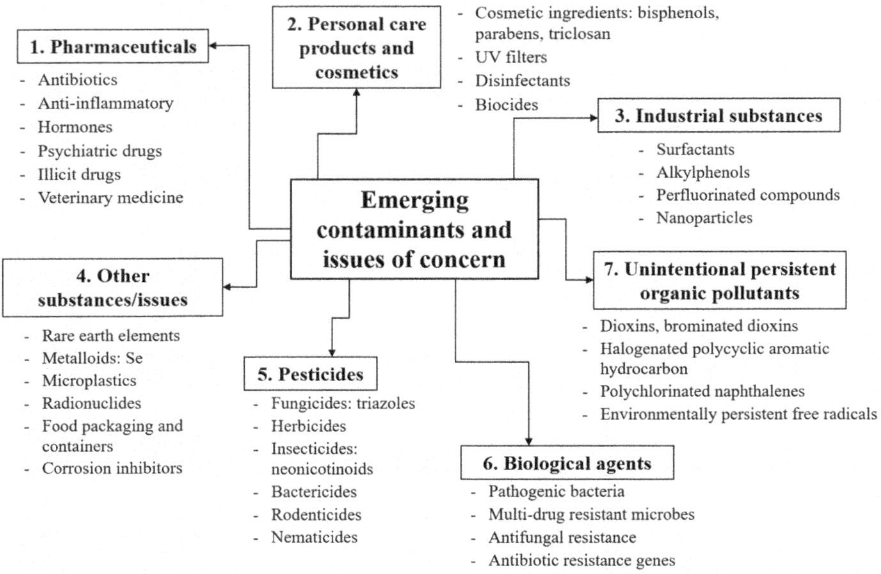

Fig. 1.1 Classification of emerging contaminants and issues of concern in water compartments

internationally recognized general classification for emerging substances. Nevertheless, many researchers agree with the definition given by the US EPA (Gogoi et al. 2018). Figure 1.1 provides a general classification of emerging contaminants and issues of concern in water compartments into seven categories, regardless of their risk to health or the environment.

1.2.3 Examples of Emerging Contaminants of Recent Concern

Polyfluoroalkyl or perfluoroalkyl, known by the abbreviation PFASs, are a family of emerging contaminants that have recently become a concern (Melzer et al. 2010; Gallo et al. 2012; Barry et al. 2013; Vestergren and Cousins 2013; DeWitt 2015; Xiao 2017; Dong et al. 2018; Lorenzo et al. 2018; Brusseau 2019; Ng et al. 2019; Klemes et al. 2019; Pan et al. 2020; Salthammer 2020). Perfluoroalkyl substances is a term used for a family of highly fluorinated chemicals (the term perfluorinated compounds can also be found in the literature). These substances contain fluorinated molecular chains that degrade very slowly in the environment and have a high bioaccumulation potential. They include subclasses such as perfluoroalkyl sulphonic acids (PFSAs) and perfluoroalkyl carboxylic acids (PFCAs). The list is particularly long as the Chemical Abstracts Service registers more than 4000 compounds.

These synthetic substances have been produced and widely used since the 1950s before their harmful health effects and persistence came to light. Today, they are still widely used in consumer products and for industrial uses, finding targeted applications in a wide variety of fields. For example, perfluoroalkyl substances (perfluoroalkyl sulfonates and perfluoroalkyl carboxylates) are widely used as polymer constituents in food containers and packaging (fast-food packaging, pizza boxes), processing aids in non-stick cookware (Teflon®), surface-active agents in waterproof clothing and stain-resistant carper (stain-resistant textiles, water-resistant textiles), and cleaning products. They are also used as individual ingredients in firefighting foams used in many military installations and airports, and in painting materials (Xiao et al. 2013; DeWitt 2015; Brusseau 2019; Pan et al. 2020). Other applications include coatings and metal plating, hydraulic fluids, medical devices, cosmetics (shampoos), pesticides, and photo imaging. The best known are perfluorooctanoic acid (PFOA) used as a repellent coating for textiles, paper products and cookware and perfluorooctanesulfonic acid (PFOS) used in fire fighting foams, carpet treatments, and mist suppressants in metal plating operations.

In 2009, perfluoroalkyl substances were listed in Annex B of the Stockholm Convention on Persistent Organic Pollutants (Stockholm Convention 2009: CN524-TREATIES 4, UNEP, Geneva) as substances of concern, restricted but not completely banned. Many substances are currently being considered for restriction under the 2015 Rome Directive (Stockholm Convention 2015: Decision POPRC-11/4, UNEP, Rome) and will be banned under the EU Chemicals Regulation from 2020. Indeed, perfluoroalkyl substances are chemicals considered to be persistent organic pollutants and long-range transported anthropic contaminants found at high concentrations in human and wildlife populations around the World, including in Arctic marine mammals. Their ubiquous persistence in the environment and their toxicity in animal models raise concerns about the effects on human health of chronic low-level exposure (Melzer et al. 2010; DeWitt 2015; Pan et al. 2020). Perfluoroalkyl substances have been linked to cancer, liver damage and thyroid disease (Melzer et al. 2010; Gallo et al. 2012; Barry et al. 2013; Vestergren and Cousins 2013). Klemes et al. (2019) recently reported that more than 16 million people in the United States were exposed to drinking water with perfluorinated alkyl substances concentrations exceeding 10 ng/L. Melzer et al. (2010) previously reported that higher concentrations of serum perfluorooctanoate and perfluorooctane sulfonate are associated with current thyroid disease in the general adult population in the United States. Gallo et al. (2012) demonstrated an association between concentrations of perfluorooctanoate and perfluorooctane sulfonate and serum levels of alanine of transaminase, a marker of hepatocellular damage. In another study, Barry et al. (2013) reported that exposure to perfluorooctanoic acid is associated with renal, pancreatic and testicular cancer in rats. Other examples can be found in the review by Vestergren and Cousins (2013).

The presence of perfluoroalkyl substances in tap water has become a priority. However, due to their chemical stability and composition, these substances are poorly or even not eliminated by conventional water treatment methods. Because they have no carbon-carbon double bonds and no functional groups, they are also resistant to advanced oxidation processes. Innovation is therefore required to find removal solutions.

Another topical issue concerns plastics/microplastics (Wagner et al. 2014; Avio et al. 2017; Lambert and Wagner 2018; Peeken et al. 2018). Plastics are synthetic polymers composed of different materials that range from cellulose to crude oil and are used in a broad ever-growing spectrum of applications. They are separated into thermoplastics and thermosets and consist of different synthetic polymers of which polypropylene, polyethylene, polyvinyl-chloride, polyurethane, polyethylene terephthalate and polystyrene were the most commercially demanded in 2017 (Plastics Europe 2017). Moreover, tire wear may be included in a larger definition of microplastics (Wagner et al. 2018) (although ISO does not cover it; ISO 2013). Plastics are important materials that are abundant in our economy (because they are lighter and cheaper than other materials) and are part of our daily lives, making our lives easier in many ways. Since the 1950s, the plastic production rose exponentially to 359 million tons in 2018 (Plastics Europe 2019). However, if they are not properly disposed of or recycled after use, they are often released into the environment where they can persist for long periods of time and break down into smaller and smaller pieces, called microplastics, which are of concern (Lambert and Wagner 2018). Figure 1.2 shows accumulation of plastic debris on a shoreline in France.

Microplastics are indeed widespread contaminants worldwide, from the Arctic to the Antarctic (Peeken et al. 2018) *via* tropical islands (Imhof et al. 2017), present not only in all the aquatic compartments (rivers, sediments, marine environment), but also in soils and in the atmosphere (Avio et al. 2017; Prata et al. 2019; Möller et al. 2020; Xu et al. 2020). In fact, plastics have also been detected in remote mountain lakes and deep-sea sediments (Cauwenberghe et al. 2013; Free et al. 2014), and

Fig. 1.2 Accumulation of plastic debris on a shoreline in France. (Source: Florence Carreras, INRA, France)

even in the air we breathe (Gasperi et al. 2018; Allen et al. 2019). They are considered as emerging contaminants. However, the first studies on the description of polystyrene spherules and pieces in the environment and their effects date back to the 1970s with the works of Carpenter and co-workers (Carpenter et al. 1972; Carpenter and Smith 1972). Microplastics were already identified as persistent in the environment and as potentially to accumulate in different marine ecosystems.

Special attention is paid to plastic pollution of the marine environment. Indeed, this issue is a growing environmental concern worldwide, as shown by the recent explosion of literature on their identification, quantification and behavior (*e.g.*, Avio et al. 2017; Juliano and Magrini 2017; Barboza et al. 2018; Lin et al. 2018b; da Costa et al. 2019; Mishra et al. 2019; Ng et al. 2019; Prata et al. 2019; Stock et al. 2019; Guo et al. 2020; Iroegbu et al. 2020; Wong et al. 2020). The first studies have been conducted in the marine environment, mainly about macroplastics, (e.g., UNEP 2009; Hidalgo-Ruz et al. 2012), and started only some years ago in freshwater environments (*e.g.*, Li et al. 2018b; Wagner and Lambert 2018). Plastics have also been investigated in rivers (Lin et al. 2018b; Rodrigues et al. 2018b), lakes (*e.g.*, Eriksen et al. 2013; Turner et al. 2019), beaches (Turra et al. 2014) and in the deep sea (Kanhai et al. 2019).

The majority of plastic debris or microplastics (particles <5 mm) in the seas comes from the fragmentation of larger products (*e.g.*, fishing net, household practices, bottles). There is growing evidence that microplastics also enters the water compartment directly from other sources, including cosmetics (nanoplastics used as ingredients), clothing and industrial processes (Juliano and Magrini 2017). Microfibers present in aquatic systems may also be released from the textile industries and from wastewater treatment plants (baths and washing clothes). Plastic microbeads are indeed present as abrasive scrubbers in personal care and cosmetic products (hand cleansers, toothpastes, facial scrubs, shampoos, etc.). Several studies have demonstrated the negative impacts of plastics and plastic debris on wildlife and the evidence of seafood contamination by microplastics, and therefore their potential effects on human health (Avio et al. 2017; Juliano and Magrini 2017; Barboza et al. 2018; da Costa et al. 2019; Mishra et al. 2019; Prata et al. 2019; Stock et al. 2019; Guo et al. 2020). From a scientific point of view, there is an urgent need for knowledge on the sources, fate and environmental concentrations over time (Wong et al. 2020).

The review by Barboza et al. (2018) showed the negative effects of plastics and plastic debris on marine fauna, and thus a potential effect on human health, due to the evidence of contamination of marine species (*e.g.*, seafood) by microplastics, for example through the ingestion of microplastics as they are confounded with prey. However, data on the presence and occurrence of microplastics in marine species (and seafood in general) used for human consumption are scarce. This information is necessary for providing a basis for a sound risk assessment. Another important challenge is to understand the processes and mechanisms involved in the entry and assimilation of plastic debris into human tissues and their potential effects on human health. From a water engineering point of view, plastic-based debris are too small to be trapped in wastewater treatment plants. Therefore, they enter into water sources through domestic drainage systems and are transported to seas and oceans. Wastewater treatment plants should adopt innovative ideas to deal with this contamination.

A third example of substances of recent concern are nanoparticles, which are organic or inorganic particles that are smaller than 100 nm, widely used in consumer products (cosmetics, sunscreens, electronics, food), pharmacy, medicine, and chemical industry (nano-plasticizers, coatings, textiles, catalysis). Other sectors are also concerned, such as automotive parts, building materials, sports equipment, and energy (Mearns et al. 2018; Zhao et al. 2018; de Marchi et al. 2019; Matich et al. 2019; Sousa and Teixeira 2020). Indeed, research and development in nanotechnology (including nanoscience, nanomaterials, and nanoparticles) is booming. However, the development of a regulatory framework and environmental impact assessment tools concerning the potential toxicity of nanoparticles for the environment and/or health has not yet been completed. Unfortunately, this lack of regulatory control means that the increasing environmental exposure to nanoparticles is occurring without awareness of the general public about the exposure and potential environmental toxicity of these emerging contaminants. In recent years, the term nanotoxicity has been coined. Likewise, the terms nanopollution and nanopollutants are also increasingly being used, but there is still very little research on the presence and behavior of nanoparticles and nanoproducts in aquatic compartments, as well as on their impact on wastewater and waste treatment systems, and on the production of drinking water. These are important areas for future research by local scientific institutions. Like other emerging substances, existing wastewater treatment and water purification technologies are not primarily designed to deal with nanopollution/nanocontamination (Sousa and Teixeira 2020). Nanoparticles have already been detected in wastewater, surface water, and treated drinking water, with concentrations ranging from ng/L to μg/L.

The scientific community is also increasingly concerned about the global contamination of water by other pollutants (Table 1.1), such as cosmetic ingredients (parabens, triclosan, bisphenol A) and UV filters (organic filters such as 2-ethylhexyl-methoxycinnamate, inorganic filters such as TiO_2 and ZnO), alkylphenols (detergents), rare earth elements (*e.g.*, used in the production of electronic devices, mobile phones, super-capacitors, or as contrast agents in medical imaging), radioactive elements or radionuclides (some such as radium-226, radium-228 and uranium are natural groundwater contaminants; and thus contamination of drinking water is a significant, emerging issue), anti-retroviral drugs (used for human immunodeficiency virus), psychiatric drugs (antiepileptics such as carbamazepine used for the treatment of epilepsy and bipolar disorder, antidepressants such as selective serotonin reuptake inhibitors, anxiolytics such as benzodiazepines), cardiovascular drugs (beta-blockers such as atenolol), hypocholesterolaemic drugs, and biological agents such as pathogenic bacteria and antibiotic-resistance genes (Table 1.1). These substances are problematic because most of them: (1) are currently not regulated in humans and the environment; (2) are not systematically monitored as part of environmental and public health programs; and (3) have their mechanism of environmental and human toxicity poorly understood. There is an urgent need for scientific knowledge from a chemical, biological, toxicological and ecotoxicological point of view (Gwenzi et al. 2018; Pedrazzani et al. 2019; Xu et al. 2019; Zhang et al. 2019; Olaniyan et al. 2020).

Another recent area of concern, not only for the public but also for the scientific community, is indoor pollutants (Slezakova et al. 2012; Salthammer 2014, 2020; Cretescu et al. 2019). It is recognized that indoor air is highly polluted: 5–10 times more polluted than outdoor air. There are many sources of indoor air pollution (at home as well as in offices), such as combustion and tobacco products (gas appliances, tobacco smoke, candles), cooking processes, building materials and furnishings, cleaning and personal care products (perfumes), paints, central heating and cooling systems, humidity, and other sources (indoor and outdoor particulate matter, biological agents, etc.). Chemical (volatile organic compounds, polycyclic aromatic hydrocarbons, polychlorinated biphenyls, polybrominated biphenyl ethers, particulate matter, radioelements such as radon, pesticides, etc.) and biological (pollens, bacteria, and mold) pollutants can accumulate to levels that can cause health, nuisance and comfort problems (Slezakova et al. 2012). The comprehensive review recently published by Salthammer (2020) can be consulted on emerging indoor pollutants.

1.2.4 Biological Agents

In recent years, cases of fatal emerging/re-emerging viral infections have significantly affected human health despite extraordinary advances in biomedical knowledge. The recurrent epidemics of dengue and chikungunya in tropical and subtropical regions, the Zika epidemic in the Americas and the Caribbean, and the global SARS epidemics and Covid-19 pandemics are dramatic examples.

Like chemical hazards, biological threats are part of our societies. Indeed, biological agents are ubiquitous in the environment and in our daily lives. Since they are invisible, it is difficult to assess the risks they present. According to the European Directive 2000/54/EC, the four main groups of biological agents are: bacteria, viruses, parasites, and fungi. In this Directive, biological agents are also defined as "the micro-organisms, including those which have been genetically modified, cell cultures and human endoparasites, which may be able to provoke any infection, allergy or toxicity" (European Directive 2000/54/EC). A microorganism is defined as a microbiological entity, cellular or non-cellular, which is capable of replication or of transferring genetic material. A cell culture is defined as the *in-vitro* growth of cells derived from multicellular organisms. Like any other organism, a biological agent simply tries to live and procreate (Parvez and Parveen 2017). Pathogenic microorganisms are present in the secretions, feces, urines of patients or asymptomatic carriers, subsequently retrieved in the wastewater, the wastewater treatment plants, and the receiving surface water. Hence, pathogenic biological agents can be considered as emerging pollutants.

Pathogens are organisms/microorganisms able to cause diseases. A minor part of organisms/microorganisms are pathogens. They include viruses, bacteria, fungi, and parasites, which can be found anywhere in the environment, including water, air, food, soil, and surfaces we are in contact with at home. Every pathogen needs a host to grow and survive. Once the pathogen has settled in a host, it must avoid the host immune responses, and thrive before being transmitted to a new host. Pathogens can be transmitted in several ways depending on their type: through skin contact, body fluids, airborne particles, water and food, contact with feces, and contact with a contaminated surface.

Viruses are responsible for a number of infections, many of which are contagious. Examples of viruses causing viral diseases include human immunodeficiency virus (HIV), dengue virus, rotavirus, and coronavirus. They are made up of a piece of genetic code (DNA or RNA) and protected by a coating or protein. Depending on the type of virus, they can cause a wide range of acute or chronic respiratory, gastrointestinal, or genital infections. However, viruses are unable to thrive and reproduce independently. Hence, viruses use the components of the host cell to replicate, damage or destroy the infected cells, before being released. There are efficient vaccines and antiviral treatments for some, but not for all, viral diseases.

Bacteria are microorganisms made of a single cell that can live almost everywhere on Earth. They are more complex organisms than viruses that can thrive independently of a host. Only a few bacterial species cause infections. Indeed, some species are beneficial (*e.g.*, those present in dairy products or in human gut), others are harmful (*e.g.*, those causing infections such as tuberculosis are called pathogenic bacteria). Examples of bacterial infections are gastroenteritis, meningitis, tuberculosis, and Lyme disease. Some bacterial diseases can be prevented by vaccination. Bacterial infections are treated by antibiotics.

Fungi can be found naturally in the environment, and also on the human skin. Some of them cause infections and the structure of their cell walls can make them resistant to antifungal agents. Some new species of fungi, such as *Candida auris,* are especially pathogenic. Parasites are organisms that behave like tiny animals, living in or on a host and feeding from or at the expense of the host. Three main types of parasites (*i.e.*, protozoa, helminths and ectoparasites) can cause diseases (*e.g.* malaria) in humans. They can be spread after an insect bite or contamination by soil, water, or air.

Antibiotic-resistant bacteria and antibiotic resistance genes have recently been identified as the greatest threats to public health by the World Health Organization. They are a potential hazard to public safety and water quality. The contamination of wastewater systems by antibiotic-resistance genes and the global spread of antibiotic-resistant superbugs is of great concern. Indeed, an area of great interest to the scientific community is the phenomenon of bacterial resistance (Zad et al. 2018; Zhao et al. 2018; Kumar et al. 2019; Liu et al. 2019; Ng et al. 2019; Reichert et al. 2019; García et al. 2020). Some bacterial clones have become resistant to most of the antibiotics, even to the last-resort compounds carbapenems or colistin, making the associated infections difficult to treat. This is due to the overuse of antibiotics by humans and animals. However, studies have also linked the use of antibiotic in agriculture with infections in humans caused by antibiotic-resistant microbes. The great majority of pharmaceuticals consumed by humans and animals end up in feces and urine, which reach sewage treatment plants and/or soil. In addition, the gut of animals and human is the major reservoir of antibiotic-resistant bacteria. Since wastewater treatment plants are not designed to eliminate pharmaceuticals or antibiotic-resistant bacteria, treated effluents still contain pharmaceuticals (*e.g.*, antibiotics, hormones) and antibiotic-resistant microorganisms. In addition, the manure produced by animals receiving antibiotics that is commonly used as fertilizer, can become a source of antibiotics and antibiotic-resistance determinants for soils and crops. Besides, in areas affected by water scarcity, wastewater reuse is becoming a common practice in agriculture, potentially introducing these emerging pollutants into the food chain. In fact, there is still very little information on antibiotic resistant organisms in water sources, drinking water treatment plants and drinking water distribution systems, and the impact of their presence on human health. There is also an urgent need to improve understanding of the mechanisms associated with the spread and development of these organisms, not only in treatment plants and the environment (water, sediment, and soil), but also in the human body and in clinical and veterinary settings (see the two chapters devoted to these topics in this book).

1.3 Analysis and Detection of Emerging Contaminants

The proliferation of chemical substances, both known and unknown, that can have adverse effects on human health and the environment has increased the need for early detection for the safe identification of these contaminants. High-performance analytical tools and validated protocols now exist to identify and quantify contaminants in various environmental matrices. All water stakeholders concerned with water pollution issues, whether they are university researchers, government agencies, regulatory bodies or environmental interest groups, have tools at their disposal that allow them to collect and evaluate data from a variety of sources. These data are important for advancing knowledge about emerging substances and for preparing future regulatory or monitoring requirements.

1.3.1 Chromatographic Methods Coupled to Mass Spectrometry for the Analysis of Emerging Contaminants

The increase of the production, as well as consumption, of compounds designed to improve health, leads to their increased release into the environment. The governments try to regulate this impact into the environment with emission limits for different chemicals and the societies try to minimize, with the available technology, the amount of these pollutants that end up in aquatic ecosystems (IOM 2014). The complexity of the chemical mixtures that are present in the environment demands that the scientific community works on increasingly sensitive methods for the analysis of these compounds, as shown in Table 1.1.

When planning to study contaminants in environmental matrices, it is necessary to develop an entire study design. Within this study design, the following points need to be considered: (1) sampling, (2) sample preparation, (3) chemical analysis, as well as (4) data visualization and analysis.

The collection of samples requires a previous study to make them representative. The previous study should depend on a great extension of the target chemicals or the expected concentrations, as well as on the information to be extracted from the results (García-Córcoles et al. 2019). For example, for a study involving pesticide analysis, the areas to be selected should be those where the major economic activity is agriculture, or in the case of a study of pharmaceuticals, it should be based on population density and age.

Sample Collection and Preparation For the study of contamination/pollution, water is usually the most often investigated environmental matrix due to its physicochemical characteristic and since it is relatively easy to analyse and is in nature an important reservoir of pollutants in Nature. These samples are usually collected from surface water, wastewater treatment plants, groundwater or even drinking water. There are two predominant methods of sample collection: grab samples or composite samples. When performing grab sampling, a certain volume of water is collected at a specific time and place. The sampled amount of water is usually relatively small. This type of sampling is usually performed only when contaminant concentrations can be expected to be high and hence easy to detect with analytical methods. The second method is more complex since the sampling is performed with a time window that can last from hours to days. It is often used to observe a wider range of contaminants or when relatively low concentrations are expected and are more difficult to detect with simple techniques. An example is the so-called large volume solid phase extraction in which the selected site is sampled for a certain time and the sample is filtered into solid phase extraction cartridges. The contaminants are accumulated and can subsequently be extracted in the laboratory (Schulze et al. 2017).

Not many methods perform direct analysis of samples. Most of the studies in water perform pre-concentration techniques using solid phase extraction (Carmona and Picó 2018; López de Alda et al. 2003; Dimpe and Nomngongo 2016) since they are considered relatively cheap, fast, reliable and reproducible methods (Tang et al. 2019).

Solid phase extraction is the most widely used extraction technique for environmental samples (Dimpe and Nomngongo 2016; Carmona and Picó 2018; Lorenzo et al. 2018; García-Córcoles et al. 2019) and is an effective tool for isolating and purifying compounds. The retention of emerging contaminants was improved thanks to the development of new polymeric sorbents, which are mainly made by hydrophilic-hydrophilic material (Wille et al. 2012). Nowadays, most of the studies with solid phase extraction techniques use Oasis® HLB cartridges (divinyl benzene and vinylpyrrolidone copolymer) followed by StrataX® cartridges (polymeric reversed phase that provides a strong retention of neutral, acidic and basic compounds). Another widely used method for the extraction of emerging pollutants is the liquid-liquid extraction, although this can be time consuming (due to its complex automation), difficult and expensive, since a large volume of organic solvents is needed. Furthermore, this procedure requires a subsequent clean-up and a pre-concentration phase prior to the chromatographic analysis (Wang et al. 2013; García-Córcoles et al. 2019).

In the case of solid matrices, such as sediments, soils or biota, the Quick, Easy, Cheap, Effective, Rugged and Safe (known as QuEChERS) method developed by Anastassiades et al. (2003) is frequently used for the extraction of emerging contaminants (Calatayud-Vernich et al. 2016). This method was optimized for the extraction of pesticides from fruits and consists mainly of two phases: extraction and clean-up. The extraction phase uses salts, such as $MgSO_4$, to reduce the water content, and others such as NaCl, sodium acetate or citrate salts for all other applications. For the clean-up step, the use of primary/secondary amine stands out for the elimination of, among others, organic acids, fatty acids, sugars and atrazine pigments, or the sorbent C18 that eliminates fats, sterols and other non-polar interferences. Other methods have been optimized over the years for the extraction of a greater number of compounds, such as solid-liquid extraction. A method using solid-liquid extraction with a McIlvaine-EDTA buffer (pH = 5.5) accompanied by solid phase extraction was described by Carmona et al. (2017) as an improvement of the traditional QuEChERS. The method using McIlvaine-EDTA buffer reduces matrix effects to less than 20% and provides recoveries above 50%. In addition, more extracted compounds have been detected in studies with screening analysis. Other methods for the extraction of organic contaminants in solid matrices are based on the technique of pressurized liquid extraction which requires solvents at high temperatures and pressures that may transform some analytes into different transformation products.

Instrumental Analysis In recent years, the number of scientific publications that focused on the study of emerging contaminants in aquatic ecosystems has increased exponentially. Most of these publications base their analyses on a multitude of chromatographic techniques coupled with mass spectrometry as shown in recent reviews (Fig. 1.3) (Carmona and Picó 2018; Lorenzo et al. 2018; García-Córcoles et al. 2019; Tang et al. 2019; Álvarez-Ruiz and Picó 2020). Chromatography is a physical method of separation in which the components are separated and distributed between two phases, a stationary and a mobile phase. The Russian botanical Mikhail Tsvett thus conceived it in 1903 based on the ability of a column of finely powdered solid materials such as Al_2O_3, to adsorb substances from a solution that drips through it. Tsvett used this method to separate coloured components of plant pigments. The column developed its colour bands, and this technique was called chromatography separation. The Tsvett method remained unknown for a long time until switched back to biochemists in the 1930s.

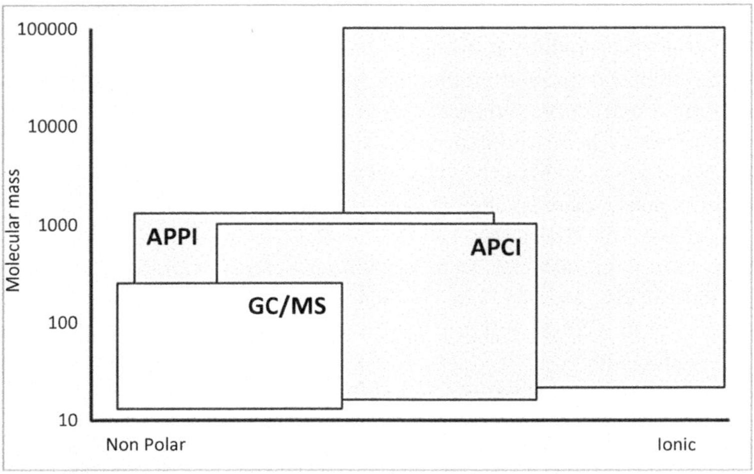

Fig. 1.3 Relationship between gas chromatography/mass spectrometry (GC/MS) and different interfaces of liquid chromatography/mass spectrometry (LC/MS) (*APPI* atmospheric pressure photoionization, *APCI* atmospheric pressure chemical ionization, source: Eric Carmona, Leipzig, Germany)

The first type of chromatography is known as adsorption chromatography. In 1942, Archer Martin and Richard Synge co-developed the partition chromatography by using a liquid absorbed on a solid as the stationary phase and the other immiscible liquid stationary phase as mobile phase. In this method, called liquid-liquid chromatography, the compounds are distributed into two liquid phases. This achievement inspired the development of other separation methods, namely gas chromatography. The effectiveness of liquid-liquid chromatography was increased in the 1970s by the use of small particles for support and by pumping fluid, in order to enhance the mobile phase through the stationary phase under pressure. This method is known as high Pressure liquid chromatography. With the various chromatography techniques, it is possible to separate and detect compounds quantities of components (parts per billion and even parts per trillion). Chromatography effectiveness is even higher when combined with other instruments such as with mass spectrometry for analysis.

One of the main reasons for the delay in the application of liquid chromatography-mass spectrometry was the difficulty involved with the coupling liquid chromatography and mass spectrometry. From the 70s several liquid chromatography-mass spectrometry interfaces have been developed, but it was only after commissioning of atmospheric pressure interfaces (electrospray ionization and atmospheric pressure chemical ionization) when liquid chromatography-mass spectrometry began to be a robust alternative to the methods of gas chromatography-mass spectrometry.

Advances in chromatographic techniques have led to increased detection of analysis of emerging contaminants (in the order of ng/L or μg/L). The most used techniques for the separation of these compounds are the gas chromatography and the liquid chromatography. The selection of one of them depends on the physical-chemical properties of the compounds to be detected. Volatile and semi volatile compounds are analysed in the gas chromatograph, such as phenols or phtalates and non-volatile ones in the liquid chromatograph (Tousova et al. 2017; Lebedev et al. 2018). According to most of the literature reviews consulted, an average of 80% of the studies use liquid chromatograph versus gas chromatograph.

Both techniques require the use of columns for the compounds separation. In Liquid Chromatography, the most common mode is reverse phase separation which is characterized by a stationary phase that is less polar than the mobile phase (Rennie 2016). Columns that are characterized by this type of separation have stationary phases with long chains of aliphatic carbons attached to silica with pores of the micrometric order. The silica columns most used in liquid chromatography are those with 8 or 18 hydrocarbon chains (C8 and C18), being the C18 column the most used due to the robustness offered through the retention of both polar and non-polar analytes. In the case of gas chromatography, the most used columns are capillary ones between 30 and 60 m, due to their efficiency, high resolution and speed (Zhang et al. 2016b). These types of columns usually have a diameter between 0.1µm and 0.5µm. These columns can be classified into three different types: support coated open tubular, porous layer open tubular and wall coated open tubular column, being the last one the most used (Dettmer-Wilde and Engewald 2014; Baduel et al. 2015).

When designing a method for analysis and depending on both the equipment available and the compounds to be identified, a robust and stable mobile phase selection is important. The mobile phase is used to transport the analytes to the stationary phase. It can also play an important role in the separation of compounds in the case of liquid chromatography. The mobile phase in liquid chromatography analysis generally consists of two solvents. One is an aqueous type (A) and the other is an organic solvent (B) and generally includes an acid or salt to improve separation. The mixture ratio of both constituents usually changes over time of analysis to allow a better elution of the compounds in the column. The most common organic solvents used are methanol or acetonitrile. These are supplemented by a salt, in the case of the negative ionization mode for further mass spectrometry detection, such as ammonium acetate, ammonium formate or ammonium fluoride. The latter has been demonstrated to improve chromatographic signals, compared to other salts and also to improve the capacity and efficiency of electrospray (Carmona et al. 2014) for the study of some emerging pollutants such as pharmaceuticals, illicit drugs and personal care products. For the positive ionization mode, the most commonly used acids are acetic acid and formic acid (Carmona et al. 2017; Sadutto et al. 2020; Álvarez-Ruiz and Picó 2020). In gas chromatography the mobile phase is usually an inert gas such as helium or another non-reactive gas such as nitrogen. Helium is usually the most used gas (about 90%) (Li et al. 2019b) for the transport of analytes. Chromatographs have been evolving, including increasing their pressure or increasing their capacity for the use of more than two mobile phases at the same time. Chromatography is a technique that is usually coupled to tandem with mass spectrometry for environmental analysis.

Mass spectrometry is a universal and specific instrumental technique, highly sensitive, which enables the unequivocal identification of a substance. The principle of the mass spectrometry is the production of ions from neutral compounds and observing the subsequent decomposition of these ions in "product ions". These ions decomposed (fragments also possess charge) move quickly and are classified according to their ratio m/z (ratio between mass and charge number of the ion). The mass spectrometer not only classifies the fragments, but also measures the amount that is formed. When a molecule is supplied with a given energy, the molecule is decomposed in a certain pattern, which fragments are always obtained at the same intensity ratio. This particular pattern is plotted in the mass spectrum, which, for this reason, is called the fingerprint of the substance. Thus, the mass spectrum allows the unequivocal identification of the molecules.

Mass spectrometers are generally sophisticated instruments, which comprise: (A) sample introduction system, that may be, in this case, a liquid chromatograph; (B) an ionization source, where the molecular fragmentation characteristic of each compound occurs; (C) a mass analyser or filter which separates the ion fragments generated according to their mass/charge ratio and (D) a detector, which collects and characterizes the fragment ions leaving the analyser. This instrumental technique is the most used for the determination and identification of emerging contaminants. This is a method with many different uses, ranging from forensics to health research, as well as for the study of the environment, since it is highly sensitive and reliable. There are different types of mass spectrometers depending on the objective of the study. Among them, for the study of emerging contaminants, the triple quadrupole (QqQ), the time of flight (TOF) and the orbitrap stand out. The most common mode for the study of emerging contaminants is electrospray ionization.

In studies on emerging pollution, the most widely used technique with mass spectrometry is the so-called target analysis or target screening, i.e., studies to identify and quantify a series of known compounds in complex matrices (Krauss et al. 2010; Ccanccapa-Cartagena et al. 2019). This technique requires a series of reference standards for the quantification and confirmation of the compound. The triple quadrupole consists of three quadrupoles connected in series. Quadrupole 1 (Q1) and the quadrupole 3 (Q3) operate as two quadrupole analysers connected in series. The quadrupole 2 (q2) or collision cell is placed in between Q1 and Q3. It is a special quadrupole in which energy is applied. The collision energy, that allows fragment ions, is obtained in the ionization source. This energy can take different values, which allows different fragments and/or different intensity relationships between them in mass spectra. This instrumental technique is highly sensitive and selective.

Another widely used technique is suspect screening. The suspect screening method implements an identification of expected compounds using databases (libraries) for their identification. This method, which has the so-called known unknowns approach, starts with a list of expected substances. Unlike the target analysis, it is not based on reference standards for confirmation and quantification, but on the specific information available for the compound, such as the molecular formula (which allows the calculation of the m/z of the expected ion). In contrast to this technique, *non-target screening* starts without any information about the compounds to be detected. The number of chemically significant structures that can be assigned to a particular peak is limited to those structures that show a close relationship with the original compound (Krauss et al. 2010). The most common mass spectrometers for this type of analysis are the QToF (time-of-flight mass spectrometry) or the orbitrap (ion trap), the latter being the least used (Andrés-Costa et al. 2016). In the QToF, which is the most used for the realization of these techniques, after the ions have crossed the quadrupole, they reach an ion button in which a high voltage pulse is applied that accelerates the ions in the so-called flight tube. An ion mirror at the end of the tube reflects the ions which sends them to the detector and records their time of arrival. The time of flight for each mass is unique and is determined by the energy at which the ion is accelerated, the distance and the m/z.

For the analysis of the obtained data, although each device has its specific software, nowadays there are new open-source softwares, such as MZmine. This can identify the compounds as well as integrate them for their later quantification. This type of software can be used for target, suspect and non-target screening.

1.3.2 Detection of Microplastics in Water and Sediment

Plastics in environmental samples are usually size categorized in macroplastics (>25 mm; MSFD Technical Subgroup on Marine Litter, MSFD 2013), mesoplastics (5–25 mm) and microplastics (<5 mm). Microplastics may further be differentiated in size classes of 1–5 mm (large microplastics) and <1 mm (small microplastics; Galgani et al. 2013) or different size fractions (1–5 mm, 0.5–1 mm, 100–500μm, 50–100μm, 10–50μm, 5–10μm and 1–5μm; Braun et al. 2018). Furthermore, different forms of microplastics are distinguished (e.g. fragments, fibers, films, spheres). Figure 1.4 shows different microplastic forms from the Elbe River, Germany. Primary microplastics mostly consist of small-sized industrially manufactured pellets (microbeads) which are used in personal care products (Fendall and Sewell 2009) or for further processing (Thompson 2015). Secondary microplastics refer to a fragmentation of larger pieces of plastics during use or because of weathering in the environment (Barnes et al. 2009; Thompson 2015). It includes fragments of plastic litter in aquatic environments and on land (Ryan et al. 2009), films from agricultural use (Astner et al. 2019) or fibers from textiles (Carr et al. 2016). Although many researchers intensively studied microplastics in the water and sediment, the results are difficult to compare. A first suggestion for standardized sampling has been published by Frias et al. (2018) for sediments. ISO TC 61/SC 14 (ISO/CD 23187; ISO 2020) is currently also working on a standardization for microplastic preparation and analytics.

Studying environmental samples for microplastic analysis is complex as the polymers are not only influenced by their manifold characteristics (Hartmann et al. 2019), but may also be altered by the development of biofilms (Oberbeckmann et al. 2015), hetero-aggregates or sorbed pollutants (Guo and Wang 2019). For identifying microplastics in water and sediment, the samples most often have to be treated

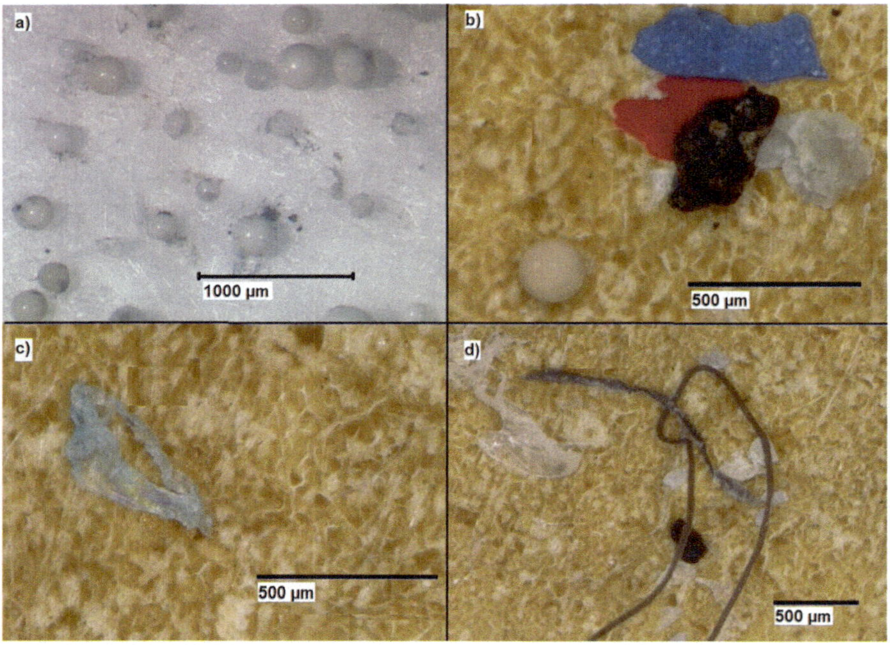

Fig. 1.4 Different microplastic forms from the Elber River, Germany: (**a**) white spheres; (**b**) blue, red, white and black fragments and one white sphere; (**c**) light blue film; and (**d**) brown and black fibers. (Source: Friederike Stock, Koblenz, Germany)

beforehand. Microplastics are relatively small and difficult to detect in environmental samples. Sediment and soil usually contain a high content of organic matter which hampers the identification by means of e.g. Raman spectroscopy or Fourier transform infrared (FTIR) spectroscopy. Therefore, the sample volume has to be reduced (for sediments *e.g.*, by means of sieving and a size fractionation; water and sediment samples: digestion of organics, density separation). The most common procedure for water and sediment samples is organic digestion with acids, bases, oxidizers or enzymes before measuring the samples (Stock et al. 2019). These solutions may however affect synthetic polymers, *e.g.*, degrade, discolorate or dissolute them (Cole et al. 2011). Therefore, samples only have to be treated when a high amount of suspended matter is present.

Before measuring the samples, most researchers separate minerals (*e.g.*, >2.5 g/cm^3 for quartz, feldspar and calcite) from organics and polymers (<0.01 and >1.4 g/cm^3, *e.g.*, foamed polystyrene and polyvinyl-chloride; Hidalgo-Ruz et al. 2012). Microplastics will generally float and minerals sink to the ground. Different density solutions are usually used: sodium chloride, zinc chloride, sodium iodide, sodium bromide, sodium polytungstate or potassium formate (Stock et al. 2019). Depending on the density of the solution (>1.2 g/cm^3), only parts or all polymers are recovered from environmental samples. The same holds true as for organic digestion; density separation is only necessary when the sample contains a larger part of inorganic material. Contamination of samples also has to be taken into account. During preparation and analysis, especially air-borne fibers may contaminate the samples (Dris et al. 2017). Therefore, a blank sample for checking the contamination during preparation and measuring is absolutely needed in order to detect external contamination.

For identifying polymers, different analytical methods have been developed. Depending on the specific research question, one of these methods is chosen rendering information about quantity, quality or mass. Visual identification of plastic particles allows a first preliminary evaluation, description, photograph and measurement of larger plastic particles (Löder and Gerdts 2015). Digital, visual or light microscopes are usually used to identify particles. Visual identification is based on polymer-specific characteristics such as color, form, absence of organic structures or equal thickness of fibers (Norén 2007). Hidalgo-Ruz et al. (2012) recommends a size of >500μm or >1 mm as smaller particles may be confounded with inorganic or organic material. However, as this method is subjective and only larger particles are taken into account, all particles have to be verified later by means of FTIR, Raman or pyrolysis.

The most common methods for identifying microplastics are FTIR spectroscopy, Raman spectroscopy and pyrolysis gas chromatography-mass spectrometry (GC-MS). Furthermore, research is conducted with thermal extraction and desorption-gas chromatography mass spectrometry (Dümichen et al. 2017), scanning electron microscopy coupled with energy dispersive X-ray spectroscopy (Gniadek and Dąbrowska 2019) or quantitative nuclear magnetic resonance spectroscopy (Peez et al. 2019).

FTIR spectroscopy allows identifying single particles of a sample by absorbing IR radiation in reflection or transmission mode. Each infrared spectrum of a sample is compared to a reference spectrum and can thus be identified (Löder and Gerdts 2015). Selected particles >500μm can be measured with an ATR-FTIR (attenuated total reflectance), smaller ones (>10μm) with a μFTIR which is very time-consumptive. μFTIR measurements in transmission mode are feasible with transparent filters (e.g. aluminum oxide) and thin samples. If water is present, particles are black or too thick (50–100μm), the IR radiation is totally absorbed and thus no identification is possible. In order to measure an entire filter, a focal plane array may be coupled to the FTIR which identifies automatically all particles present.

Raman spectroscopy detects the molecular composition by investigating its vibrational and rotational frequencies with lasers and wavelengths from 500 nm to 1064 nm. The spectral fingerprint of a chemical structure allows the determination of the specific polymer. The basis of a Raman is the Raman shift which is the difference of Rayleigh photons and inelastically scattered photons (McNesby and Pesce-Rodriguez 2006; Kuhar et al. 2018). Such as the μFTIR, μRaman is used for measuring small particles (>1μm) and is also time-consumptive when analyzing singe particles. Raman and FTIR are complementary methods as the signals are stronger in Raman and weaker in IR. The advantage of Raman spectroscopy is that water does not influence the measurement and that a higher depth can be reached, i.e. more particles can be measured and identified. However, samples need to be well prepared as organic/inorganic particles may lead to fluorescence influencing the spectra and Raman measurements are still time-consuming (Dris et al. 2015).

Pyrolysis gas chromatography mass spectrometry renders qualitative information about one single particle or quantitative information by calculating the mass of selected plastic polymers in environmental matrices. Moreover, plastic additives might be measured. Molecules are thermally degraded to determine their chemical composition by using a gas chromatograph for separation and a mass spectrometer for identification (Käppler et al. 2018; Dierkes et al. 2019). The measured pyrograms are compared to a library with the most common polymer types. The disadvantage of pyrolysis gas chromatography mass spectrometry is that particles can only be used once, as they are destroyed during the measurement.

In summary, microplastics are small plastic/polymer fragments in debris of various sizes that have accumulated in the global environment, particularly in all major marine habitats from the poles to the equator, from the sea surface and coastline to the depths. Techniques and methodologies for evaluating the detection of microplastics in aqueous environments are available. However, the detection and quantification of plastics in environmental matrices is currently an analytical challenge due to the laborious and varied analytical procedures currently used (Barboza et al. 2018). Indeed, the comparability of data on microplastics is currently hampered by a wide variety of different methodologies, resulting in the production of data of widely differing quality and resolution. Recently, da Costa et al. (2019) and Prata et al. (2019) also highlighted the difficulty comparison of data obtained in different studies due to the lack of standardized sampling protocols, methodologies and techniques for their identification and quantification, because of the many different characteristics of polymers. A critical evaluation of the methodologies used for the detection and identification of microplastics is necessary. Environmental research on these materials is also not consensual and definitions are unclear (Barboza et al. 2018). The classification of plastics is not a trivial issue. Recently, the term nanoplastics (<100 nm) has been introduced but there is no universally agreed definition. In addition, there are no established standard methods yet to accurately establish their ecotoxicological effects. This aspect is also of interest to the scientific community.

1.4 Emerging Contaminants in the Environment

1.4.1 Occurrence, Fate and Toxicity of Alkylphenols in the Aquatic Environment

Alkylphenols are a group of chemicals used in the production of surfactants, resins, antioxidants, pesticides, construction materials and automobile supplies (Ying et al. 2002; Lamprea et al. 2018). Nonylphenol ethoxylates and octylphenol ethoxylates are the most common non-ionic surfactants. When nonylphenol ethoxylates and octylphenol ethoxylates are released to wastewater treatment plants and environmental matrices, they are subjected to incomplete biodegradation, leading to the formation of nonylphenol and octylphenol (Fairbairn et al. 2016; Xu et al. 2016). Nonylphenol and octylphenol can also be leached from concrete, elastomeric bitumen, and tires in runoff (Lamprea et al. 2018). Short-chain alkylphenols, such as 4-*tert*-butylphenol and 4-*tert*-pentylphenol, are often present in building materials such as paints, coating, and adhesives (Janousek et al. 2020). Some alkylphenols, *e.g.*, 6-di-*tert*-butyl-4-methylphenol, are also used as antioxidants in food and cosmetics (Lu et al. 2019). Produced water discharges from offshore oil and gas production can be an important source for some alkylphenols (C1-C3) in the marine environment (Lofthus et al. 2016). Alkylphenols are known as environmental endocrine disruptors, and their levels in some areas may exceed water quality criteria (Zhang et al. 2017). Alkylphenols are therefore recognized as emerging contaminants of concern and, in the European Water Framework Directive (2013/39/EU), nonylphenols (in particular 4-*para*-nonylphenol) and octylphenols (4-*tert*-octylphenol) are classified as priority substances.

Occurrence Nonylphenol and octylphenol are frequently detected in the aquatic environment worldwide due to their widespread use (Table 1.2). It should be noted that nonylphenol or technical nonylphenol is a mixture of alkylphenols with different carbon chain lengths, branching, and substitution positions. Technical nonylphenol consists mostly of 4-nonylphenol, which is theoretically composed of 211 isomers, and 4-*n*-nonylphenol is the linear isomer. Similarly, 4-*tert*-octylphenol is one of the isomers of octylphenol. Most of the recent reports on nonylphenol and octylphenol are from China (Table 1.2). The highest levels have been reported from Guangzhou, the largest city in Pearl River Delta, where maximum concentrations of nonylphenol in river water and sediment were 5050 ng/L and 14,400 ng/g, respectively (Xu et al. 2018). Levels of nonylphenol and octylphenol in other Asian countries are also high, with nonylphenol as high as 1100 ng/L in seawater in Tokyo Bay, Japan (Song et al. 2020). Levels of nonylphenol and octylphenol in European countries are much lower, with the highest concentration reported in pond sediments in Greece being 323 ng/g for nonylphenol (Bokony et al. 2018). Limited data are available for countries in Africa and North and South America (Table 1.2). The presence of other alkylphenols has hardly been reported. Liu et al. (2016) reported levels of 4-*tert*-butylphenol, 2,4-di-*tert*-pentylphenol, 4-*n*-heptylphenol, 4-butylphenol, and 4-*n*-hexylphenol in surface waters, suspended particulates, and sediment in Lake Taihu and its tributaries in China, and their concentrations were lower than those of nonylphenol and octylphenol. Saeed et al. (2017) reported C4-C9 alkylphenols in coastal waters and sediments in Kuwait, with concentrations of ng/L and subng/L in water and ng/g and subng/g in sediments, respectively.

Environmental Fate Adsorption of alkylphenol to solid particles (sediment and suspended solids) and organic matter is a critical process that dominates their fate, transport, bioavailability and toxicity in the aquatic environment. Hu et al. (2019) reported that log K_{OC} values of nonylphenol and octylphenol calculated under field conditions were 4.63–4.78 and 4.74–4.84, respectively, indicating their high affinity

Table 1.2 Occurrence of nonylphenol and octylphenol in aquatic environment (concentration in water in ng/L; concentration in sediment in ng/g dry weight; detection frequency in %)

Region, Country	Matrix	Compound	Concentration range (mean)	Sample number	Detection frequency	References
Yinma River Basin, China	River water	4-Nonylphenol	14.13–392.41 (71.77)	51	100%	Sun et al. (2019)
Yinma River Basin, China	River sediment	4-Nonylphenol	1.33–98.44 (30.68)	51	100%	Sun et al. (2019)
Bahe River, China	River water	4-*tert*-Octylphenol	<1.2–126.0 (47.43)	10	100%	Wang et al. (2018b)
Bahe River, China	River water	Nonylphenol	108.1–634.8 (252.57)	10	100%	Wang et al. (2018b)
Three Gorges Reservoir Region, China	River water	Nonylphenol	<1.7–33.3 (10.8)	14	71.4%	Wang et al. (2016c)
Three Gorges Reservoir Region, China	River water	Octylphenol	<0.3–120 (32.3)	14	71.4%	Wang et al. (2016c)
Three Gorges Reservoir Region, China	River sediment	Nonylphenol	<1.5–8.5 (5.0)	14	85.7%	Wang et al. (2016c)
Three Gorges Reservoir Region, China	River sediment	Octylphenol	<0.5–12.4 (5.30)	14	92.9%	Wang et al. (2016c)
Xiangjiang River, China	River water	4-*tert*-Octylphenol	<0.27–156.1 (16.28)	160	72.2%	Luo et al. (2019)
Xiangjiang River, China	River water	4-Octylphenol	<0.67–40.8 (7.3)	160	69.5%	Luo et al. (2019)
Xiangjiang River, China	River water	4-Nonylphenol	<0.79–12.9 (2.65)	160	74%	Luo et al. (2019)
Zunyi, Guizhou province, China	River water	Nonylphenol	174–3411 (1730)	6	100%	Yu et al. (2017)
Pearl River networks, China	River water	4-Nonylphenol	27.20–824.18 (219.33)	31	100%	Xie et al. (2020)

(continued)

Table 1.2 (continued)

Region, Country	Matrix	Compound	Concentration range (mean)	Sample number	Detection frequency	References
Pearl River networks, China	River water	Octylphenols	1.11–57.19 (13.64)	31	100%	Xie et al. (2020)
Pearl River networks, China	River sediment	4-Nonylphenol	128.59–3831.95 (1248.39)	19	100%	Xie et al. (2020)
Pearl River networks, China	River sediment	Octylphenol	4.59–79.71 (32.17)	19	100%	Xie et al. (2020)
Hong Kong, China	Sea water	Nonylphenol	76.5–1353.6 (342.3)	24	100%	Xu et al. (2018)
Hong Kong, China	Suspended particles	Nonylphenol	97.7–1492.5 (365.1)	24	100%	Xu et al. (2018)
Hong Kong, China	Marine sediment	Nonylphenol	114.6–2665 (579.0)	24	100%	Xu et al. (2018)
Guangzhou, China	River water	4-Nonylphenol	22.5–5050 (1300)	22	100%	Peng et al. (2017)
Guangzhou, China	River water	4-*tert*-Octylphenol	<0.42–165 (50.3)	22	92%	Peng et al. (2017)
Guangzhou, China	River sediment	4-Nonylphenol	10.9–14,400 (4100)	22	100%	Peng et al. (2017)
Guangzhou, China	River sediment	4-*tert*-Octylphenol	<0.3–261 (64.3)	22	79%	Peng et al. (2017)
Tainan Canal, China	Marine sediment	Nonylphenol	1.16–324 (20.34)	69	93.9%	Shiu et al. (2019)
Tainan Canal, China	Marine sediment	Octylphenol	0.36–55.81 (3.99)	10	22.5%	Shiu et al. (2019)
Kenting National Park, Taiwan	Sea water	Nonylphenol	0.04–26.5 (7.69)	10	100%	
Kenting National Park, Taiwan	Sea water	Octylphenol	<0.0024–113 (15.4)	125	20%	Kung et al. (2018)
The Yellow and Bohai seas, South Korea and China	Marine sediment	4-*tert*-Octylphenol	<0.10–2.33 (0.30)	125	56.8%	Yoon et al. (2020)

(continued)

Table 1.2 (continued)

Region, Country	Matrix	Compound	Concentration range (mean)	Sample number	Detection frequency	References
The Yellow and Bohai seas, South Korea and China	Marine sediment	Nonylphenol	<0.91–111.15 (4.55)	125	38.4%	Yoon et al. (2020)
Tokyo Bay and Nagasaki Bay, Japan	Sea water	4-*tert*-Octylphenol	4.2–290.0 (61.3)	6	100%	Song et al. (2020)
Tokyo Bay and Nagasaki Bay, Japan	Sea water	4-Nonylphenol	37–1100 (410)	42	100%	Song et al. (2020)
Nationwide, Vietnam	River water	4-*tert*-Octylphenol	<10–850	42	47.6%	Chau et al. (2018)
Sai Gon and Dong Nai river basin, Vietnam	River water	Octylphenol	<2.0–35 (12.26)	14	N/A	Minh et al. (2016)
Villages in Delta and Anambra state, Nigeria	Groundwater	NP	100–90,300 (34479.8)	5	100%	Onyekwere et al. (2019)
Vojvodina Province, Serbia	Surface water	4-*tert*-Octylphenol	6–15 (9)	18	72%	Skrbic et al. (2018)
Nationwide, Hungary	Pond sediment	Nonylphenol	70–323 (165)	12	100%	Bokony et al. (2018)
Strymonas River, Greece	Surface water	Nonylphenol	<10–119 (17)	18	78%	Tousova et al. (2017)
Strymonas River, Greece	Surface water	Octylphenol	<3–28	18	22%	Tousova et al. (2017)
Marina del Este beach, Spain	Marine sediment	Nonylphenol	146–340 (237)	9	100%	Martin et al. (2017)
Malopolska, Poland	River sediment	4-*tert*-Octylphenol	2.9–8.9 (3.87)	25	100%	Czech et al. (2016)
Po River delta, Italy	Lake sediment	Octylphenol	0.44–3.97 (1.73)	6	100%	Casatta et al. (2016)

(continued)

Table 1.2 (continued)

Region, Country	Matrix	Compound	Concentration range (mean)	Sample number	Detection frequency	References
Po River delta, Italy	Lake sediment	4-Octylphenol	0.13–1.43 (0.68)	6	100%	Casatta et al. (2016)
Po River delta, Italy	Lake sediment	Nonylphenol	19.44–202.27 (82.19)	6	100%	Casatta et al. (2016)
Po River delta, Italy	Lake sediment	4-Nonylphenol	<0.12–1.24 (0.62)	6	50%	Casatta et al. (2016)
Nationwide, Mexico	Recreational water	4-Nonylphenol	0.83–12.61 (3.61)	10	83%	Vargas-Berrones et al. (2020a)
Cuautla River, Mexico	Surface water	4-Nonylphenol	1.23–44.77 (7.53)	9	44%	Calderon-Moreno et al. (2019)
Cuautla River, Mexico	Surface water	4-*tert*-Octylphenol	0.30–28.10 (11.24)	9	100%	Calderon-Moreno et al. (2019)
São Paulo State, Brazil	Surface water	4-*n*-Nonylphenol	1–2018 (429)	205	2%	Montagner et al. (2019)
São Paulo State, Brazil	Surface water	4-*n*-Octylphenol	2–1029 (266)	205	2%	Montagner et al. (2019)
Santos Bay, Brazil	Marine sediment	4-Nonylphenol	<10–72.5 (32.2)	28	100%	Dos Santos et al. (2018)

for organic matter. Gong et al. (2016) found that in the Pearl River system in South China, log K_{OC} values for nonylphenol in suspended particulate matter and sediment were 5.05 ± 0.33 and 4.59 ± 0.26, respectively. The corresponding values for octylphenol were 4.89 ± 0.41 and 4.32 ± 0.49, respectively. Biodegradation is another major factor in the fate of alkylphenols in the aquatic environment. Fate modeling in the Yong River in China indicated that the contribution of adsorption was greater than that of biodegradation, while the opposite was true for 4-t-octylphenol (Cheng et al. 2018). The half-lives for nonylphenol in estuarine waters alone and estuarine waters with sediment would be 2.36 ± 1.76 and 1.31 ± 0.51 d (Yang et al. 2016). Another study showed that the half-lives of nonylphenol in the water/sediment system under several redox conditions, ranging from aerobic to sulfate reduction, ranged from 3.2 to 6.2 d (Koumaki et al. 2018). A variety of bacteria, fungi and microalgae have been found to degrade alkylphenols. The consortium of cyanobacteria *Arthrospira Maxima* and the green microalgae *Chlorella vulgaris* can remove

96% of 4-nonylphenol from water in 96 h (Zaytseva and Medvedeva 2019). Other microalgae can also remove nonylphenol to varying degrees (He et al. 2016). Fungi of the genera *Aspergillus* and *Penicillium* have been found to effectively degrade 4-*n*-nonylphenol within a few days (Zhang et al. 2016a; Yang et al. 2018). It should be noted that the biodegradation of nonylphenol is isomer-specific and affected greatly by the redox potential; for example, the half-lives of 19 nonylphenol isomers in oxic river sediments ranged from 0.9 to 13.2 d (Lu and Gan 2014a). Biodegradation pathways of 4-nonylphenol include *ipso*-hydroxylation in the *para* position, nitrification at the *ortho*-position, and oxidation of the alkyl-chain, in particular ω-oxidation and then β-oxidation of the nonyl chain when fungi are involved. However, the routes of action of 4-*n*-nonylphenol are different. One pathway is initiated by hydroxylation at the terminal β-carbon atom, followed by oxidation and decarboxylation. Another pathway is initiated by hydroxylation at the *ortho*-positions and followed by ring-opening reactions (Zhang et al. 2016a; Yang et al. 2018).

Toxicity The estrogenicity of nonylphenol and octylphenol has been known for decades and has been confirmed in fish (Ahmadpanah et al. 2019; Sayed et al. 2019), aquatic invertebrates (Hart et al. 2016; Guo et al. 2019) and plants (Cahyanurani et al. 2017; Cheng et al. 2019). Nonylphenol is hepatotoxic (Abd-Elkareem et al. 2018), hemotoxic (Ahmadpanah et al. 2019), histopathological (Shirdel and Kalbassi 2016), genotoxic (Sharma and Chadha 2017a, b; Sayed et al. 2018), immunotoxic (Sayed et al. 2019), and reproductive toxic (Watanabe et al. 2017; Saravanan et al. 2019) to fish. In an extended one-generation medaka test, when 3 generations of Japanese medaka (*Oryzias latipes*) were exposed to 1.27–89.4μg/L of 4-nonylphenol, the parental generation was not affected while the fecundity, fertile egg and the fertility of the first generation decreased. However, 4-nonylphenol did not affect the survival and hatching of the second generation (Watanabe et al. 2017). Vazquez et al. (2016) also reported that the reproductive capacity of *Cichlasoma dimerus* caused by 4-t-octylphenol could be recovered when transferred to octylphenol-free water. Exposure to nonylphenol may increase vitellogenin and vitellogenin receptor expression in the shrimp *Macrobrachium rosenbergii* (Guo et al. 2019), weaken the immune system of the Pacific oyster *Crassostrea gigas* (Hart et al. 2016), reduce the biomass of mussels (*Lampsilis siliquoidea* and *Villosa iris*) (Ivey et al. 2018), and disrupt lipid storage (Jordao et al. 2016), moulting, and reproduction of the water flea *Daphnia magna* (Kim et al. 2019). Nonylphenol and octylphenol can also induce oxidative responses in algae such as *Gracilaria lemaneiformis* (Zhong et al. 2017) and *Dictyosphaerium sp.* (Cheng et al. 2019) and the submersed macrophyte *Ceratophyllum demersum* (Cahyanurani et al. 2017). The Yeast Estrogen Screen Test, the Two Hybrid Yeast Test, the MVLN Cell Test and the E-Screen Test all showed that the estrogenicity of nonylphenol was isomer-specific (Lu and Gan 2014a). However, no studies were conducted in aquatic species. The co-existence of other pollutants may increase or decrease the toxicity of alkylphenols to aquatic species. The equitoxic mixture of copper and nonylphenol was antagonistic to the embryos and larvae of *Rhinella* *arenarum* toad at LC50 (Aronzon et al. 2020). Carnevali et al. (2017) demonstrated that dietary intake of bisphenol A, octylphenol, and nonylphenol by juvenile sea bream was antagonistic to its lipid metabolism. However, Cd(II) enhanced growth inhibition and superoxide induction in the alga *Chlorella sorokiniana* (Wang et al. 2018a). All these results suggest that further research is needed on the co-toxicity of nonylphenol and other pollutants.

1.4.2 Environmental Impact of Rare Earth Elements

Rare earth elements (REEs) are a group of 17 elements (fifteen lanthanides as well as yttrium and scandium) that tend to occur in the same ore deposits in Nature and exhibit similar physical and chemical properties. One of the members of the lanthanide series, promethium (Pm) which has only radioactive isotopes is usually recovered from the leftovers of uranium fission. Rare earth elements occur at trace concentration levels in the upper continental crust with total content ranging from <1 to 65μg/g (Table 1.3). These are highly electropositive and normally exists predominantly in trivalent stage (M^{3+}), with the exception of cerium which exists in quadrivalent valent state (Ce^{4+}) and europium which exists in divalent state (Eu^{2+}) under oxidizing and reducing environments respectively (Balaram 1996). Their unique electronic, optical, magnetic, and catalytic properties give them the ability to donate and accept electrons, making them suitable for integral components of many hi-tech products such as mobile phones, computers, home entertainment, medical devices, military defense systems, autocatalytic converters, petroleum refining and for polishing of optical-quality glass, and emerging clean energies. Thus, their industrial applications can be divided mainly into catalysts, ceramics, metallurgy, polishing and other applications (Fig. 1.5). A number of reviews has been published in recent times on all aspects of rare earth elements including their applications, occurrence, exploration, and their chemical analysis (*e.g.*, Mubashir 2018; Balaram 2019). The extensive applications of these elements have led to their wide distribution in all environmental compartments, raising concerns over their potential effects on human health. Because of well-known toxic effects of elements such as As, Cd, Pb, Hg and U, most of the environmental toxicity studies so far have been confined to this set of elements only (*e.g.*, Sparks 2005; Reddy et al. 2012; Duggal et al. 2017). Even platinum group elements have received considerable attention in recent years for their impact on human health due to their extensive utilization in autocatalysts (Balaram 2020). But in recent times, technological innovations coupled to modern living conditions have been contributing to the enhancing intake of significant quantities of rare earth elements by humans. Several studies demonstrated that these elements have toxic effects on human health and therefore are considered as emerging pollutants (*e.g.*, Pagano et al. 2015; Rim 2016). These elements can accumulate in the different environmental compartments and also in the biota, by many different routes such as mining, mineral processing, e-waste, agriculture practices, medicines and rudimentary recycling and disposal practices.

Table 1.3 Rare earth element (REE) concentrations (in μg/g) in different environmental compartments in comparison to the average concentrations of the upper continental crust

Rare earth element	Soils of ex-mining area[a]	Plant species[b] – *Dicranopteris dichotoma*			Fertilizer enriched with REE[c]	Runoff water from traffic area[d], Beijing, China (ng/mL)	Background concentrations in surface water[5] (ng/mL)	Background concentrations in ground water[e] (ng/mL)	Maximum permissible concentrations (ng/mL)	UCC[f]
		Leaves	Shoot	Root						
La	94.69	1568.00	59.39	1305.07	15,400	5.56	<0.08	<0.08	10.1	31
Ce	420.5	2290.33	738.63	1482.60	24,100	10.4	<0.13	0.956	22.1	63
Pr	33.79	149.77	48.53	566.70	11,800	1.40	<0.08	0.125	9.1	7.1
Nd	112.3	1499.00	112.23	687.40	1100	4.04	<0.39	<0.39	1.8	27
Sm	–	–	–	–	2000	0.97	<0.56	<0.56	8.2	4.7
Eu	59.84	29.32	12.53	161.55	200	0.25	<0.17	<0.17	–	1.0
Gd	42.26	75.08	47.49	157.63	1100	0.91	<0.33	0.198	7.1	4.0
Tb	42.26	85.37	45.85	98.40	25.8	0.12	<0.10	<0.10	–	0.7
Dy	18.66	47.78	17.11	166.30	91.6	0.57	<0.22	0.069	9.3	3.9
Ho	12.12	13.24	6.39	59.39	4.3	0.13	<0.09	<0.09	–	0.83
Er	43.5	13.24	6.56	34.58	26.9	0.011	<0.14	0.072	–	2.3
Tm	10.15	8.37	13.50	17.57	1.4	0.07	<0.07	0.175	–	0.30
Yb	38.23	18.49	15.45	151.56	5.3	0.43	<0.13	<0.13	–	1.96
Lu	9.33	7.48	6.18	16.10	0.5	0.10	<0.05	<0.05	–	0.31
Sc	39.89	33.16	33.40	177.83	–	–	<2.80	0.001	–	14.0
Y	79.2	219.07	57.70	111.65	–	3.12	<0.22	0.416	6.4	21

[a]Soils of ex-mining area (representative) (Khan et al. 2017)
[b]Plant species, *Dicranopteris dichotoma* (Khan et al. 2017)
[c]Fertilizer enriched with rare earth elements (Hu et al. 2006)
[d]Runoff water from traffic area, Beijing, China (Shajib et al. 2020)
[e]Background concentrations in surface/ground water and maximum permissible concentrations (Sneller et al. 2000)
[f]UCC – recommended average concentrations of rare earth elements in the upper continental crust (Rudnick and Gao 2003)

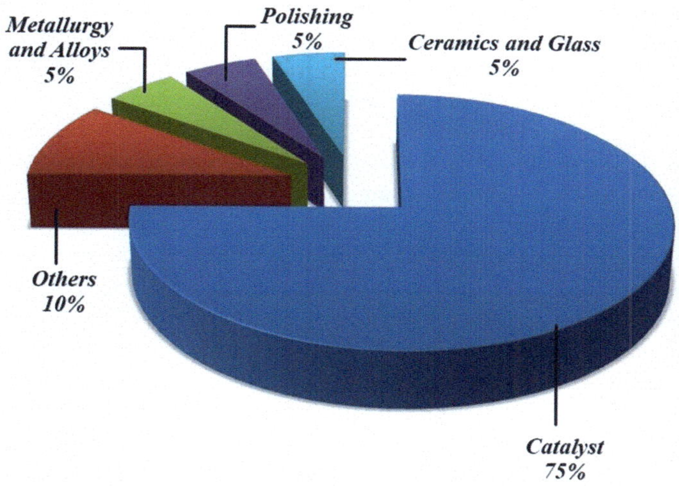

Fig. 1.5 The estimated current end use of rare earth elements. (Adapted from U.S. Geological Survey, USGS 2020, Reston, Virginia)

Impact of Rare Earth Elements Mining Though its more than a century since commercial mining of rare earth elements started, their global production and consumption have seen a significant increase during last half century, thus currently posing mounting environmental risk. Intense rare earth elements mining and production activities have led to significant environmental and health impacts already in countries such as China, US, India, Malaysia and Brazil. There are a number of potential environmental implications to mining of rare earth elements if not done properly. Mining activities such as cutting, drilling, blasting, transportation, stockpiling, and processing can release dust containing rare earth elements, other toxic metals and chemicals into air and surrounding water bodies that can impact local soil, vegetation, wildlife, and humans. Some of the rare earth element minerals contain significant amounts of radioactive elements such as uranium and thorium, which can contaminate air, soil, surface water and groundwater (Balaram 2019). According to Hurst (2010), approximately one ton of calcined rare earth ore generates 9600–12,000 cubic meters of waste gas containing dust concentrate, hydrofluoric acid, sulfur dioxide, and sulfuric acid, approximately 75 cubic meters of acidic wastewater, and about one ton of radioactive waste residue containing water. Most of that wastewater is normally discharged without being effectively treated, which can not only contaminate potable water for daily living of the neighboring communities, but also contaminates the surrounding water environment and irrigated farmlands. For example, an urban street dust of an industrial city, Zhuzhou in central China recorded very significant concentrations of rare earth elements (ΣREEs ranged from 66.1 to 237.4μg/g with an average of 115.9μg/g) which reveals the gravity of the rare earth elements pollution, particularly in industrial cities (Sun et al. 2017). More mining of rare earth elements, however, will mean more environmental degradation and human health hazards as waste disposal areas can be exposed to weathering conditions and have the potential to pollute the air, soil and water if adequate monitoring and protection measures are not applied (Barakos et al. 2018). These problems are due to insufficient environmental regulations and controls in the mining and processing areas in different countries.

Rare Earth Elements in e-Waste Toxic metals are increasingly present in water systems around the World due to human activities, such as unregulated mining and more recently, the inadequate disposal of electronic waste. Dumping of huge amounts of e-waste that facilitate the release of significant quantities of rare earth elements along with several other toxic elements is leading to severe contamination of the subsoils and ground water (Haxel et al. 2002). Each year, the electronics industry generates up to 50 million tons of e-waste, but as the number of consumers rises, and the lifespan of devices shrinks in response to demand for the updated equipment, this figure could cross 50 million tons in the coming years. Rare earth elements are known to be more mobile in solutions rich in F^-, Cl^-, HCO_3^-, CO_3^{2-}, HPO_4^{2-}, PO_4^{3-} ions (Xiong and Yusheng 1991). In addition, large quantities of rare earth elements are also getting into the agricultural soils through the phosphate-based fertilizers (Table 1.3). Under natural conditions, rare earth elements may become available only in small amounts via the groundwater and the atmosphere; however, their increased use and enhanced amounts have created several new routes for bioaccumulation (in plants, animals, and human beings). Background level of rare earth elements content in waters, both surface and subsoil, varies significantly and depends mostly on the local geology.

Recycling of Rare Earth Elements Recycling and recovering activities of the valuable rare earth elements found in advanced electronic gadgets such as mobile phones, smart television sets, magnets, computers, batteries and catalysts, have been a challenge for both manufacturers and environmental groups. Several research groups are involved in the developments of technologies that could recover these materials for reuse (Hasegawa et al. 2018; Jowitt et al. 2018). The intensity of this activity is driven by the demand for these elements and its future potential and sustainability requires a significant amount of research on these aspects. One of the biggest electronic companies, Apple has already started expanding the company's global recycling programs (Balaram 2019). Increasing the amount of rare earth elements recycling may contribute to overcoming some of the criticality issues related to the supply of these elements. However, recycling activities such as the treatment of raw materials, beneficiation, separation of rare earth elements as oxides and their purification can also contribute to the release of substantial amounts of these metals into the environment.

Adverse Effects of the Use of Rare Earth Elements in Agriculture Rare earth elements containing micro fertilizers are extensively utilized in agriculture because of the striking positive effects in terms of crop yield and quality (Hu et al. 2006), therefore leading to further increase in the concentrations of these elements in soils (Guo et al. 1988; Diatloff et al. 1995). Although several findings reported positive effects of rare earth elements on plant growth, many questions about their biological role remain unclear. Their concentrations in vegetables produced from mining area are higher and can have health implications especially in children (Zhuang et al. 2017). In general, the mineral fertilizers (*i.e.*, phosphate fertilizers) and soil conditioners contain macronutrients (Ca, Mg, N, P, and S), micronutrients (such as Fe and Si) and rare earth elements. In soils, the rare earth elements can also originate from local geological parent materials (Liu 1988). Although the accumulated concentrations of rare earth elements have been reported to be very low, the accumulation capacity of a particular plant depends upon several factors such as plant species, their growing conditions and the rare earth elements content in the substrate soil or rock (Fu et al. 2001). Table 1.3 presents the concentrations of rare earth element in different parts of plants such as roots, stem and leaves of a particular plant species, which can give an idea on how these elements can reach biota. Following excessive application of rare earth elements in agriculture, there is a raising environmental concern that these elements may enter the food chain in greater quantities though plant uptake, which might be deleterious to human health. For instance, the studies conducted by Redling (2006) confirmed very low concentrations of rare earth elements in cereal grains and no significant accumulation due to rare earth elements fertilization. Thus, grains and products made of them such as wheat flour are considered to be safe. Thomas et al. (2014) have indicated that countries like Russia and Nigeria where the natural abundance levels of rare earth elements are high in their soils will have more environmental threats arising from the increased input of these elements. However, close monitoring may be needed in countries where phosphate-based fertilizers (mined from monazite deposits) are applied in large scale.

Rare Earth Elements in Medicine Their unique properties, such as radiation emission and magnetism, allow these elements to be used in many health and medical applications, such as in anti-tumor agent, kidney dialysis medicine and surgical equipment (Giese 2018). Though, there are currently a few major applications of rare earth elements in medicine and many others are on the horizon (Lu et al. 2017). Lanthanide compounds with outstanding stability (photo- and redox-stability) and luminescent properties have found wider applications in cancer therapy and imaging (Teo et al. 2016). For example, Gadolinium has been used in a chelated form as a contrast agent in magnetic resonance imaging measurements (Raju et al. 2010). Even the pharmaceutical samples analyzed for inorganic impurities are found to contain appreciable amounts of rare earth elements when analyzed by inductively coupled plasma mass spectrometry (ICP-MS) (Balaram 2016). The implications of the presence of this range of rare earth elements concentrations in pharmaceuticals, are not clear at present. On one hand, rare earth elements have been found to disease-causing and occupational poisoning of local residents in mining areas, water pollution, and farmland destruction, etc. A body of evidence; has shown that rare earth elements are associated with antioxidant properties that are used in the treatment of many diseases (Rim 2016). Several studies (Zhang et al. 2000; Wakabayashi et al. 2016) confirmed the antibacterial and antifungal activities of rare earth elements that are comparable to those of copper ions, and thus these elements are starting to be used for several pharmaceutical applications. Rare earth elements can also be used as nematicide as they can also inhibit the formation and germination of fungal spores and thus influence large number of organisms (Zhang et al. 2000). Recently, the medical and biological properties of rare earth elements have been reviewed by Panichev (2015). New medical applications for these elements are being found at an increasing rate and emerging advancements such as nanotechnology might be used to enhance their use in medicine in the future.

Health Effects to Humans Extensive mining activities and dumping of huge amounts of e-waste are facilitating release of significant quantities rare earth elements along with several other toxic elements and radionuclides into the subsoils and ground water. In addition, the largescale application of rare earth elements in various modern technologies is continually growing despite the knowledge about the environmental concerns of these elements. In addition, there is also a widespread use of these elements in agriculture and medicine sectors (Xiong and Yusheng 1991). As a result, large quantities of rare earth elements are also getting migrated into the agricultural soils, groundwater and thereby food products. Unfortunately, there is no sufficient data available about the toxicity of rare earth elements to human health and thus, their maximum acceptable limits in drinking water were not provided yet by any international health organization. Sneller et al. (2000) have reported maximum permissible concentrations for different rare earth elements in drinking water and the validity of these values is not known (Table 1.3). Currently, there are substantial gaps in our understanding of the adverse effects of rare earth elements to human health, their anthropogenic levels and fate, their biogeochemical or anthropogenic cycling, and their individual and additive toxicological effects. Most of the harmful effects resulting from human exposure to rare earth elements exposure are reported in occupational studies of mine workers and others who regularly deal with rare earth elements or their products, where exposure is typically much higher than that which the general population would experience. There are also several reports of occupational exposure to rare earth elements that resulted in bioaccumulation and adverse effects to human health (McDonald et al. 2015; Yoon et al. 2005; Giovanni 2016; Rim 2016). More studies are required to identify various anthropogenic sources, transfer mechanisms, bioaccumulation and their environmental behavior to minimize human health risks in future.

Despite substantial increase in the number of studies, many questions about the environmental and biological effects of rare earth elements remain unanswered. As the use of rare earth elements in industrial, agricultural and health applications is expected to increase to a great extent in the future, risk assessment studies on the daily intake of rare earth elements through dust, air, water and food, and on the subsequent long-term health effects on humans including pregnant women, children, and other vulnerable populations, have to be speeded up. In order to fill in the current gaps in our understanding of the roles of different rare earth elements, multifaceted studies focusing on different organisms and exposure routes are needed. As the mining, processing and recycling of rare earth elements will be imperative in the future in order to meet the demand for these elements, it is necessary to formulate verified guidelines and monitoring systems to guarantee the welfare of workers, but also to protect the environment in general.

1.4.3 Engineered Nanoparticles in Drinking Waters Sources

In the last decade, nanotechnology has brought a large number of nanoparticles, engineered nanoparticles and nanomaterials to applications in multiple daily products and for many industrial products, due to their specific physicochemical properties (Mueller and Nowack 2008; Gottschalk and Nowack 2011). For example, titanium dioxide and zinc oxide nanoparticles can be found in sunscreens and cosmetics (Lu et al. 2015); titanium dioxide, silver, and silicon dioxide in paints and coatings (Hanus and Harris 2013); silver nanoparticles in textiles due to their antimicrobial properties (Hardin and Kim 2016); cerium oxide nanoparticles as fuel additives (Dale et al. 2017); C60-fullerenes in dermatological and skin care products (Mousavi et al. 2017); or carbon nanotubes as nanofertilizers (Iavicoli et al. 2017). This increase in use of engineered nanoparticles ends up in waste streams, and consequently in the environment and drinking water resources. The potential human risks from engineered nanoparticles in drinking water are still unknown and uncertain, due to the limited data on the occurrence and toxicity of nanoparticles in waters (Good et al. 2016; Bhattacharyya et al. 2017a, b). Potential impacts on the environment and human health have been reported (Born et al. 2006; Biskos and Schmidt-Ott 2012).

According to Gottschalk and Nowack (2011), engineered nanoparticles may enter into the environment through production, incorporation of nanoparticles into products, the use of nanoparticles containing products, and through wastewater treatment plants, waste incineration, landfills, and recycling and disposal processes. In addition, during their life cycle, the characteristics of engineered nanoparticles can change (due to physical/chemical interactions such as agglomeration, dissolution), as well as aging/degradation of the nanoparticles or nanoparticles containing products (Gottschalk and Nowack 2011). Surface water, such as lakes, rivers and streams, may therefore receive engineered nanoparticles, which may represent a threat to human health if this water is used for human consumption. In developed countries, 80% of the water supply is from surface water, while smaller communities or private residences are supplied by groundwater (Westerhoff et al. 2018).

Occurrence in Surface Waters Surface waters have many sources of engineered nanoparticles, including natural, unintentional and intentional sources (Hochella et al. 2008; Park et al. 2017). Natural sources such as mineral weathering, sea spray, volcanoes and forest fires also include nanoparticles beyond other particle sizes (Hochella et al. 2008). Unintentional release usually includes non-point sources and release from products such as cosmetics (*e.g.*, sunscreens, Weir et al. 2012), wearing or washing textiles containing engineered nanoparticles (Gottschalk and Nowack 2011), paints (Mueller and Nowack 2008), urban runoff (Baalousha et al. 2016), or combustion, wear and corrosion processes (Westerhoff et al. 2018). However, major incidental engineered nanoparticles releases into the environment occur via wastewater treatment for cosmetics, coatings, cleaning agents, and dietary supplements (Gottschalk et al. 2009). Intentional sources are usually related with the application of the engineered nanoparticles directly, for example, when used in drinking water or wastewater treatments (Brar et al. 2010), in remediation of contaminated sites because of their sorption capacities (Nowack and Bucheli 2007) or applied as agrochemicals and nanopesticides (Kah and Hofmann 2014; Iavicoli et al. 2017). Groundwater may also receive engineered nanoparticles through direct use, *e.g.*, nano zero-valent iron injected into groundwater contaminated with chlorinated solvents then directly pumped into the subsurface (Nowack and Bucheli 2007; Grieger et al. 2010). Carbon nanotubes are intentionally loaded in water and soil to detect, prevent or remove pollutants (Mauter and Elimelech 2008).

Once released into water, engineered nanoparticles can undergo various transformations, including degradation, agglomeration, dissolution and sedimentation, and their fate, mobility, and persistence in aquatic systems are largely dependent on these transformations and their stability (Nowack and Bucheli 2007). Thus, the significance of engineered nanoparticles to drinking water supplies depends upon their stability and mobility in water resources used for drinking water production, which is largely determined by engineered nanoparticles colloidal stability. Engineered nanoparticles have high surface energies, due to their high surface-to-volume ratios, so they tend to aggregate (Hu et al. 2010; Sousa and Teixeira 2013; Serrão Sousa et al. 2017). Furthermore, if repulsive electrostatic forces are dominant, an increase in ionic strength leads to reduced engineered nanoparticle stability and an increase in aggregation (Chen and Elimelech 2006; Li and Lenhart 2012; Sousa and Teixeira 2013). However, aggregation rate stops increasing with ionic strength because each collision leads to attachment and aggregation becomes diffusion-limited, at a particular electrolyte concentration known as the critical coagulation concentration (Troester et al. 2016). However, at Na^+ and Ca^{2+} concentrations typically found in surface waters, destabilization of engineered nanoparticles by Na^+ and Ca^{2+} is, respectively, too low or limited (Troester et al. 2016). The naturally occurring organic matter present in surface waters can interact with engineered nanoparticles and influence suspension stability. Organic matter surface coating of engineered nanoparticles influences the aggregation rate of particles by electrostatic stabilisation mechanisms and electrosteric repulsion (Chen and Elimelech 2007; Baalousha et al. 2008; Klaine et al. 2008; Sousa and Teixeira 2013). The extent of the stabilization is dependent of organic matter type and concentration (Chen and Elimelech 2007; Sousa and Teixeira 2013; Zhang et al. 2013). Baalousha (2009) referred that an increase of nanoparticle concentration enhances their aggregation, particularly at pH values close to the engineering nanoparticle isoelectric points. In addition, at pH equal to the engineering nanoparticle isoelectric point, particles have little or no charge, leading to a substantial weakening in the repulsive forces between nanoparticles and aggregation occurs (Baalousha 2009). For surface waters pHs, between 6.5 and 8.5, only limited electrostatic stabilization can be expected for bare cerium oxide and bare iron oxide nanoparticles, since their isoelectric points are reported within this range (Illés and Tombácz 2006; Baalousha et al. 2008; Zhang et al. 2008).

In a very simplified way, Hammes et al. (2013) evaluated the effects of the physicochemical properties of the water on the ultimate fate of nanoparticles and contaminant-colloid assemblies across Europe. These authors showed more stable conditions for contaminant-colloid assemblies in northern Europe, the alpine region and Austria, Scotland, Northern Ireland and on the north-western Iberian Peninsula. In addition, as sedimentation rates were expected to be low in these areas, they concluded that aquatic organisms may be exposed to higher to small sized nanoparticle compared to waters in Italy, southern Portugal and Spain, France, Germany, east England and eastern Europe where conditions to nanoparticle aggregate prevail.

Predicted and Measured Engineered Nanoparticles Concentrations in Surface Waters Data available on the occurrence and quantities of engineered nanoparticles in surface waters are scarce. The uncertainties about the input and production quantities, the multiple pathways for engineered nanoparticles entering into the environment, but mostly the absence of widespread adequate detection techniques for identifying low concentrations of engineered nanoparticles in natural waters contribute to these limited data (von der Kammer et al. 2012; Westerhoff et al. 2018). Rather models have been developed to predict engineered nanoparticles concentrations.

Gottschalk et al. (2013) reviewed published data on the distribution of engineered nanoparticles in surface waters, both modelled and experimental data. Troester et al. (2016) made an actualisation of that review. According to Gottschalk et al. (2013), modelled nano-titanium dioxide concentrations in rivers ranged from 0.003 to 1.6μg/L, at high temporal and geographical resolution, and this range included other modelled concentration ranges. Nano-silver presented high variations between the highest and lowest values, approximately a factor of $1x10^5$ for the difference, which might reflect the differences between initial estimates of nanosilver production (Gottschalk et al. 2013). Modelled nano-zinc oxide varied between *ca.* 1×10^{-4} and 1×10^{-1}μg/L, but only two studies have reported concentrations in aqueous systems; carbon nanotube varied between 1×10^{-5} and 1×10^{-3}μg/L; fullerenes ranged from 1×10^{-5} to 1×10^{-4} μg/L; and cerium oxide nanoparticles were in the range 1×10^{-2} to 1×10^{-1}μg/L (Gottschalk et al. 2013). More recent modelling studies revised in by Troester et al. (2016), showed concentrations ranging for titanium dioxide nanoparticles between 0.40 and 1.4 μg/L; for silver nanoparticles between 0.51 ng/L and 0.94 ng/L; for zinc oxide nanoparticles between 0.05 ng/L and 0.29μg/L; for C60 between 0.07 ng/L and 0.28 ng/L; and for carbon nanotubes between 0.17 ng/L and 0.35 ng/L for percentile 15 and 85%, respectively (Sun et al. 2014). In addition, some studies predicted engineered nanoparticles concentrations at different distances from emission source, showing a decrease in concentrations justified by hetero-aggregation and sedimentation of the engineered nanoparticles (Troester et al. 2016). Experimental studies revealed similar concentrations of nano-titanium dioxide from the ones predicted (Gottschalk et al. 2013; Troester et al. 2016). Table 1.4 presents an overview of recent published data on

Table 1.4 Overview of the measured engineered nanoparticle concentrations in surface waters (*SP-ICP-MS* single particle inductively coupled plasma mass spectrometry, *ICP-OES* inductively coupled plasma optical emission spectroscopy, *ET-AAS* electrothermal atomic absorption spectrometry)

Material	Location	Analytical technique	Results	References
Titanium dioxide	Clear Creek, Colorado collected hourly over a 72 h period	SP-ICP-MS	5 to 28 μgTi/L; metals naturally co-occurring with Ti (Al, Fe) also showed an increase during bathing periods, suggesting the Ti increase may be partially of sediment resuspension	Reed et al. (2017)
Silver	Dutch rivers	SP-ICP-MS56–60	Silver nanoparticles (average size 15 nm) at 0.3–2.5 ng/L	Peters et al. (2018)
Cerium oxide	Dutch rivers	SP-ICP-MS56–60	Cerium oxide nanoparticles (average size 19 nm) at 0.4–5.2 ng/L	Peters et al. (2018)
Titanium dioxide	Dutch rivers	SP-ICP-MS56–60	Titanium dioxide particles (300 nm) at 0.2–8.1μg/L	Peters et al. (2018)
Titane total	North East Italy	SP-ICP-MS	3.7–11.2 μgTi/L	Piccoli et al. (2018)
Cerium total	North East Italy	SP-ICP-MS	0.2–0.4 μgCe/L	Piccoli et al. (2018)
Zinc total	North East Italy	SP-ICP-MS	0.12–7.9 μgZn/L	Piccoli et al. (2018)
Silver total	North East Italy	SP-ICP-MS	0.64–0.80 μgAg/L	Piccoli et al. (2018)
Silver	Melana River, Malaysia	ICP-OES	0.13 ± 0.06 to 10.16 ± 1.32 mg/L	Syafiuddin et al. (2018)
Silver	Sekudai River, Malaysia	ICP-OES	0.10 ± 0.00 to 9.63 ± 0.80 mg/L	Syafiuddin et al. (2018)
Titanium dioxide	Salt River, Arizona	SP-ICP-MS	Maximum concentration of 659 ng/L	Venkatesan et al. (2018)
Silver-*b*-nanoparticles (silver-based nanoparticles)	River Isar, southern Germany	ET-AAS after cloud point extraction	1–2 ng/L; load peak of 20.88 ng/L near the wastewater treatment plant Decrease in silver-*b*-NP concentration at 1.5 km downstream of discharge wastewater treatment plant points	Wimmer et al. (2019)
Ti	Lake Taihu, China	SP-ICP-MS	0.09 to 10.2μg/L, in October 2016 0.11 to 4.16μg/L, in April 2018	Wu et al. (2020a)

measured engineered nanoparticles in surface waters that are dependent on the nanoparticles type and ranges within with the ealier studies presented in the table. Some studies demonstrate a remarkable decrease in measured engineered nanoparticles concentration with the distance of the point source, namely wastewater treatment plants (Syafiuddin et al. 2018; Wimmer et al. 2019).

The data currently available suggest that engineered nanoparticles may be present in certain surface waters, but not at high concentrations. However, the amount of analytical data available remains too small. There is a continuing need to detect and quantify these emerging contaminants in water. The low concentrations result from the removal mechanisms occurring in surface waters, particularly in waters with moderate or high ionic strengths, and organic matter high concentrations. Nevertheless, an expected increase of engineered nanoparticles production and use might increase the concentration of engineered nanoparticles in surface waters and the risk of water contamination. Furthermore, localized high-input sources may exist and engineered nanoparticle releases that may impact drinking water treatment plants intake should be carefully analysed and monitored.

1.5 Substances of Emerging Global Concern

1.5.1 Persistent Organic Pollutants in Lake and Ocean Ecosystems

The term "persistent organic pollutant" describes a wide range of organic compounds of anthropogenic origin present in the environment. This family of substances is subdivided into four main categories, namely polycyclic aromatic hydrocarbons (PAHs), chlorinated aromatic compounds, pesticides (organochlorine pesticides) and brominated flame retardants. The best known substances are, for example, 2,4-dichlorophenoxyacetic acid herbicide (well known as 2,4-D), hexachlorobenzene (another chlorinated fungicide used to treat seeds, especially wheat), polychlorinated biphenyls commonly known as PCBs used as lubricants in transformers, or PAHs from combustion or pyrolysis. The list of substances is particularly long due to the large number of congeners and isomers for the same class of substances. For example, polychlorinated biphenyls represent a group of 209 different chemical compounds (Puzyn and Mostrag-Szlichtyng 2012; Vestergren and Cousins 2013; DeWitt 2015; Eggen and Vogelsang 2015; Harmon 2015; Kallenborn et al. 2015; Zeng 2015; Xiao 2017; Dong et al. 2018; Lorenzo et al. 2018; Brusseau 2019; Ng et al. 2019; Klemes et al. 2019; Pan et al. 2020; Salthammer 2020).

Persistent organic pollutants represent a family of substances that is not strictly speaking "new" as it has been known for more than 50 years to date, but this group of pollutants is still of concern. Indeed, their presence in the aquatic environment is not a recent phenomenon, but this problem has become more widely evident over the last decade thanks to the constant improvement of analytical techniques and more regular monitoring of aqueous compartments, including sediments, and organisms. Persistent organic pollutants in the Arctic atmosphere have been investigated since the 1970s when the first atmospheric measurements revealed their presence in this pristine polar environment (Kallenborn et al. 2015). Nevertheless, this family of substances is regularly updated, for example, the emerging perfluoro alkyl and polyfluoro alkyl substances have been classified as persistent organic pollutants (Puzyn and Mostrag-Szlichtyng 2012; Vestergren and Cousins 2013; Zeng 2015). Only limited information is available for these "newly" identified substances.

Persistent organic pollutants are transported from primary and secondary sources (industry, agriculture, urbanized areas, transport, etc.) through the atmosphere, and they can be transported over long distances to areas of the World through natural atmospheric and oceanic processes where there are no anthropogenic activities, *e.g.*, in Arctic soils (Gioia et al. 2011; Cabrerizo et al. 2018). Indeed, they are found all over the World, even in areas where these substances have never been used or produced (Puzyn and Mostrag-Szlichtyng 2012; Kallenborn et al. 2015).

The majority of persistent organic pollutants identified so far are banned or restricted worldwide due to concerns about their harmfulness to human health and ecosystems. They are indeed substances of global concern due to their toxicity, persistence, bioaccumulation and long-range transport (Puzyn and Mostrag-Szlichtyng 2012; Datta et al. 2018). Once released into the environment, persistent organic pollutants, that remain intact for exceptionally long periods of time (even some long-banned substances persist in the environment), are widely distributed in all environmental compartments (water, sediment, soil and atmosphere). They accumulate in the fatty tissue of living organisms (due to their low solubility and high lipophilicity, the molecules are subject to bioaccumulation in human and animal fatty tissues, and then biomagnify along food chains), being toxic to humans and wildlife (Mortimer 2013; Harmon 2015). Indeed, due to their special intrinsic physical and chemical properties, they are resistant to environmental degradation through physical, chemical and biological processes (Harmon 2015; Zeng 2015). Exposure to persistent organic pollutants can lead to serious health effects, including certain cancers, birth defects, immune and reproductive system dysfunction, increased susceptibility to disease and damage to the central and peripheral nervous systems. Some substances are also known as disrupting endocrine compounds.

In order to eliminate or reduce the release of persistent organic pollutants into the environment, several international treaties have been established, which call for efforts and actions by the international community, such as the Protocol to the UNECE Regional Convention on Long Range Transboundary Air Pollution (CLRTAP) and the Stockholm Convention on persistent organic pollutants. The Stockholm Convention (signed in 2001; Puzyn and Mostrag-Szlichtyng 2012), which has 184 countries/regions member signataries, listed 12 original persistent organic pollutants in 2004 and has updated it over the last decades. These instruments have established strict international regimes for lists of persistent organic pollutants, including 16 chemicals in the UNECE Protocol and 12 original chemicals (the "dirty dozen" including DDT, aldrin, endrin, heptachlor, polychlorinated biphenyls, polychlorinated dibenzofurans and polychlorinated dibenzo dioxins) and 16 newly added chemicals in the Stockholm Convention.

Lake ecosystems, composed of the water body, sediments, organisms, surrounding soils and the atmosphere, are one of the major sinks for persistent organic pollutants and are also an important part of surface water systems. Persistent organic pollutants in lakes can accumulate in the fatty tissue of living organisms through the food chain and pose potential risks to humans (Puzyn and Mostrag-Szlichtyng 2012; Duedahl-Olesen 2013; Schrenk and Chopra 2013; Mortimer 2013). Legacy persistent organic pollutants, such as polycyclic aromatic hydrocarbons, organochlorine pesticides, polychlorinated biphenyls, polychlorinated dibenzo-p-dioxins and dibenzofurans and polybrominated diphenyl ethers, have been commonly detected and reported in various lake ecosystems, while emerging persistent organic pollutants such as perfluorinated compounds, polychlorinated naphthalenes, tetrabromobisphenol A and hexabromocyclododecane have been relatively less detected or reported, but these chemicals are also of great concern. Figure 1.6 shows the average concentrations of various persistent organic pollutants in the Chinese Taihu lake ecosystem. Lake Taihu, located in the southern Jiangsu province and northern Zhejiang province, is the third largest freshwater lake in China and has provided local communities with valuable fisheries for centuries. There are 38 cities and more than 40 million people living around the lake, Taihu Basin being one of the most developed regions in China (Li et al. 2009; Wang et al. 2009).

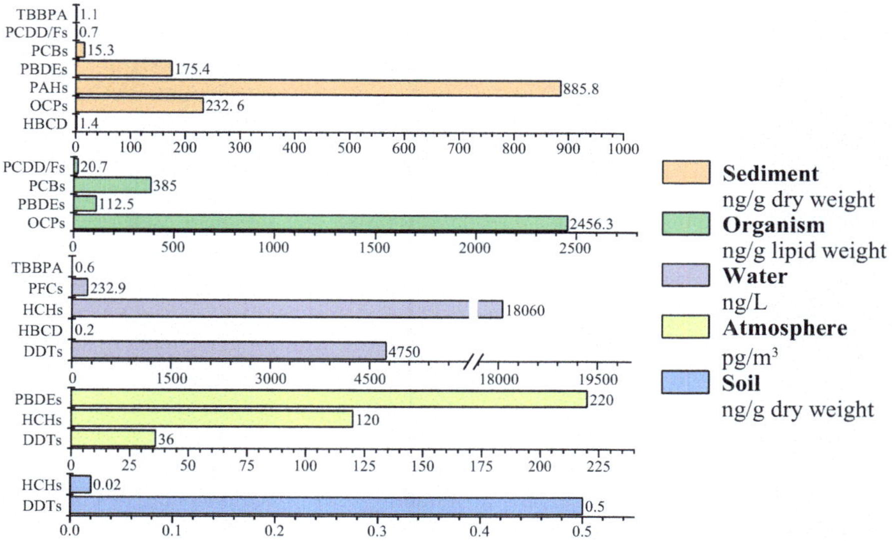

Fig. 1.6 Average concentrations of various persistent organic pollutants in the Chinese Taihu lake ecosystem (*TBBPA* tetrabromobisphenol A, *PCDD/Fs* polychlorinated dibenzo-p-dioxins and dibenzofurans, *PCBs* polychlorinated biphenyls, *PBDEs* polybrominated diphenyl ethers, *PAHs* polycyclic aromatic hydrocarbons, *OCPs* organochlorine pesticides, *HBCD* hexabromocyclododecane, *DDT* dichlorodiphenyltrichloroethane, *DDTs* DDT-related compounds)

Various carcinogenic persistent organic pollutants from the past are commonly detected in typical lake ecosystems in Europe and Asian. Organochlorine pesticides, polychlorinated biphenyls, and polycyclic aromatic hydrocarbons are the main persistent organic pollutants in water bodies (Javedankherad et al. 2013), sediments (de Mora et al. 2004; Boehm et al. 2005; Nemirovskaya and Brekhovskikh 2008; Varnosfaderany et al. 2015; Baniemam et al. 2017), organisms (Rajaei et al. 2011; Mashroofeh et al. 2015), and the surrounding soils and atmosphere (Aliyeva et al. 2012, 2013; Shahbazi et al. 2012) of the Caspian Sea located between Europe and Asia, the World's largest lake.

The levels of these persistent organic pollutants in the Caspian Sea ecosystems are at a scale of approximately ng/g, with polycyclic aromatic hydrocarbons having a relatively higher concentration. However, persistent organic pollutants newly added to the Stockholm Convention, such as polychlorinated naphthalenes, perfluorinated compounds, hexabromocyclododecane, etc., are rarely studied in the Caspian Sea at present. Persistent organic pollutants, including organochlorine pesticides, polychlorinated biphenyls, polychlorinated dibenzo-p-dioxins and dibenzofurans, polycyclic aromatic hydrocarbons, polybrominated diphenyl ethers, polychlorinated naphthalenes, hexabromocyclododecane, perfluorinated compounds and tetrabromobisphenol A, are also present in different environmental compartments and biota in Chinese lake ecosystems (Qiu et al. 2004; Zhang and Jiang 2005; Qiao et al. 2006; Wang et al. 2007; Wen et al. 2008; Yang et al. 2011; Yu et al. 2012; Zhang et al. 2012; Zhou et al. 2012a, b; Xu et al. 2013, 2015; Wang et al. 2016a, b; Cui 2018; Liu et al. 2018; Wei et al. 2019; Xiang et al. 2019).

Recently, the presence of perfluorinated compounds in the environment has been of great concern. Levels of perfluorinated compounds in the water body of Chinese lakes have averaged between 30 and 50 ng/L and perfluorooctanoic acid and perfluorooctane sulfonate were the most frequent perfluorinated compounds (Yang et al. 2011; Gao et al. 2016). Mean concentrations of organochlorine pesticides, polycyclic aromatic hydrocarbons, polychlorinated biphenyls, polybrominated diphenyl ethers, polychlorinated dibenzo-p-dioxins and dibenzofurans, perfluorinated compounds and hexabromocyclododecane in samples of biota from various Chinese lakes are usually at ng/g levels (Liu et al. 2018). In summary, European and Asian lake ecosystems are contaminated to varying degrees by various persistent organic pollutants. There are large differences in the levels of persistent organic pollutants between different lake ecosystems and other environmental media. Further studies on newly added persistent organic pollutants are needed in the future to better understand their environmental fate and ultimate control.

Persistent organic pollutants, old and new, are also widely distributed in the water bodies, sediments, organisms, surrounding soils and atmosphere of lake ecosystems typical of North America and Africa. Lake Superior is the largest freshwater lake in the World and also the largest of the Great Lakes in North America. Polychlorinated biphenyls and organochlorine pesticides occuring in the Lake Superior water body in the 1990s were at ng/g range and by 2018 they were at pg/g levels, which is lower than in lakes in Europe and Asia (Jeremiason et al. 1994; Ruge et al. 2018). Polychlorinated dibenzo-p-dioxins and dibenzofurans, polybrominated diphenyl ethers and polychlorinated biphenyls were found in Lake Superior sediments at ng/g magnitude, which was comparable to that of European and Asian lakes (Pearson et al. 1998; Shen et al. 2009).

In addition to organochlorine pesticides and polychlorinated biphenyls, some emerging persistent organic pollutants such as polybrominated diphenyl ethers, hexabromocyclododecane, polychlorinated naphthalenes and perfluorinated compounds were also present in Lake Superior organisms (Kannan et al. 2000; Luross et al. 2002; Dykstra et al. 2005; Fernie and Letcher 2010; Furdui et al. 2015). Polycyclic aromatic hydrocarbons, polybrominated diphenyl ethers, organochlorine pesticides, polychlorinated naphthalenes and polychlorinated biphenyls were present in the surrounding atmosphere (Hillery et al. 1997; Strandberg et al. 2001; Fernandez et al. 2002; Helm et al. 2003). Lake Tanganyika is a freshwater lake located in Central Africa. Reports of persistent organic pollutants in Lake Tanganyika are mainly related to lake organisms. Organochlorine pesticides, polychlorinated biphenyls, polybrominated diphenyl ethers and hexabromocyclododecane were detected in fish samples from Lake Tanganyika (Mahugija et al. 2018; Polder et al. 2014). Among these reported persistent organic pollutants, the concentration of polychlorinated biphenyls in fish samples ranged from 36 to 167 ng/g, which is generally lower than in Asian lakes. In addition, the contamination of persistent organic pollutants in Lake Tanganyika was lower or comparable to that in other parts of Africa (Manirakiza et al. 2002).

Polybrominated diphenyl ethers have raised new concerns in recent decades. Some studies have focused on polybrominated diphenyl ether contamination in the North Pacific Ocean, where the World's largest fishery, that of Hokkaido, is being lacerated. It is the main source of seafood for people living in countries along the North Pacific coast, especially those living in Japan, South Korea and China. However, this coast is also the World's largest producer and consumer of polybrominated diphenyl ethers. The use and emissions of polybrominated diphenyl ethers in this region are significant, and pollution by polybrominated diphenyl ethers is becoming an increasingly serious problem. Xiang et al. (2019) showed that the current concentration of polybrominated diphenyl ethers in the atmosphere, water, sediments and organisms in the North Pacific Ocean is 0.14 to 896 pg/m^3, 3 to 4360 pg/L, 0.24 to 7397.7 ng/g dry weight and 0.56 to 3930 ng/g lipid weight, respectively. The predicted concentration of polybrominated diphenyl ethers in the Hokkaido fishery could be 4.6 to 9 pg/m^3 (air), 106 to 341 pg/L (water), 3.4 to 248 ng/g dry weight (sediment) and 5 to 64.7 ng/g lipid weight (biota). The concentration of polybrominated diphenyl ethers in the atmosphere, water bodies, sediments and organisms in the North Pacific has a characteristic near-shore > ocean or low latitude > high latitude distribution (Xiang et al. 2019).

From a global perspective, lake ecosystems are contaminated to varying degrees by several persistent organic pollutants, both from past and recent contamination. Although many studies have been conducted on these substances in lake ecosystems, these studies have mainly focused on traditional persistent organic pollutants such as organochlorine pesticides, polychlorinated biphenyls, polychlorinated dibenzo-p-dioxins and dibenzofurans, while emerging persistent organic pollutants such as polychlorinated naphthalenes, perfluorinated compounds, hexabromocyclododecane, etc. are relatively less reported. The extent and significance of all these substances are still largely unknown. Environmental risk assessing of persistent organic pollutants remains particularly difficult, not only in lake ecosystems, but also in freshwater and marine ecosystems.

These ecosystems are fragile, living at the rate of anthropogenic disturbances (pollution, artificialization, withdrawals), increasingly vulnerable for example to climate change, but essential to life in general. Indeed, lake ecosystems are important freshwater resources closely linked to human life, but our understanding of pollution by persistent organic pollutants in lake ecosystems is not yet sufficient. A better understanding of these substances, in terms of their degradation, behavior, accumulation, transfer, toxicity and impact, still requires research in the different environmental compartments and living organisms of these systems. More comprehensive international health policies and standards are also needed in the future.

1.5.2 Monitoring of Emerging Contaminants in Portuguese Rivers

Portugal is on the top three of countries with more studies in the field of surface water even by normalizing the amount of reports by the number of inhabitants per country, according to a recent review on the environmental monitoring of water organic pollutants dedicated to the literature published between 2012 and 2018 (Sousa et al. 2018). That review paper provides an overview of the worldwide occurrence of contaminants identified by European Union guidelines, namely priority substances of European Directive 39/2013/EU and contaminants of emerging concern of Watch Lists of European Decisions 495/2015/EU and 840/2018/EU, and presenting also comprehensive data about other specific organic pollutants not considered in those European Union documents and with reported high concentrations and frequencies. Another review paper covering the period between 2001 and 2015 systematically overviewed the published data on the monitoring of industrial compounds, natural and synthetic estrogens, phytoestrogens and phytosterols, pharmaceuticals, and pesticides in the Portuguese aquatic environment (Ribeiro et al. 2016b). The data collected in that review show that some compounds were determined at high concentrations in the studies performed between 2001 and 2015, namely industrial and household compounds (*e.g.*, bisphenol A, alkylphenols and alkylphenol polyethoxylates) found from few ng/L to dozens of μg/L, in agreement with other European countries such as Spain, Switzerland, Netherlands, and France (Ribeiro et al. 2016b). In the case of natural and pharmaceutical estrogens, they were detected always at ng/L levels in Portuguese surface waters in the reviewed literature, similar or higher than those concentrations reported in European countries (Ribeiro et al. 2016b). While phytoestrogens are described as frequently studied and detected at μg/L range, the authors pointed out that most publications about pharmaceuticals in Portuguese surface waters published before 2016 referred only few therapeutic classes in the north of Portugal and the majority of them consisted in the application of new analytical methods rather than consistent monitoring data (Ribeiro et al. 2016b). A wide range of classes of pesticides including priority substances of European Directive 39/2013/EU were compiled in that review from 2001 to 2015, some of them up to μg/L, including atrazine that was already banned in Europe (Ribeiro et al. 2016b). The present section aims to update the information about the occurrence and distribution of emerging contaminants in Portuguese rivers (Fig. 1.7) and results from a literature survey comprising reports published since 2015 in Scopus database, using as keywords: "river water", "surface water" or "estuarine water" and "Portugal" or "Portuguese", and "monitor" or "occurrence".

In the last years, natural organic compounds have been studied in Portuguese rivers, namely phytoestrogens and phytosterols that are found in many plants and mycotoxins produced by fungi. One study monitoring these classes in the Douro River estuary reported the maximum concentration of daidzein (up to 277.4 ng/L) in the summer (Ribeiro et al. 2016a), whereas the highest value in another seasonal study performed by the same research group in the Ave River was determined in the spring (up to 404.0 ng/L) (Ribeiro et al. 2016c). In that river, a seasonal trend for phytoestrogens and phytosterols was verified and the largest level of coumestrol (up to 165.0 ng/L) was observed in the summer. Interestingly, this compound was not detected in any sample collected in the Douro monitoring campaign (Ribeiro et al. 2016a). Both reports showed that the mycotoxin deoxynivalenol was ubiquitously found up to 373.5 ng/L in the Douro (Ribeiro et al. 2016a) and between 59.5 to 642.4 ng/L in the Ave (Ribeiro et al. 2016c), whereas the micotoxin zearalenone and its metabolite α-zearalenol were only detected in spring and summer in the Douro (Ribeiro et al. 2016a). This last finding is in agreement with another monitoring study in seven Portuguese rivers and one creek, where the highest concentration of

Fig. 1.7 River pollution by emerging contaminants and related challenges for their management. Ave River (left) and Leça River (right). (Source: Ana Rita Lado Ribeiro, Porto, Portugal)

zearalenone was reported during spring. In that survey, it was estimated a contamination frequency of 23.7% for this mycotoxin, with concentrations ranging between 5.6 and 82.6 ng/L (Laranjeiro et al. 2018). In general, the highest values found in the last years are quite lower than those reviewed in the previous period 2001–2015 by Ribeiro et al. (2016b). Industrial and household compounds have been also monitored in Portuguese rivers. In a monitoring program developed in Mira River over a 1-year period, including also natural and pharmaceutical estrogens, phytoestrogens and the phytosterol sitosterol, both estrogens and industrial/household pollutants were found at high concentrations in all sampling locations, including some positioned in environmental protected areas (Rocha et al. 2016). Specifically, the annual average of industrial/household pollutants was of ca. 1.3μg/L, whereas values at ng/L levels were described for phytoestrogens and sitosterol (Rocha et al. 2016). Another recent study targeting endocrine disruptor compounds in Minho, Ave and Mondego River estuaries, reported quite high concentrations of alkylphenols and alkylphenol ethoxylates (up to 4855 ng/L) despite those of the natural estrogens were markedly lower than previous surveys (Rocha et al. 2016).

Regarding personal care products, one study performed in northern Portuguese rivers in the summer (2018) targeted 43 compounds and detected 28 substances (Celeiro et al. 2019): fragrances up to 200 ng/L (citronellol), limonene, α- isomethylionone, tonalide and galaxolide in all samples, the latter up to 379 ng/L; phthalates up to 92 ng/L (benzylbutylphthalate), including the priority substance diethylhexylphthalate determined at 88 ng/L in one sampling point; the priority substance antioxidant butylhydroxytoluene between 8 and 25 ng/L in half of the samples; and UV filters up to 254 ng/L (benzophenone 3).

In the last years, pharmaceuticals have been extensively monitored by some research groups. A wide group of 66 human and veterinary pharmaceuticals belonging to seven therapeutic groups was monitored in Tejo estuarine waters and the therapeutic families with an overall frequency of detection higher than 90% were antibiotics, β-blockers, antihypertensives, lipid regulator and anti-inflammatories (Reis-Santos et al. 2018). The highest concentrations of each class were 304 ng/L for antidepressants (sertraline), 51.8 ng/L for non-steroidal anti-inflammatories (diclofenac), 77.0 ng/L for lipid regulators (gemfibrozil), 161.9 ng/L for antihypertensives (ibersartan) and 128.0 ng/L for antibiotics (doxycycline) (Reis-Santos et al. 2018). Another one-year occurrence study published before targeting 23 pharmaceuticals in Tejo and Mondego Rivers (Pereira et al. 2017a), showed one site in Tejo River with comparable concentrations of diclofenac, gemfibrozil and bezafibrate. In that study, pharmaceuticals were found in 27.8% of the samples at ng/L levels, with the following therapeutic families having the highest detection frequencies and mean concentrations: selective serotonin reuptake inhibitors, anti-inflammatories and antibiotics (Pereira et al. 2017a). The authors confirmed the impact of wastewater treatment plants by an increase of the mean concentrations downstream and the impact of lower river flow rates on increasing detection frequencies and concentrations (Pereira et al. 2017a). In this sense, despite the risk quotients were higher than one only for two pharmaceuticals, the authors stressed the ecotoxicological pressure especially due to water scarcity in drought periods, which may raise the risk in rivers located in the Iberia region (Pereira et al. 2017a). A study carried out in the Lisbon's drinking water supply system included the analysis of 31 pharmaceuticals in surface waters from Tejo and Zêzere Rivers, where 15 and 10 pharmaceuticals were detected at ng/L levels, respectively (de Jesus Gaffney et al. 2015). Caffeine, erythromycin, acetaminophen, sulfadiazine, sulfapyridine, sulfamethoxazole, carbamazepine and atenolol were quantified in both rivers; propranolol, sulfamethazine, gemfibrozil, indomethacin, ibuprofen, diclofenac, and naproxen were present in the Tejo River; and nimesulide was only quantified in the Zêzere River (de Jesus Gaffney et al. 2015). In that study, the risk assessment was performed for human health and it was suggested that the exposure to residual pharmaceuticals in drinking water would be particularly improbable. One environmental relevant therapeutic class is the cytostatics as the cancer incidence has raised worldwide and thus their consumption has evidently increased over the last years. From the seven cytostatics (cyclophosphamide, capecitabine, mycophenolic acid, imatinib, bicalutamide, prednisone and 5-fluorouracil) determined in river water samples in a very recent study encompassing also wastewaters, only mycophenolic acid was detected in Uíma, Douro and Leça Rivers, at average concentrations of 210, 211 and 541 ng/L, respectively (Santos et al. 2018). Twenty-seven pharmaceuticals including antibiotics and psychiatric drugs were also analysed in Douro and Leça Rivers (Fernandes et al. 2020). In that study, higher frequency and concentrations of pharmaceuticals were found in the Leça River, where six antibiotics (azithromycin, ciprofloxacin, clarithromycin, moxifloxacin, ofloxacin, and trimethoprim) were found, the highest concentration was registered for azithromycin (2819 ng/L), and fluoxetine was the most detected pharmaceutical (Fernandes et al. 2020). Interestingly, in the same report, none of the studied antibiotics was detected in river samples of Douro River (Fernandes et al. 2020).

Some occurrence studies have been developed in the last years, by simultaneously targeting pharmaceuticals and other classes of compounds, namely pesticides. One recent monitoring program performed recently in four rivers (Ave, Antuã, Cértima, and Leça) described the spatial and temporal variations of 39 priority substances and contaminants of emerging concern, some belonging to European Directive 39/2013/EU, European Decisions 840/2018/EU and 495/2015/EU (Barbosa et al. 2018). From the 13 pollutants detected in all the rivers, various pharmaceuticals were detected (up to 396 ng/L) and the priority pesticide isoproturon was found up to 92 ng/L (Barbosa et al. 2018). The industrial priority substance perfluorooctanesulfonic acid was detected below its quantification limit in the Antuã, Cértima, and Leça (Barbosa et al. 2018). Atrazine, a priority substance of European Union Directive 39/2013 already banned, was also detected in Ave, Antuã and Leça (up to 41 ng/L), whereas simazine was found in the Cértima and Leça (Barbosa et al. 2018), both pesticides being found at higher concentrations than those reported in the Arade River by Gonzalez-Rey et al. (2015). The neonicotinoids acetamiprid and imidacloprid of the European Decision 840/2018/EU were only detected during the dry season in the Ave (Barbosa et al. 2018).

In the same report, higher concentrations were observed downstream of industrial areas and urban wastewater treatment plants (Barbosa et al. 2018). In another study published by the same group, all the contaminants of emerging concern preconized in the Watch List in force at the time of that monitoring program (European Decision 495/2015/EU) were monitored during four seasons in two Portuguese rivers located in the northern region, Ave and Sousa (Sousa et al. 2019). In that study, 13 contaminants of emerging concern were found, being diclofenac, 2-ethylhexyl 4-methoxycinnamate, and azithromycin the most frequent compounds and diclofenac, 2-ethylhexyl 4-methoxycinnamate and imidacloprid those determined at the highest concentrations (Sousa et al. 2019). The same authors reported the first spatial distribution of 37 pollutants in the entire Portuguese coast, including the ocean shore and the nearest river discharging on it (Sousa et al. 2019). This monitoring program included some priority substances and contaminants of emerging concern of the European Watch List (European Decisions 495/2015/EU and 840/2018/EU; EU 2015, 2018). In general, the authors found high concentrations of diclofenac, tramadol and carbamazepine, the latter being ascribed as having medium to high risk for algae (Sousa et al. 2020). While some pharmaceuticals and perfluorooctanesulfonic acid were largely distributed, atrazine and alachlor were also determined in most samples, with the concentrations of alachlor in some locations being considered as medium to high risky (Sousa et al. 2020). Interestingly, higher levels were described in some seawater samples in comparison to that of the nearest estuarine site, which was attributed to the possible direct discharge into the sea (Sousa et al. 2020). The first study on the presence of pharmaceutical compounds and pesticides in the Arade River estuary included 19 pharmaceuticals and 47 pesticides, some of which defined as priority substances of European Directive 39/2013/EU or included in the European Decision 495/2015/EU (EU 2013; Gonzalez-Rey et al. 2015). In that study, the detected compounds varied temporally and seasonally, except the stimulant caffeine (804 ng/L), which occurred without any remarkable differences. Other study targeting caffeine as anthropogenic indicator in the Lis River over an 11-month period, reported its presence in all river samples ranging from 25.3 to 321 ng/L, with its highest levels found downstream of effluent discharge points, confirming that it could be an effective indicator of human-born pollution (Paíga et al. 2019). Besides caffeine, those compounds found at highest concentrations in the Arade River were the antiasthmatic theophylline (184 ng/L), the analgesic paracetamol (88 ng/L) and the fungicide carbendazim (45 ± 18 ng/L) (Gonzalez-Rey et al. 2015). In turn, atrazine, diuron, isoproturon and simazine were found at few ng/L and below the environmental quality standards.

Some monitoring surveys focusing on pesticides have been also published. Surface waters of the Lezíria do Tejo agricultural area were monitored and the measured environmental concentrations for the 19 pesticides were compared with their environmental quality standards for risk assessment, being the risk demonstrated in 100% of the samples containing insecticides, which represented 60% of the total risk identified (Pereira et al. 2017b). One occurrence study was performed between 2002 and 2004 in the river basins Mondego, Sado and Tejo in order to assess the risk of 29 pesticides and metabolites, including priority substances and other compounds used in dominant crops of a number of agricultural areas within the catchment of these Mediterranean river basins (Silva et al. 2015a, b). In that study, 20 substances were detected, from which seven are priority substances (alachlor, atrazine, chlorfenvinphos, chlorpyrifos, endosulfan, simazine and terbutryn) and remarkably all exceeded their individual environmental quality standards values. Other specific pollutants namely molinate, oxadiazon, pendimethalin, propanil, terbuthylazine, and the metabolite desethylatrazine, had non-acceptable aquatic risks (Silva et al. 2015a). The authors highlighted the need to monitor such substances not only at European Union, but also at local/river-basin/national level, allowing to describe the exceedances of the environmental quality standards (Silva et al. 2015a). The most frequently found compounds in a previous study reporting 29 pesticide compounds (21 herbicides, 5 insecticides and 3 metabolites) in surface waters of Mondego, Sado and Tejo from 2002 and 2008 were: chlorfenvinphos (mean 0.16μg/L), propanil (mean 0.007μg/L) and the metabolite 3,4-dichloroaniline (mean 0.33μg/L) in all rivers; molinate (mean 1.03μg/L) and oxadiazon (mean 0.006μg/L) in Mondego and Sado Rivers; and atrazine (mean 0.16μg/L), simazine (mean 0.08μg/L) and metolachlor (mean 0.06μg/L) in Mondego and Tejo Rivers (Silva et al. 2015a). Endosulfan was also detected in all rivers, but less frequently (3–7%) and with a quite lower mean value of 0.0008μg/L (Silva et al. 2015a). In that study, the authors predicted the aquatic risk of the measured pesticide mixtures for primary producers, arthropods and fish in the three Portuguese river basins and proposed a ranking per taxonomic group and river basin for the relative contribution of individual compounds or groups sharing the same mode of action, where the highest toxicity role was observed for oxadiazon on primary producers and for the organophosphorus insecticides chlorfenvinphos and chlorpyrifos on arthropods and fish, respectively. It is noteworthy that a study conducted between 2010 and 2011 in the Mondego estuary, comprising 56 priority pesticides (insecticides, herbicides and fungicides), reported more than half of the quantified pesticides above the maximum values set by the European Directives (98/83/EC and 2013/39/EU), a potential risk for the pesticide mixture at the maximum concentrations found, and a significant toxic effect for *Artemia salina*, with observed affected locomotion (Cruzeiro et al. 2016).

In another spatial and temporal monitoring study encompassing the pesticides atrazine, azoxystrobin, bentazon, λ-cyhalothrin, penoxsulam and terbuthylazine in the Mondego estuary, azoxystrobin was found as the most frequently detected pesticide (Rodrigues et al. 2018a), which had been also reported in all aqueous samples collected seasonally during 2012–2013 in a wetland of worldwide interest (Ria Formosa Lagoon) (Cruzeiro et al. 2015a), where pesticide concentrations surpassed those recommended by European Directive 2013/39/EU in two monitoring studies (Cruzeiro et al. 2015a, b). In the Mondego (Rodrigues et al. 2018a), atrazine was the second most detected, reinforcing that this priority substance should be monitored despite its ban. Although the concentrations of pesticides in estuarine water were estimated to lead to low risk to estuarine organisms, all pesticides were bioaccumulated by bivalves and triazine pesticides were found also in macroalgae (Rodrigues et al. 2018a). In the same study, the pesticides were quantified typically during summer concurring with the pesticide usage period and no severe contamination occurred during a flood event that was expected to promote the runoff of pesticides from the adjacent agricultural areas (Rodrigues et al. 2018a). The quantified residues of pesticides were also more frequent during the summer and thus in the pesticide application period in another spatial and temporal monitoring program performed in 2017 in the Sado estuary (Rodrigues et al. 2019a). That report suggested a long-term aquatic exposure for five herbicides, namely alachlor, bentazon, metobromuron, metribuzin and triclopyr, which were determined in the water samples before and after the production season (Rodrigues et al. 2019a). Regarding the potential adverse effects of the application of agricultural pesticides on the aquatic organisms in this estuary, it was not found any severe effect, even considering the potential mixture effect of pesticides (Rodrigues et al. 2019a).

The problematic of microplastics in river waters has received little attention in comparison to marine system. Two recent studies performed in Antuã River (Rodrigues et al. 2018b) and Douro river estuary (Rodrigues et al. 2019b) targeted this environmental issue. The first study on the occurrence of microplastics in freshwater systems performed in Portugal was developed in the Antuã River during March and October 2016, showing a correlation of their occurrence with the urbanization and industrial activities, with an abundance in water up to 193 items/m^3 in March and up to 1265 items/m^3 in October (Rodrigues et al. 2018b). It was verified a higher abundance in October than in March for almost all sampling sites (one exception) and a decrease from upstream to downstream areas (Rodrigues et al. 2018b). Generally, the most frequently found plastic type was foam (>50%) and the least was pellets, but fibers were the dominant type during March (Rodrigues et al. 2018b). In the Douro report performed between from 2016 and 2017, the authors found a total of 2152 particles, with an average density of 17.06 items/100 m^3, including fibers, soft and hard plastic, colourful and transparent plastic, being hard fragments and fibers the most frequent (Rodrigues et al. 2019b). In the same study, an average ratio of fish larvae to microplastics was found to be greater than one in some months, highlighting the possible ingestion by fish and the resulting impact in these communities (Rodrigues et al. 2019b).

All these results show that we still need to mobilise to protect water resources in order to guarantee the sustainability of the ecosystems that are essential to Portugal's economy. Indeed, the spatial distribution of pollutants along the entire Portuguese mainland coast shows the need for mitigation measures to enhance the elimination of pharmaceuticals in wastewater treatment plants and measures to control the use and discharge of pesticides and industrial compounds.

1.5.3 Presence of Emerging Contaminants in the Aquatic Environment in Latin America

With about one-third of the World's water resources and 24,400 m^3 of water per capita per year, Latin America is a region with abundant freshwater resource availability. This resource is nevertheless fragile, particularly impacted by increasing economic development (Peña-Guzmán et al. 2019). Indeed, in many Latin America countries, there is a high population density in areas suffering from poor sanitary conditions, which contributes to the contamination/pollution of rivers and reservoirs that supply water (Caldas et al. 2019; Peña-Guzmán et al. 2019; Starling et al. 2019). The chemical and physicochemical properties (*e.g.*, water solubility, high adsorption capacity and low biodegradability) of many substances in daily use facilitate their flow without resistance reaching water sources, which can trigger problems not only for the environment but also for the health of living organisms (Janssens et al. 1997; Knepper et al. 1999; Kuster et al. 2008; Hernández et al. 2011a, b; Peña-Guzmán et al. 2019).

In a recent comprehensive review by Peña-Guzmán et al. (2019), it was addressed the water quality involving various aqueous compartments of Latin American countries (Argentina, Bolivia, Brazil, Chile, Colombia, Costa Rica, Ecuador, Guatemala, Mexico, Uruguay, and Venezuela), such as water reservoirs for human consumption, drinking water treatment plants, groundwater, surface waters, rivers and wastewater treatment plants. The authors of this survey concluded that the most frequently detected emerging substances are pharmaceuticals (mainly antibiotics), followed by personal care products, and that the most common molecules are 17-β-estradiol, bisphenol A, estrone and caffeine. Most of these chemical substances are not yet regulated by any environmental law. Peña-Guzmán et al. (2019) indicated that a total of 51 emerging pollutants were found in studies related to wastewater, with Ecuador being the country with the largest number of samples, followed by Mexico, Brazil and Colombia. In wastewater, caffeine had the highest concentrations (5597μg/L), followed by the cocaine metabolite benzoylecgonine (1065μg/L) and carbazepine (830μg/L). In the drinking water supply, the substance with the highest concentration (625μg/L) was ibuprofen (Brazil, Colombia, and Venezuela). For surface water sources, the contaminants with the highest concentrations were cholesterol (301μg/L), caffeine (106μg/L), stigmasterol (85.5μg/L) and bisphenol A (64.2μg/L) (Brazil, Costa Rica, Colombia, Bolivia, Chile, Mexico and Argentina). Mexico is the only country that has recorded cases of emerging contaminants in groundwater in the urban water cycle, with the highest concentrations for naproxen (2μg/L), sulfasalazine (0.78μg/L), ibuprofen (0.51μg/L), and salicylic acid (0.464μg/L). For drinking water treatment plant effluents (Brazil only), the substances identified were caffeine (4.083μg/L), nonylphenol (0.228μg/L), 17-α-ethinylestradiol (0.798μg/L), bisphenol A (0.0005μg/L), estriol (0.0003μg/L), and estrone (0.00003μg/L).

Other recent studies have also shown that pharmaceuticals, personal care products, industrial substances and pesticides are widely detected at concentrations of the same order of magnitude (Machado et al. 2016; Hernández-Padilla et al. 2017; Botero-Coy et al. 2018; Gilabert-Alarcón et al. 2018; Sodré et al. 2018; Montagner et al. 2019; Starling et al. 2019). All studies concluded that these types of chemicals are persistent in the environmental compartments of Latin America; at the same time, their continuous introduction into rivers yields a complex mixture with potential synergistic effects that can result in a serious ecological risk.

As in many other countries around the World, emerging substances enter the aquatic environment of Latin America by many pathways, with the main one being the conventional wastewater treatment plants not designed to remove these type pollutants, and their discharge containing organic substances and some metabolites (Dougherty et al. 2010; Hernández-Padilla et al. 2017; Starling et al. 2019). Besides, it is estimated that only 20% of municipal wastewater is treated, and in some Latin America countries, many water treatment plants are abandoned due to the high operating costs or low performance (Hernández-Padilla et al. 2017). In Mexico, less than 50% of wastewater is treated and wastewater treatment plants are not financially self-sustaining, resulting in uneven water quality (Gilabert-Alarcón et al. 2018). In Brazil, a country of more than 200 million inhabitants, the percentage of households with access to the wastewater treatment system does not exceed 50%, and a large part of the collected wastewater does not receive appropriate treatment before being discharged into rivers (Machado et al. 2016). In Colombia, raw wastewater is sometimes discharged directly into surface waters in certain Colombian regions such as Florencia and Tumaco (Botero-Coy et al. 2018). When untreated water is released, this contamination further reduces the availability of freshwater for domestic purpose and other uses, as well as damage to ecosystems is caused.

However, this is not just a domestic water problem. Indeed, treated industrial wastewater from the pharmaceutical, chemical, textile or metal industries, as well as water from agricultural activities including irrigation, livestock and aquaculture, can also contain a cocktail of substances, which end up in rivers. All these water uses can also pollute freshwater resources. Before being discharged, like municipal wastewaters, industrial effluents must nevertheless comply with regulations concerning a number of "conventional" substances such as metals and water parameters, namely suspended solids, chemical and biochemical oxygen demand, but organic substances are rarely registered in industrial wastewater (Hernández-Padilla et al. 2017; Gilabert-Alarcón et al. 2018).

The assessment of the occurrence of new contaminants in rivers is one of the major challenges for the scientific community. This task is important because there is a lack of information in most Latin American countries (Hernández-Padilla et al. 2017; Peña-Guzmán et al. 2019; Starling et al. 2019). Unlike Europe and other regions, many Latin American countries do not have an established list of priority pollutants or laws controlling their production or release into the environment. However, regulations are beginning to evolve, particularly in Mexico, Colombia and Brazil, for example with regard to the treatment of municipal and industrial wastewater before discharge. Treatment plants and industries in Latin America do not have a formal structure to detect, assess, control and manage new compounds. Researchers and industrialists in Mexico, Colombia and Brazil draw on the experience of other developed countries in order to protect the aquatic environment of their respective regions, which have different conditions, environmental culture and pollutions levels (Hernández-Padilla et al. 2017).

1.5.3.1 Pharmaceuticals, Personal Care Products and Pesticides in the Aquatic Environment in Mexico

Based on data of the Federal Agency of Consumer (PROFECO) of Mexico, this country is one of the main economies in Latin America, which has registered a strong growth in the personal care products market during the last decade, where around 40% of the demand is related to hair and skin care products. On the other hand, according to the Ministry of Health (Mexico), it is estimated that more than 80% of the Mexicans self-medicate, which means that they use drugs regardless medical prescription. As an example, diabetes is one of the main health issues in Mexico, and high pharmaceuticals consumption is associated to this disease. Unfortunately, similar to other countries, the uncontrolled use of drugs and personal care products, coupled to the discharge of untreated domestic wastewater or the inefficient removal of this type of substances in conventional wastewater treatment plants, can increase the contamination levels in the aquatic environment of various Mexican regions (Gilabert-Alarcón et al. 2018; Lesser et al. 2018; Pérez-Alvarez et al. 2018).

In some States, the reuse of wastewater has been implemented without integrative planning and assessment of the impact on ecosystems and public health (Gilabert-Alarcón et al. 2018). Despite emerging contaminants are rarely monitored in Mexico, the occurrence of pharmaceuticals and personal care products in effluents and surface waters has been investigated. With that aim, measurements of trimethoprim (0.11–0.15μg/L, 0.28–0.32μg/L), clarithromycin (0.37–0.45μg/L, 1.10–1.40μg/L), naproxen (6.2–6.74μg/L, 2.84–3.16μg/L), and diclofenac (0.25–0.34μg/L, 0.42–0.50μg/L) were determined in wastewater from Mexico City-Mezquital Valley (Pumping station "Gran Canal" and Emisor Profundo "El Salto"). The concentrations of these substances were found to be similar to levels reported for sewage in Europe, Japan, and USA. Levels of metoprolol (0.21–0.25μg/L) in water from the Gran Canal were comparative to the concentrations detected in wastewater in Switzerland (0.14–0.29μg/L), but these levels were higher than those found in sewage in Spain (< 0.005–0.09μg/L). The concentration of ibuprofen had been previously reported close to 0.3μg/L for both sampling points (Siemens et al. 2008).

Lesser et al. (2018) carried out analyses of untreated wastewater used for agricultural irrigation, from different points in Mezquital Valley (located in Central Mexico), reporting that from the group of 7 reproductive hormones and 118 pharmaceuticals measured, 3 hormones and 65 drugs were detected. Some examples include acetaminophen (39,900 ng/L), ciprofloxacin (1190 ng/L), sulfamethoxazole (5360 ng/L), oxytetracycline (134 ng/L), naproxen (11,800 ng/L) and metformin (82,900 ng/L).

In the same way, hospital wastewater from Toluca City (the capital city of the State of Mexico) was studied by the quantification of different pharmaceuticals, where antidiabetics (*e.g.*, 1.92μg/L and 1.31μg/L for glibenclamide and metformin, respectively), β-blockers (0.20μg/L for atenolol and 2.02μg/L for metoprolol), β-lactams (3.77μg/L for penicillin G and 0.42μg/L for penicillin V), hormones

(0.08μg/L for 17-β-estradiol), and anti-inflammatory drugs (0.59μg/L for diclofenac, 0.62μg/L for ibuprofen, 1.79μg/L for naproxen and 2.66μg/L for acetaminophen), were detected (Pérez-Alvarez et al. 2018).

Similarly, the presence of some emerging contaminants was also investigated in five effluents of domestic wastewater produced in the coastal zone of Cihuatlan, Jalisco (a State of Mexico on the edge of the Pacific Ocean) (Arguello-Pérez et al. 2019). These authors specified that diclofenac, ibuprofen and ketorolac were determined in all the analyzed effluents at concentration considered as "toxic", whilst estradiol was found in two of the effluents at concentrations that fall into a very toxic range.

Díaz and Peña-Alvarez (2017) previously reported an analytic method for the detection of ibuprofen, 2-benzyl-4-chlorophenol, naproxen, triclosan, ketoprofen, diclofenac, bisphenol A and estrone in river sediments. Their analyses showed that sediment samples from the Tula River (Hidalgo, in central Mexico) had ibuprofen, naproxen and triclosan at concentrations in the order of ng/g. The authors remarked that the water sample with higher naproxen concentration (240 ng/L) was taken from the riverside where wastewater from Ixmiquilpan City was discharged.

Others monitoring studies exhibited the occurrence of pharmaceuticals in influent/effluent of the wastewater treatment plant "Acapantzingo" and surface water (Apatlaco River in Cuernavaca, the capital of the State of Morelos in south-central Mexico) (Rivera-Jaimes et al. 2018). The total concentrations of pharmaceuticals were higher in surface water and influent wastewaters as compared to those detected in effluent wastewaters, showing the important role of wastewater treatment plants. Rivera-Jaimes et al. (2018) indicated that the discharge of untreated wastewater increased the concentration of pharmaceuticals in the river, where naproxen (732–4880 ng/L), acetaminophen (354–4460 ng/L), diclofenac (258–1398 ng/L) and bezafibrate (286–2100 ng/L), were the most abundant.

Recently, Pérez-Coyotl et al. (2019) analyzed water samples from an urban reservoir (Madín dam, located in the municipalities of Naucalpan and Atizapán in the metropolitan area adjacent to Mexico City) that is a source of drinking water and represents an area for recreational activities, including carp fishing. Their results showed different concentrations of personal care products (especially sunscreens) and pharmaceuticals. As example, 2,2′,4,4′-tetrahydroxybenzophenone (13.7 ng/L), 2-ethylhexyl 4-(dimethylamino) benzoate (29 ng/L), 4-methylbenzylidine camphor (134 ng/L) and methyl benzotriazol (250 ng/L), antidiabetic agents glibenclamide and metformin (2148 ng/L and 9557 ng/L, respectively), and the analgesic acetaminophen (9156 ng/L) were detected. Hence, it is notably essential to implement a monitoring program of these emerging contaminants in drinking water sources of this region.

Another important problem in Mexico, as in other Latin American countries, is the presence of pesticides in water resources (García-de la Parra et al. 2012; Arellano-Aguilar et al. 2017). The high availability and convenient cost of agrochemicals, as well as flexible environmental laws have led to an overuse of pesticides in Mexico, in order to satisfy the global demand of agricultural products. Nevertheless, residual agrochemicals in agricultural runoff are rarely monitored by farmers, and some of their leachates drain these pollutants to lakes or rivers (sources of drinking water), which eventually contaminate bays and coasts (Fig. 1.8) (Osuna-Flores and Riva 2002; Arellano-Aguilar et al. 2017). In particular cases, people of various villages are in contact with this type of hazardous leachates; furthermore, private wells for drinking water, which is untreated (used directly), are located close to agricultural fields (Picos-Corrales et al. 2020). In this regard, some pesticides are regulated in Mexico by the Federal Commission for the Protection against Sanitary Risk, an organ of the Health Secretary of Mexico. Based on this environmental regulation (NOM-127-SSA1-1994), the maximum limits for the main

Fig. 1.8 Contamination sequence involving river pollution in Sinaloa from agricultural wastewater (Sinaloa is an important agricultural region of Mexico; source: Lorenzo A. Picos-Corrales, Sinaloa, Mexico)

organochloride pesticides for drinking water are: 30μg/L for 2,4-D (2,4-dichlorophenoxyacetic acid), 20μg/L for methoxychlor, 2μg/L for γ-lindane, 1μg/L for DDT (dichlorodiphenyltrichloroethane, total) and hexachlorobenzene, 0.2μg/L for chlordane, 0.03μg/L for heptachlor, aldrin and dieldrin (separated or combined).

Sinaloa is an example of an industrialized agricultural region in Northwestern Mexico that is characterized by a variety of crops and high production. In this State of Mexico, pesticides have been found in sediments of agricultural drainages (García-de la Parra et al. 2012). From this assessment, concentrations of organochlorine compounds (15 substances) and organophosphorus compounds (8 substances) were within the ranges 0.1–20.19 ng/g and 0.03–1294 ng/g, respectively, being diazinon the compound with the highest concentration. Arellano-Aguilar et al. (2017) also reported that this industrialized agricultural region has a documented pesticides usage of 700 t/year, being at least 17 of the molecules classified as moderately to highly toxic.

With the same focus, Leyva-Morales et al. (2017) assessed the quality of river water from central Sinaloa (Humaya, Tamazula and Culiacan rivers) by monitoring residual pesticides. Their results showed that the following chemicals were present: lindane (0.0041–0.0104μg/L), endosulfan (0.0198–0.03601μg/L), heptachlor, dichlorodiphenyltrichloroethane (0.0525μg/L), diazinon (0.0211–0.0403μg/L), chlorpyrifos, methyl parathion, cyromazine (0.1961μg/L), permethrin (0.2351μg/L), ethion, carbofenothion, lambda cyhalotrine, pyrimicarb, aldrin (0.0099–0.1023μg/L) and malathion. Leyva-Morales et al. (2017) suggested that pesticide contamination derived mainly from runoffs of polluted soil toward rivers.

Besides, the connection between agrochemical loading in drains/rivers and pollution in coastal lagoons of the Mexican Pacific has been studied (Arellano-Aguilar et al. 2017). The authors indicated that the organochlorine chemicals were clearly accumulated in the coastal lagoons from the drains and rivers, with hexachlorocyclohexanes (organochlorine) showing the highest concentration. In another work regarding pesticides accumulation, levels of dichlorodiphenyltrichloroethane (∑DDTs ~0.6–137μg/kg) were quantified in sediments samples taken from Texcalac River in Tlaxcala (Mexico), exceeding the guidelines for protection of aquatic life according to the authors (García-Nieto et al. 2019).

From the literature, Yucatan (southeastern Mexico) is a state with high levels of groundwater pollution, where different pesticides have been detected (3.2 mg/L for endrin, 10.86 mg/L for δ-lindane, 5.23 mg/L for γ-lindane, 6.53 mg/L for α-lindane and 12.54 mg/L for heptachlor) (Polanco-Rodríguez et al. 2018). This type of hazardous chemicals has been also identified in the drinking water source "Madín dam" of Naucalpan and Atizapan (Mexico), namely diazinon up to 12.592 ng/L and fenthion sulfoxide up to 2.465 ng/L (Pérez-Coyotl et al. 2019). Based on the current literature, it is recommended to increase the monitoring work regarding pesticides content in surface water in Mexico.

In addition to monitoring, examples related to the impact of emerging pollutants on human health and ecosystems in Mexico have also been reported. Researchers developed a strategy for the determination of some residual personal care product ingredients in hydroponic French lettuce from a local supermarket in Mexico

(Cabrera-Peralta and Peña-Alvarez 2018). The results indicated that the use of wastewater to irrigate crops can result in contaminated food since the absorption by plants is favored in the presence of compounds with intermediate polarity.

Martinez-Gomez et al. (2015) previously assessed the effect of pharmaceuticals and personal care products on aquatic life using the rotifer *Plationus patulus*, which is a member of riverine food networks in Mexico. Results indicated that the continuous exposure to emerging contaminants could reduce rotifer population growth. Hence, the constant discharge of treated wastewater to aquatic ecosystems could represent a risk for the zooplankton community.

In another work, the toxicological risk of a hospital effluent from Toluca (the capital of the State of Mexico) was evaluated using two bioindicator species, i.e. *Xenopus laevis* and *Lithobates catesbeianus* (Pérez-Alvarez et al. 2018). This research group suggested that the hospital effluents represent a latent risk for the ecological balance and diverse organisms, being necessary a prior treatment for the removal of the pharmaceuticals (Pérez-Alvarez et al. 2018).

Based on hazard quotients for pharmaceuticals detected in the Apatlaco River in Cuernavaca (the capital of the State of Morelos in south-central Mexico), Rivera-Jaimes et al. (2018) found that contamination levels of this river could represent a risk for the aquatic ecosystem, due to the high concentrations of naproxen, ibuprofen, diclofenac and sulfamethoxazole. Similarly, experiments performed using samples from the drinking water source "Madín dam" (Naucalpan-Atizapán, State of Mexico) revealed that residual pharmaceuticals, personal care products and pesticides originated embryotoxicity, embryolethality, congenital abnormalities and oxidative stress on the common carp embryos. This was a rational explanation regarding the reduction in the population of *Cyprinus carpio* species of this ecosystem (Pérez-Coyotl et al. 2019). Also, in past years in the Bay of Ohuira (Topolobampo, a port on the Gulf of California in northwestern Sinaloa), the decrease in shrimp production was associated, among other causes, with the high coastal pollution by pesticides (Osuna-Flores and Riva 2002). In Chiapas (Southern Mexico), concentrations of organochlorine pesticides were found in a greater amount in cow's milk and forage than in water samples. Despite the content of contaminants was below the limits established by Mexican government, a monitoring program was recommended by the authors (Murga et al. 2016).

It is also estimated that in Yucatan (southeastern Mexico) around 30% of the residents drink water from contaminated sources, and investigations reveled high levels of pesticides (7.352 mg/L for endosulfan I, 2.336 mg/L of 4,4′-dichlorodiphenyldichloroethane, 3.695 mg/L for aldrin and 1.434 mg/L for heptachlore) in the blood of women having cervical uterine cancer. Additionally, other studies showed the high levels of these chemicals in breast milk (18.436 mg/L for heptachlor epoxide and 2.10 mg/L for dieldrin) (Polanco-Rodríguez et al. 2018).

In conclusion, Mexico is facing the problem of emerging substances in its aqueous compartments. This is not a novelty itself, but this problem is accentuated by the diversity of the substances found and the levels of quantified concentrations. Table 1.5 summarizes examples of contamination by emerging substances reported in water compartments in Mexico, comparing them with those reported by

Table 1.5 Contamination by emerging substances reported in water compartments in Columbia, Mexico and Brazil (from 2013 onward)

Country	Water type	Emerging substance	Concentration (ng/L)	References
Columbia (La Fe)	Drinking water supply reservoir	Ibuprofen	7–39	Aristizabal-Ciro et al. (2017)
Columbia (Rio Grande)	Drinking water supply reservoir	Ibuprofen	5–62	Aristizabal-Ciro et al. (2017)
Mexico (Mexico City)	Surface water	Ibuprofen	15–45	Félix-Cañedo et al. (2013)
Mexico (Toluca City)	Hospital wastewater	Ibuprofen	620	Pérez-Alvarez et al. (2018)
Mexico (Morelos)	Surface water	Diclofenac	258–1398	Rivera-Jaimes et al. (2018)
Mexico (Mexico City)	Surface water	Diclofenac	28–32	Félix-Cañedo et al. (2013)
Mexico (Morelos)	Wastewater	Diclofenac	600–2500	Rivera-Jaimes et al. (2018)
Mexico (Cuernavaca)	Wastewater	Diclofenac	258–1398	Rivera-Jaimes et al. (2018)
Mexico (Toluca City)	Hospital wastewater	Diclofenac	590	Pérez-Alvarez et al. (2018)
Mexico (Morelos)	Surface water	Naproxen	732–480	Rivera-Jaimes et al. (2018)
Mexico (Tula River)	Surface water	Naproxen	240	Díaz and Peña-Alvarez (2017)
Mexico (Mexico City)	Surface water	Naproxen	52–186	Félix-Cañedo et al. (2013)
Mexico (Mezquital Valley)	Wastewater	Naproxen	11,800	Lesser et al. (2018)
Mexico (Cuernavaca)	Wastewater	Naproxen	732–4880	Rivera-Jaimes et al. (2018)
Mexico (Morelos)	Wastewater	Naproxen	800–4200	Rivera-Jaimes et al. (2018)
Mexico (Toluca City)	Hospital wastewater	Naproxen	2660	Pérez-Alvarez et al. (2018)
Brazil (São Paulo)	Surface water	Paracetamol	30,421	Campanha et al. (2015)
Brazil (São Paulo)	Surface water	17-α-ethynylestradiol	777	Montagner et al. (2019)
Brazil (São Paulo)	Surface water	Caffeine	19–127,000	Montagner et al. (2019)
Brazil	Surface water	Caffeine	40–19,000	Machado et al. (2016)
Brazil (São Paulo)	Surface water	Caffeine	129.585	Campanha et al. (2015)
Brazil	Drinking water	Caffeine	1.8–2000	Machado et al. (2016)

(continued)

Table 1.5 (continued)

Country	Water type	Emerging substance	Concentration (ng/L)	References
Brazil (São Roque)	Drinking water	Caffeine	4083	Peña-Guzmán et al. (2019)
Brazil (São Paulo)	Drinking water	Caffeine	121	Montagner et al. (2019)
Brazil (Rio Grande)	Drinking water supply reservoir	Caffeine	18,828	Machado et al. (2016)
Brazil (São Paulo)	Surface water	Bisphenol A	2–13,016	Montagner et al. (2019)
Brazil (Rio Grande)	Drinking water supply reservoir	Bisphenol A	11	Machado et al. (2016)
Brazil (São Paulo)	Drinking water	Bisphenol A	23	Montagner et al. (2019)
Brazil (São Roque)	Drinking water	Bisphenol A	0.5	Peña-Guzmán et al. (2019)
Brazil (São Paulo)	Surface water	Cocaine	10	Montagner et al. (2019)
Brazil (São Paulo)	Surface water	Benzoylecgonine	133	Montagner et al. (2019)
Colombia (Bogotá)	Wastewater	Benzoylecgonine	4000	Bijlsma et al. (2016)
Colombia (Bogotá)	Wastewater	Amphetamines	12–68	Bijlsma et al. (2016)
Brazil (São Paulo)	Drinking water	Triclosan	2–37	Montagner et al. (2019)
Brazil (Rio Grande)	Surface water	Atrazine	5–49	Caldas et al. (2019)
Brazil (São Paulo)	Drinking water	Atrazine	687	Montagner et al. (2019)
Brazil (Rio Grande)	Drinking water	Atrazine	5–37	Caldas et al. (2019)
Brazil	Drinking water	Atrazine	2–6000	Machado et al. (2016)
Mexico (Yucatan)	Groundwater	Endrin	3.2 mg/L	Polanco-Rodríguez et al. (2018)
Mexico (Yucatan)	Groundwater	γ-Lindane	10.86 mg/L	Polanco-Rodríguez et al. (2018)
Mexico (Yucatan)	Groundwater	Heptachlore	12.54 mg/L	Polanco-Rodríguez et al. (2018)

Columbian and Brazilian colleagues. It is necessary to mobilize all the stakeholders and the population to protect aquatic ecosystems in order to ensure the production of quality water. To achieve this goal, a thorough monitoring of surface waters is necessary and indispensable as part of a vast program at several spatial and temporal scales, not only at the national level but also as part of a more comprehensive water policy in collaboration with Mexico neighboring countries. In addition, attention should also be paid to treatment plants and their purification performance, as part of a more organizational water management at national and regional levels.

1.5.3.2 Emerging Contaminants in Colombian Rivers

Colombia is a country in South America with an area of 1,141,748 km^2, a large marine area covering 928,660 km^2 and a population of 46,581,823 (*Departamento Administrativo Nacional*, DANE 2011). Colombia is a privileged country because of its geographical position in the continent, which gives it direct access to both ocean Atlantic and Pacific. Colombia has many water sources throughout its territory and the most important biodiversity not only in the region but also in the World. The country has indeed many rivers, the most important of which are the following: Amazon (6992.6 km, shared with Peru and Brazil), Caquetá (Japura, 2816.3 km, shared with Brazil), Negro (2230.5 km, shared with Brazil and Venezuela), Orinoco (2140.4 km, shared with Venezuela), Putumayo (1609.3 km, shared with Peru and Brazil), Guaviare (1496.6 km, shared with Venezuela), Arauca (1049.2 km, shared with Venezuela), Cauca (965.6 km), Goal (804.6 km, shared with Venezuela), Magdalena (528.8 km) and its affluent Rio Bogotá. Concerning the latter river, it originates in the northeastern region of Cundinamarca, in the lagoon valley of Paramo Guacheneque, municipality of Villa Pinzon, Eastern Cordillera of Colombia, at 3300 m above sea level. The waters of the Bogotá River flow for 380 kilometers to Girardot, where they flow into the Magdalena River. Its basin is located in the center of the country, in the department of Cundinamarca, with an area of approximately 6.000 km^2 (Hernández et al. 2015; Bedoya-Ríos et al. 2018; Bedoya-Ríos and Lara-Borrero 2018).

Colombia bases its income mainly on the production of raw materials to export and to manufacture consumer products for the domestic market. Its main activities are the production of oil for exportation (fourth in Latin America and sixth in the continent) and mining, particularly carbon, gold, emeralds, sapphires and diamonds (*Ministerio de Ambiente y Desarrollo Sostenible*, 2016). The strongest industrial sectors in Colombia are petrochemicals, textiles, automobiles, and chemicals.

Colombia, like most countries in the World, is not immune to the problem of emerging compounds and Colombian developing regions are concerned (Martínez Vidal et al. 2006; Barceló and Petrovic 2008; Ibáñez et al. 2008, 2009; Klamerth et al. 2009; Teijón et al. 2010; Tobón-Marulanda et al. 2010; Gil et al. 2012; Díaz-Casallas et al. 2019; Reichert et al. 2019). As in many other developing countries, Colombia's indiscriminate industrial growth has led to severe pollution of its waterbodies. Indeed, numerous studies have shown that several of its water sources are heavily contaminated, including by emerging contaminants. The sources of contamination are industrial, agricultural and domestic, the latter often uncontrolled. Once entered into the environment, chemicals and their metabolites can produce subtle effects on aquatic and terrestrial organisms, especially on the former since they are exposed to long-term continuous influx of wastewater effluents. Recently, Díaz-Casallas et al. (2019) proposed a comprehensive analysis of the water quality at the upper basin of the Bogotá River between 2008 and 2017, pointing out an insufficient quality of water. This study highlighted the necessity for further efforts on the continuous monitoring of Colombian rivers basins.

Indeed, Colombia is currently focusing its efforts on the detection and the quantification of emerging substances, in order to compile a list of priority substances adjusted to the needs and characteristics of Latin America. Research programs are also underway to provide recommendations for future monitoring, prevention and elimination programs for these chemicals. It is indeed important not only to monitor discharge waters from municipal treatment plants and industrial wastewaters but also to encourage the improvement of their systems and treatment techniques in order to achieve better removal efficiencies.

The list of substances that have been found in effluents from wastewater treatment plants, and consequently in Colombian waters (surface water, groundwater, etc.) is extremely varied, ranging from pharmaceuticals, hormones and steroids to surfactants, cosmetics and pesticides (Hernández et al. 2005, 2012, 2015; Martínez Vidal et al. 2006; Ibáñez et al. 2008, 2009; Tobón-Marulanda et al. 2010; Bijlsma et al. 2016; Sarria-Villa et al. 2016; Portilla et al. 2017; Botero-Coy et al. 2018; Serna-Galvis et al. 2019; Reichert et al. 2019). These studies also helped to improve the understanding about the dynamics and behavior of the target compounds.

The substances most frequently found in Colombian water sources were pharmaceuticals such as antihypertensives, antibiotics, analgesics and psychiatric products, and illicit substances, *e.g.*, cocaine and its main metabolite, heroine, and amphetamines (methamphetamine). They are detected at trace levels, usually between ng/L and μg/L, depending on the molecule. Raw wastewater from the cities of Bogotá, Medellín and Florencia contain levels of several pharmaceuticals above 1μg/L, while the concentrations for antibiotics are commonly above 5μg/L in raw hospital wastewater. A survey of the Latin American literature, including Colombian data, shows that all these substances are often identified and quantified by gas chromatography coupled to mass spectrometry (Peña-Guzmán et al. 2019). Continual improvements in analytical equipment and methodologies enable the determination of pharmaceuticals and their metabolites at lower and lower concentration levels in water samples and also in other environmental matrices.

Colombia is facing the problematic of the high use of pharmaceutical products and cosmetics (Fig. 1.9). Pharmaceuticals include antihypertensives (valsartan, irbesartan), antibiotics (ciprofloxacin, norfloxacin), anticonvulsants (carbamazepine), diuretics (furosemide), psychiatric products (gabapentin, benzodiazepine derivatives, antiepileptics), and psychoactive agents (hallucinogens). Benzophenones (benzophenone-1, benzophenone-2, benzophenone-3, benzophenone-4), parabens (methylparaben, ethylparaben, butylparaben, propylparaben), triclosan and bisphenols are examples of cosmetic ingredients considered as endocrine disruptors. Among these substances, antihypertensives (valsartan, irbesartan), antibiotics (ciprofloxacin, norfloxacin) and endocrine disruptors are of great concern and their hazardous effects are already reported (Darbre 2015; Gee et al. 2015). Triclosan is also a substance of concern. This hydrophobic substance can accumulate in fatty tissues, such as fish and human samples (urine, breast milk and serum). Moreover, triclosan

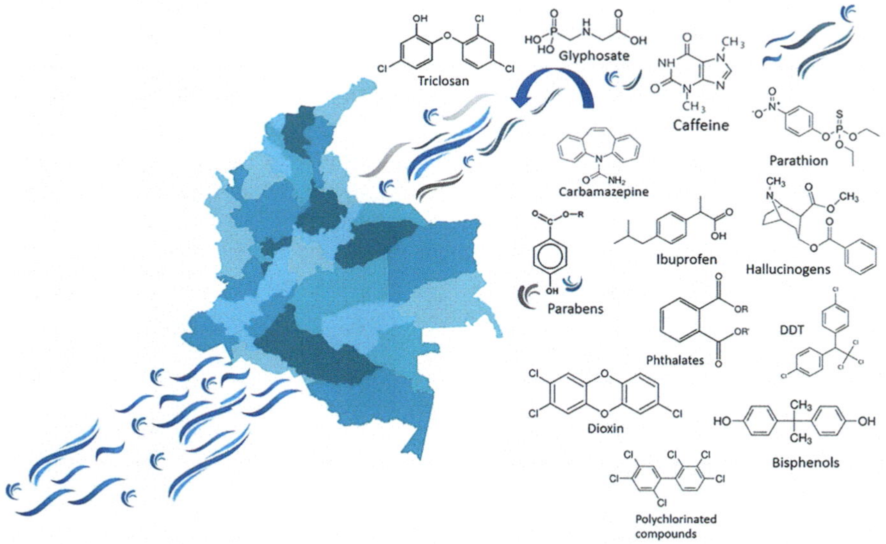

Fig. 1.9 Example of emerging contaminants in Colombian Rivers. (Source: Juan Carlos Moreno-Piraján, Bogotá, Colombia)

has been shown to have toxic effects in aquatics organisms at environmental concentrations and its degradation products can be more toxic and persistent than the parent compound (Kosma et al. 2014; Juliano and Magrini 2017; Machado et al. 2016; Montagner et al. 2019).

A first typical example is the presence of losartan, an antihypertensive widely used in Colombia (Portilla et al. 2017; Serna-Galvis et al. 2019). Another typical example is ciprofloxacin that is among the top five most common antibiotics prescribed in Colombian hospitals. Botero-Coy et al. (2018) and Serna-Galvis et al. (2019) have reported its frequent presence in hospital effluents. However, the load from hospital wastewater varied from study to study. Another problem is the presence of endocrine disruptors, which are capable of blocking and altering every hormonal function, disrupt the endocrine system. Endocrine disruptors are compounds that mimic the hormones produced by living organisms and interfere with various functions (Darbre 2015). These chemical compounds include the natural (17-β-estradiol) and synthetic (17-α-ethynylestradiol) hormones used in birth control pills, fertility treatments and hormone replacement therapy. All these endocrine disruptors end up in water via domestic, industrial and/or hospital discharges. However, conventional municipal wastewater treatment plants are not designed to deal with these substances, and consequently such molecules and their metabolites can reach the aquatic environment. Endocrine disrupting effects such as intersexuality and reproductive disorders have been reported in fish and other aquatic organisms living downstream from wastewater treatment plant outfalls (Vajda et al. 2008; Gagné et al. 2011; Tetreault et al. 2011).

Cocaine is a highly consumed illicit drug in populated Colombian cities such as Bogotá, and consequently, its main metabolite (benzoylecgonine) has a high prevalence in wastewaters (Bijlsma et al. 2016; Hernández et al. 2015; Serna-Galvis et al. 2019). Bijlsma et al. (2016) conducted for the first time a wastewater-based epidemiology in Colombia, reporting that cocaine was the most consumed illicit drug, particularly in Medellin (department of Antioquia). Benzoylecgonine, a cocaine metabolite, exceeded 4000 ng/L in all samples from wastewater treatment plants. Amphetamines were also found in all samples at concentrations between 12 ng/L and 68 ng/L (Bijlsma et al. 2016). Illicit substances, together with pharmaceuticals, are commonly measured in Colombian effluents as both groups of substances are not completely removed with conventional treatments (Serna-Galvis et al. 2019).

Another important issue is the presence of pesticides, such as glyphosate, dichlorodiphenyltrichloroethane (DDT), parathion and clofibric acid, in water, since Colombia is a major agricultural country that uses phytosanitary products. A typical example is the presence of glyphosate used to protect illicit crops. The frequent detection of this substance suggests that it is a widespread environmental contaminant. Another typical example is the insecticide DDT which has well know adverse health consequences, namely an established link with malformations in newborns. Studies have also shown significant impacts on fish and reptiles (deformities, reduced fertility, and abnormal behavior).

There are also industrial products such as phthalates, bisphenol A and its analogues (bisphenol F and bisphenol S) used in the production of resins and plastics, paints/lacquers and binding materials, or cleaning agents (nonylphenol). Another example is triclosan used as a broad spectrum antibacterial agent into many consumer products and as preservative in products of domestic use. It is also found in cosmetics, textiles and antibacterial fibers. These substances have been identified and quantified in industrial wastewaters and in rivers.

In Colombia and in other Latin America countries, the situation is worrying not only from a water pollution point of view but also from a health point of view, with the appearance of resistance phenomena for certain bacteria. Antibiotics are antimicrobial drugs that kill or reduce the growth of bacteria. They have been used in large quantities for several decades around the World and antibiotic resistance of pathogens has been for long a focus of research in clinical settings and, more recently, in environmental research. Incorrect use of antibiotics increases resistance to them, a process called antimicrobial resistance, the antibiotic being ineffective on a bacterial infection. The more antibiotics are consumed, the greater the risk of developing resistant bacteria (García et al. 2020). One widely reported case is that of *Staphylococcus aureus*, which has developed resistance to methicillin (MRSA). This bacterium is a clear example of hospital infections originated when a certain patient stays in a hospital for long periods of time, creating a big problem in terms of hospital hygiene. There are three main routes of entry of antibiotics into freshwater: (1) effluents from wastewater treatment plants, (2) chemical manufacturing plants, and (3) livestock, agricultural and aquaculture sites. Indeed, antibiotics can bypass water treatment processes and end up directly in the environment. They are detected in Colombian rivers at very low concentrations and are diluted more than a million times compared to concentrations in the human body. However, these residual concentrations are highly likely to generate antimicrobial resistance, having direct and indirect effects on the microbial component of aquatic communities. Indeed, even at low concentrations, these antibiotics could have significant consequences for ecosystems and human health.

In conclusion, the presence of emerging substances in Colombian waters is a concern that is becoming increasingly important in society. The demand for water, whether for domestic, agricultural or industrial use, is constantly increasing, with the consequences of water pollution and ecosystem degradation, as a significant amount of wastewater is either untreated or poorly treated. This scenario is complicated by a climate context that is affecting the predictability of our most precious resource. All water stakeholders are concerned by this issue and mobilized. There is a growing consensus that the challenges can be tackled by adopting an integrated approach for water resources management. In particular, special efforts need to be made in the operation of municipal wastewater treatment plants and in the protection of the ecosystems and water resources.

1.5.3.3 Analysis of Emerging Contaminants in Brazilian Water Resources

As in Mexico and Colombia, the presence of the same vast group of emerging contaminants (pharmaceuticals, personal care products and cosmetics, industrial substances and pesticides) in Brazil's water resources, including rivers, lakes, drinking water reservoirs and tap water, has been reported at the same orders of magnitude of concentrations, showing that the conventional wastewater treatments are not effective for their removal, even in regions that have adequate sanitation. Indeed, numerous studies indicated that the conventional applied wastewater treatments are not efficient enough to remove these substances from wastewater and sludge, and as a result they find their way into the environment (Barbosa et al. 2015; Campanha et al. 2015; Albuquerque et al. 2016; Machado et al. 2016; Caldas et al. 2019; Starling et al. 2019).

The collection, treatment and reuse of wastewater from households and industry, the reduction of diffuse pollution and the improvement of water quality are major challenges for Brazilian water stakeholders. Freshwater quality is indeed under threat. Its pollution is widespread and increasing in many regions. In Brazil, regulations concerning emerging contaminants in both natural waters and drinking water require a commitment among researchers and regulatory authorities, since this issue has not been considered as a priority by the government (Machado et al. 2016).

The substances that have been analyzed in Brazilian waters include mainly, hormones, antibiotics, endocrine disruptors, caffeine (as licit drug), illicit drugs (methamphetamine, cocaine, heroin), industrial compounds (bisphenols, triclosan), and pesticides. Among them, pesticides in particular have been monitored, as Brazil is the World's largest consumer of herbicides, fungicides and insecticides (Barbosa et al. 2015; Albuquerque et al. 2016; Caldas et al. 2019; Starling et al. 2019), which means that these substances may end up in the environment, especially in aquatic systems and reservoirs used for drinking water production. There are 381 active ingredients authorized by the Brazilian Ministry of Agriculture for pesticides used on crops and 1670 formulated plant protection products in the market (Albuquerque et al. 2016). Brazil has regulations on the use of pesticides and their presence in water. For example, the Brazilian drinking water norm (Ordinance 2914/2011 of the Ministry of Health) includes 64 chemical substances, of which 27 are pesticides that must be monitored every 6 months. However, this number represents <10% of the current pesticide active ingredients approved for use in the country (Barbosa et al. 2015). Of the 27 pesticides, 21 are banned in Europe due to the risks they offer to health and the environment. A comprehensive review on the presence of pesticides in Brazilian freshwaters was published by Albuquerque et al. (2016). They reported that data in peer-reviewed literature are scarce and often incomplete, concluding that, whitout the implementation of a nationwide pesticide freshwater monitoring program and a clear definition of proper water quality standards, it is not possible to evaluate the risks posed by pesticides to the aquatic life in Brazil.

Among the most widely used pesticides in Latin America, glyphosate and atrazine occupy prominent positions as the best-selling active ingredients in Brazil (Brazil 2011; Barbosa et al. 2015; Marin et al. 2019). Since its introduction as an herbicide active ingredient in 1971, glyphosate became and remains the market leading herbicide worldwide (glyphosate-based formulations are mainly represented by Roundup). The limit of glyphosate in drinking water in Brazil is 0.5μg/L (Brazil 2011; Barbosa et al. 2015), while in Europe only 0.1μg/L is permitted (0.9μg/L for World Health Organization, 0.8μg/L for Canada). The 0.1μg/L value is also the maximal residue level of glyphosate in European surface and ground water. The maximum contaminant level in the U.S. in drinking water is 700μg/L (sum of glyphosate and its residues) and the health-based guideline value in Australia is 1000μg/L (Székács and Darvas 2018). The contamination of the environment by glyphosate is worrying since it impairs enzymatic activity, causes cytoxicity and DNA data in human cells (Marin et al. 2019).

In drinking and surface waters, acceptable limit for atrazine (herbicide of the triazine class) and methyl parathion (organophosphate insecticide) are 9 and 2μg/L, respectively (Ordinance 2914/2011 of the Ministry of Health). These limits are based on toxicological and neurotoxicological trials, but they do not take into account possible estrogenic effects that atrazine can cause when human and wildlife are chronically exposed to low concentrations (Machado et al. 2016). In Canada, the maximum acceptable concentration for atrazine in drinking water is 5μg/L (this value is applicable to the sum of atrazine and its metabolites) while for the United States Environmental Protection Agency, the maximum concentration in water is 3μg/L. In the European Union, a limit of 0.1μg/L has been set for all pesticide residues in drinking water and groundwater. This value of 0.1μg/L for each substance represents a standard for water quality, not for the health risk. If more than one pesticide is found in drinking water, the sum of all pesticides has a legal limit of 0.5μg/L (source: French Agency for Food, Environmental and Occupational Health & Safety).

Sodré et al. (2010) investigated the occurrence of endocrine disruptors in drinking water of the city of Campinas. Six emerging contaminants (stigmasterol, cholesterol, bisphenol A, caffeine, estrone, and 17-β-estradiol) were found in twelve samples. Estrone and 17-β-estradiol were detected only during the dry season, with concentrations below quantification limits. Stigmasterol showed the highest average concentration (0.34 ± 0.13μg/L), followed by cholesterol (0.27 ± 0.07μg/L), caffeine (0.22 ± 0.06μg/L), and bisphenol A (0.16 ± 0.03μg/L). Alarmingly, the level of bisphenol A detected in drinking water (0.0005μg/L) was higher than the average effluent from effluent treatment plants (0.00002μg/L) (Sodré et al. 2010). Ten years later, Peña-Guzmán et al. (2019) also reported similar levels of concentrations. Sodré et al. (2010) concluded that, in Campinas, where surface drinking water supplies receive large amounts of raw sewage inputs, the emerging contaminants levels in drinking waters were higher than median values compiled for drinking and finished water samples around the World. Montagner and Jardim (2011) also reported that among endocrine disruptors, bisphenol A (a compound used as a manufacturing intermediate) is distinguished by its widespread presence in the environment. In the State of São Paulo, this substance was detected in all surface water sampling campaigns with concentrations between 204 and 13,016 ng/L (Montagner and Jardim 2011).

A special substance in Latin America is caffeine. It is consumed by people of all ages, cultures and socio-economic status (Campanha et al. 2015; Sodré et al. 2018; Peña-Guzmán et al. 2019; Starling et al. 2019). As caffeine is present in everyday beverages such as coffee, tea, chocolate and cola drinks, a large amount of caffeine waste is generated, which enters the aquatic environment. Many drugs also contain caffeine, namely analgesics, antihistamines, diet pills, cold remedies, and stimulants of psychophysical activity. Campanha et al. (2015) reported a three-year study on the occurrence of pharmaceuticals and hormones in surface waters of a central urban region of São Paulo State of Southeast Brazil (the Monjolinho River in São Carlos). The most frequently detected compounds at highest concentrations were caffeine, paracetamol, and atenolol (maximum concentrations 129.585μg/L, 30.421μg/L, and 8.199μg/L, respectively), while the hormones estrone and 17-β-estradiol were the least detected, at levels up to 0.0148 ng/L. The results also showed that there was an increasing trend in concentrations of most of the compounds along the river course, especially downstream of the river where there are discharges of both wastewater treatment plant effluent and raw sewage from a particular region of São Carlos city. The concentrations of contaminants were higher during dry periods as a result of the decline in the water levels. A decrease of concentrations occurred near the river mouth at different extents for each compound, being high for caffeine and atenolol and very low for carbamazepine and diclofenac.

The first nationwide survey of emerging contaminants in Brazilian waters has been published by Machado et al. (2016). One hundred drinking water samples were investigated in 22 Brazilian State capitals during two campaigns between July and September of 2011 and 2012. In addition, seven source water samples from two of the most populous regions of the country were evaluated. The analytical monitoring showed that caffeine, triclosan, atrazine, phenolphthalein and bisphenol A were found in at least one of the samples collected in the two sampling campaigns. Caffeine and atrazine were the most frequently detected substances in both drinking and source water, in agreement with the results published by Campanha et al. (2015). Caffeine concentrations in drinking water ranged from 1.8 ng/L to values above 2.0μg/L while source-water concentrations varied from 40 ng/L to approximately 19μg/L. The frequency of detection of caffeine corresponded to 93%, corresponding to a higher frequency than those found in similar Spanish and Chinese studies (<88% in the two cases). For atrazine, concentrations were found ranging from 0.002 to 0.006μg/L in drinking water and at concentrations of up to 0.15μg/L in source water. The levels of atrazine were between two and three orders of magnitude below the maximum limit established by the Brazilian regulations (2.0μg/L is the authorized value in drinking and surface water). Atrazine was found in 75% of the drinking water samples collected in both campaigns. Only two other substances, triclosan and phenolphthalein, were detected in drinking water samples. Emerging contaminants such as pharmaceuticals, hormones, and industrial products were not detected in the samples. Caffeine, atrazine, and bisphenol were detected in all seven surface samples. The highest concentrations of caffeine and bisphenol A detected were 18.828μg/L and 0.011μg/L in a reservoir connected to the Rio Grande River. Machado et al. (2016) concluded that in both drinking and source waters, caffeine reached concentration values from a few ng/L to μg/L, depending on the region of collection. The detection frequency of caffeine in drinking water was higher than other values reported in the literature. Its widespread presence in samples of treated water (detected in 93% of samples) suggested the presence of domestic sewage in the source water, considering that caffeine is a compound of anthropogenic origin, exhibiting great residence time in the environment and low susceptibility to degradation. Caffeine can be proposed as an indicator of contamination by sewage (Campanha et al. 2015; Machado et al. 2016).

More recently, Montagner et al. (2019) summarized the results of 10 years of analyses carried out in the State of São Paulo that has one of the highest population densities in Brazil and intense agricultural and industrial activities. In this work, 58 compounds (9 hormones, 14 pharmaceuticals and personal care products, 8 industrial compounds, 17 pesticides and 10 illicit drugs) were determined in 708 samples between 2006 and 2015. Samples were collected in 13 cities in the State of São Paulo (drinking water), in 10 rivers (surface waters) and 4 reservoirs, and in 5 municipal wastewater treatment plants (wastewaters). The frequency of detection of each substance strongly varied among samples. The average concentration for the synthetic contraceptive 17-α-ethynylestradiol was 777 ng/L. The highest frequencies of detection in surface water were in the following order: caffeine (97%), atrazine (69%), triclosan (43%), estriol (31%), estrone (28%) and testosterone (13%). Caffeine levels varied from 19 to 127,000 ng/L, reflecting the differences between sampling points according to their anthropogenic impact, similar to the results reported by Campanha et al. (2015) in the Monjolinho River, the main water body of São Paulo. Bisphenol A was quantified in 145 samples in a wide range of concentrations, between 2 and 13,016 ng/L. Cocaine ant its metabolite (benzoylecgonine) were detected in 53% (average concentration: 10 ng/L) and 84% (average concentration: 133 ng/L) of the samples, respectively. For drinking water (289 samples between 2007 and 2015), bisphenol A was the most frequently analyzed substance

(258 samples, average concentration 23 ng/L and a maximum concentration of 178 ng/L) followed by caffeine (231 samples, average concentration of 548 ng/L), testosterone (215 samples, average concentration of 3 ng/L), 4-n-nonylphenol (189 samples, average concentration of 114 ng/L), triclosan (186 samples with concentrations ranging from 2 to 37 ng/L), and atrazine (179 samples, average concentration of 36 ng/L). All hormones were analysed in more than 100 samples, but the frequencies of detection were between 0 and 4%, the maximum concentration being 125 ng/L for estriol. Other substances such as dioctylphthalate and octylphenols were found in drinking water. From the 33 groundwater samples collected, the most prevalent substance was atrazine (86% of samples 6/7) followed by caffeine (55%, 17/31) and bisphenol A (50%, 16/32). Finally, a preliminary risk assessment for aquatic life protection identified potential risks for caffeine, paracetamol, diclofenac, 17-α-ethynylestradiol, 17-β-estradiol, estriol, estrone, testosterone, triclosan, 4-n-nonylphenol, bisphenol A, atrazine, azoxystrobin, carbendazim, fipronil, imidacloprid, malathion and tebuconazole. As expected, raw wastewaters had high levels of concentrations, with the highest values reported for caffeine, benzoylecgonine and cocaine. Peña-Guzmán et al. (2019) reported similar findings. However, even after treatment, some emerging contaminants were still detected at relatively high concentrations in the treated wastewaters. Montagner et al. (2019) suggested that the wastewater treatment technology used in São Paulo State (Brazil) was not effective for the elimination of most of the target compounds. Campanha et al. (2015) also indicated that the main source of emerging contaminants in the river was related to the discharges from wastewater treatment plants. Drinking water criteria were available only for 22 compounds of those studied by Montagner et al. (2019) and for those substances no adverse effects were expected at the concentrations found, except for 17-β-estradiol and atrazine. Montagner et al. (2019) concluded that, among the 58 studied contaminants, caffeine, estrone, 17-α-ethynylestradiol, 17-β-estradiol, bisphenol A, atrazine, carbendazim, fipronil, malathion and imidacloprid were underlined as substances of priority concern, due to the high frequency of detection above water quality criteria. Another important conclusion was the fact that caffeine and benzoylecgonine were pertinent anthropogenic indicators, especially for regions highly inhabited and with poor sanitation structure. Since the cocaine metabolite can persist in the aquatic environment even after 2 weeks from the discharge, this substance may be a reliable marker of contamination of surface water and also for drinking water.

Caldas et al. (2019) studied the presence of 33 pesticides in surface and drinking water in Southern Brazil (data of four-year monitoring; 48 samples) and reported that 30 substances were identified and quantified, whereas four compounds (tebuconazole, atrazine, azoxystrobin, clomazone) were identified in more than half of the samples. Five class of pesticides, *i.e.*, triazines, triazoles, carbamates, strobilurins and imidazolinones, were identified. Among them, the herbicide atrazine was the most frequently detected pesticide, with concentrations ranging from 0.005μg/L to 0.049μg/L in drinking water and from 0.005μg/L to 0.037μg/L in surface water. In this study, atrazine was identified as a priority substance amongst those that pose a significant risk to the aquatic environment. The authors hypothesized a relationship with agricultural activities because atrazine is widely used in Brazil as a weed-killer in the cultivation of crops (rice, soybeans, sugarcane, corn). Other pesticides detected in the whole 4-year monitoring period were the fungicide azoxystrobin (concentrations ranging from 0.041μg/L to 0.233μg/L and from 0.1μg/L to 0.192μg/L in surface and drinking water, respectively), the herbicide clomazone (identified in more than 50% of the samples at concentrations up to 0.164μg/L) and the triazole molecules such as cyproconazole, difenoconazole, epoxiconazole, propiconazole and tebuconazole (identified in more than 80% of the samples at concentrations up to 0.46μg/L). Triazole substances are used not only as fungicides in agriculture and as biocides but also as antifungal agents in human and veterinary pharmaceuticals. Caldas et al. (2019) concluded that the concentrations detected (ranging from 0.004 to 1μg/L) were of the same order of magnitude as those detected in other regions of Brazil. The concentrations of the target pesticides detected in drinking water in the city of Rio Grande were below the limits established by the Brazilian legislation and by the World Health Organization. However, considering the values established by the European Union for individual pesticides in drinking water of 0.1μg/L, the maximum values detected for some pesticides (*e.g.*, azoxystrobin, cyproconazole, clomazone, quinclorac and tebuconazole) exceeded this limit.

Numerous Brazilian studies come to the conclusion that, in many Brazilian regions that suffer from poor sanitary conditions, there is an urgent need to study the occurrence of emerging contaminants in natural waters. Those reports show the importance of continuous monitoring of this type of substances and other water parameters for proper water management. The use of pesticides should be minimized, and if necessary, agricultural practices should be modified. Another important action should focus on improving the treatment processes applied in the municipal wastewater treatment plants (Sodré et al. 2018; Montagner et al. 2019; Peña-Guzmán et al. 2019; Caldas et al. 2019; Starling et al. 2019).

In conclusion, the extent of industrial pollution is not yet well known in Latin America, as discharges are poorly monitored and rarely aggregated at the national level. Although some domestic and industrial wastewater is treated on-site, few data are available and aggregated for national and regional assessments, unlike in Europe. Many countries lack the capacity to collect and analyze the data needed for a comprehensive assessment. Reliable monitoring of water quality is nevertheless essential to guide investment priorities. It is also important for assessing the status of aquatic ecosystems and the need to protect and/or restore them.

1.6 Conclusion

Emerging substances are a diverse group of chemical and biological agents, which possible consequent health and ecological effects are of increasing concern. Many chemical substances are non-biodegradable or refractory, toxic or carcinogenic, some are regulated, and others have been banned for several years but are still found in the environment. As described in this chapter, emerging substances are present in water resources around the world, from all the 5 continents, with specific examples given from China, Portugal, Mexico, Colombia and Brazil, raising concerns for human health and the environment. The list of substances and products is exceptionally long: pesticides, pharmaceuticals, hormones and steroids, cosmetics, personal care products, illicit drugs, surfactants, cleaning products, industrial formulations and chemicals, food additives, food packaging, metalloids, rare earth elements, nanomaterials, microplastics and pathogens. Due to insufficient chemical treatment abatement, the discharge of effluents from municipal wastewater treatment plants is an important route for the appearance of chemical substances in the aquatic environment. One of the levers for action concerns the improvement of the operation of wastewater treatment plants, such as, for example, the installation of tertiary treatment systems in order to move towards zero pollution discharge. However, attention must be paid to all sources of water pollution, such as those resulting from the intensive use of pesticides.

We are all both concerned and worried about water pollution. The multifaceted pollution resulting from our lifestyles and consumption patterns and the worrying synergies that can occur in wastewater due to the multiplicity of pollutants (pathogens, emerging substances, metals, etc.) are major threats to this vital element that is water, with consequences also for the environment and human health. However, the behavior of emerging substances in the environment and their effects on all living organisms remain largely unknown. We must (continue) to mobilize to protect this resource.

Acknowledgements Nadia Morin-Crini and Grégorio Crini (Besançon, France) thanks the FEDER (*Fonds Européen de Développment Régional*) for its financial support (NIRHOFEX Program) and the *Région Bourgogne Franche-Comté* (France) for the research grant awarded to Guest Research Director G. Torri. Ana Rita Lado Ribeiro (Porto, Portugal) acknowledges the support from Projects: PTDC/QUI-QAN/30521/2017 – POCI-01-0145-FEDER-030521 – funded by FEDER funds through COMPETE2020 – *Programa Operacional Competitividade e Internacionalização* (POCI) and by national funds (PIDDAC) through FCT/MCTES and Base Funding – UIDB/50020/2020 of the Associate Laboratory LSRE-LCM – funded by national funds through FCT/MCTES (PIDDAC).

References

Abd-Elkareem M, Abou Khalil NS, Sayed AH (2018) Hepatotoxic responses of 4-nonylphenol on African catfish (*Clarias gariepinus*): antixoidant and histochemical biomarkers. Fish Physiol Biochem 44:969–981. https://doi.org/10.1007/s10695-018-0485-1

Acir IH, Guenther K (2018) Endocrine-disrupting metabolites of alkylphenol ethoxylates – a critical review of analytical methods, environmental occurrences, toxicity, and regulation. Sci Total Environ 635:1530–1546. https://doi.org/10.1016/j.scitotenv.2018.04.079

Ahmadpanah K, Soltani M, Islami HR, Shamsaie M (2019) Effects of nonylphenol on hematological parameters and immune responses in immature rainbow trout (*Oncorhynchus mykiss*). Mar Freshw Behav Physiol 52:151–165. https://doi.org/10.1080/10236244.2019.1661779

Albuquerque AF, Ribeiro JS, Kummrow NAJA, Montagner CC, Umbuzeiro GA (2016) Pesticides in Brazilian freswaters: a critical review. Environ Sci Proc Impacts 18:779. https://doi.org/10.1039/c6em00268d

Aliyeva G, Kurkova R, Hovorkova I, Klanova J, Halsall C (2012) Organochlorine pesticides and polychlorinated biphenyls in air and soil across Azerbaijan. Environ Sci Pollut Res 19:1953–1962

Aliyeva G, Halsall C, Alasgarova K, Avazova M, Ibrahimov Y, Aghayeva R (2013) The legacy of persistent organic pollutants in Azerbaijan: an assessment of past use and current contamination. Environ Scie Pollut Res 20:1993–2008

Allen S, Allen D, Phoenix VR, Le Roux G, Durántez Jiménez P, Simonneau A, Binet S, Galop D (2019) Atmospheric transport and deposition of microplastics in a remote mountain catchment. Nat Geosci 12:339–344. https://doi.org/10.1038/s41561-019-0335-5

Álvarez-Ruiz R, Picó Y (2020) Analysis of emerging and related pollutants in aquatic biota. Trends Environ Anal Chem 25:e00082. https://doi.org/10.1016/j.teac.2020.e00082

Amato-Lourenço LF, dos Santos Galvão L, de Weger LA, Hiemstra PS, Vijver MG, Mauad T (2020) An emerging class of air pollutants: potential effects of microplastics to respiratory human health? Sci Total Environ 749:141676. https://doi.org/10.1016/j.scitotenv.2020.141676

Amin MM, Bina B, Ebrahimi A, Yavari Z, Mohammadi F, Rahimi S (2018) The occurrence, fate, and distribution of natural and synthetic hormones in different types of wastewater treatment plants in Iran. Chinese J Chem Eng 26:1132–1139. https://doi.org/10.1016/j.cjche.2017.09.005

Anastassiades M, Lehotay SJ, Stajnbaher D, Schenck FJ (2003) Fast and easy multiresidue method employing acetonitrile extraction/partitioning and "dispersive solid-phase extraction" for the determination of pesticide residues in produce. J AOAC Int 86:412–431. https://doi.org/10.1093/jaoac/86.2.412

Anastopoulos I, Pashalidis I, Orfanos AG, Manariotis ID, Tatarchuk T, Sellaoui L, Bonilla-Petriciolet A, Mittal A, Núñez-Delgado A (2020) Removal of caffeine, nicotine and amoxicillin from (waste)waters by various adsorbents. A review. J Environ Manag 261:110236. https://doi.org/10.1016/j.jenvman.2020.110236

Andrés-Costa MJ, Carmona E, Picó Y (2016) Universal method to determine acidic licit and illicit drugs and personal care products in water by liquid chromatography quadrupole time-of-flight. MethodsX 3:307–314. https://doi.org/10.1016/j.mex.2016.04.004

Annamalai J, Namasivayam V (2015) Endocrine disrupting chemicals in the atmosphere: their effects on humans and wildlife. Environ Int 76:78–97. https://doi.org/10.1016/j.envint.2014.12.006

Archer E, Petrie B, Kasprzyk-Hordern B, Wolfaardt GM (2017) The fate of pharmaceuticals and personal care products (PPCPs), endocrine disrupting contaminants (EDCs), metabolites and illicit drugs in a WWTW and environmental waters. Chemosphere 174:437–446. https://doi.org/10.1016/j.chemosphere.2017.01.101

Archundia D, Boithias L, Duwig C, Morel MC, Flores Aviles G, Martins JMF (2018) Environmental fate and ecotoxicological risk of the antibiotic sulfamethoxazole across the Katari catchment (Bolivian Altiplano): application of the GREAT-ER model. Sci Total Environ 622-623:1046–1055. https://doi.org/10.1016/j.scitotenv.2017.12.026

Arellano-Aguilar O, Betancourt-Lozano M, Aguilar-Zárate G, Ponce de Leon-Hill C (2017) Agrochemical loading in drains and rivers and its connection with pollution in coastal lagoons of the Mexican Pacific. Environ Monit Assess 189:270. https://doi.org/10.1007/s10661-017-5981-8

Arguello-Pérez MA, Mendoza-Pérez JA, Tintos-Gómez A, Ramírez-Ayala E, Godínez-Domínguez E, Silva-Bátiz FA (2019) Ecotoxicological analysis of emerging contaminants from wastewater discharges in the coastal zone of Cihuatlán (Jalisco, Mexico). Water 11:1386. https://doi.org/10.3390/w11071386

Aristizabal-Ciro C, Botero-Coy AM, López FJ, Peñueal GA (2017) Monitoring pharmaceuticals and personal care products in reservoir water used for drinking water supply. Environ Sci Pollut Res 24:7335–7347. https://doi.org/10.1007/s11356-016-8253-1

Aronzon CM, Peluso J, Coll CP (2020) Mixture toxicity of copper and nonylphenol on the embryo-larval development of *Rhinella arenarum*. Environ Sci Pollut Res 27:13985–13994. https://doi.org/10.1007/s11356-020-07857-7

Astner AF, Hayes DG, O'Neill H, Evans BR, Pingali SV, Urban VS, Young TM (2019) Mechanical formation of micro- and nano-plastic materials for environmental studies in agricultural ecosystems. Sci Total Environ 685:1097–1106. https://doi.org/10.1016/j.scitotenv.2019.06.241

Avio CG, Gorbi S, Regoli F (2017) Plastics and microplastics in the oceans: from emerging pollutants to emerged threat. Marine Environ Res 128:2–11. https://doi.org/10.1016/j.marenvres.2016.05.012

Baalousha M (2009) Aggregation and disaggregation of iron oxide nanoparticles: influence of particle concentration, pH and natural organic matter. Sci Total Environ 407:2093–2101. https://doi.org/10.1016/j.scitotenv.2008.11.022

Baalousha M, Manciulea A, Cumberland S, Kendall K, Lead JR (2008) Aggregation and surface properties of iron oxide nanoparticles: influence of pH and natural organic matter. Environ Toxicol Chem 27:1875–1882. https://doi.org/10.1897/07-559.1

Baalousha M, Yang Y, Vance ME, Colman BP, McNeal S, Xu J, Blaszczak J, Steele M, Bernhardt E, Hochella MF (2016) Outdoor urban nanomaterials: the emergence of a new, integrated, and critical field of study. Sci Total Environ 557–558:740–753. https://doi.org/10.1016/j.scitotenv.2016.03.132

Baduel C, Mueller JF, Tsai H, Gomez Ramos MJ (2015) Development of sample extraction and clean-up strategies for target and non-target analysis of environmental contaminants in biological matrices. J Chromatogr A 1426:33–47

Balaram V (1996) Recent trends in the instrumental analysis of rare earth elements in geological and industrial materials. Trends Anal Chem 15:475–486. https://doi.org/10.1016/S0165-9936(96)00058-1

Balaram V (2016) Recent advances in the determination of elemental impurities in pharmaceuticals e status, challenges and moving frontiers. Trends Anal Chem 80:83–95. https://doi.org/10.1016/j.trac.2016.02.001

Balaram V (2019) Rare earth elements: a review of applications, occurrence, exploration, analysis, recycling, and environmental impact. Geosci Front 10:1285–1303. https://doi.org/10.1016/j.gsf.2018.12.005

Balaram V (2020) Environmental impact of Pt, Pd and Rh emissions from autocatalytic converters – a brief review of the latest developments. In Hussain CM (ed) Handbook of environmental materials management. Springer. https://doi.org/10.1007/978-3-319-58538-3_194

Baniemam M, Moradi AM, Bakhtiari AR, Fatemi MR, Khanghah KE (2017) Seasonal variation of polycyclic aromatic hydrocarbons in the surface sediments of the southern Caspian Sea. Marine Pollut Bull 117:478–485

Barakos G, Mischo H, Gutzmer J (2018) A forward look into the US rare-earth industry. How potential mines can connect to the global REE market? Min Eng 70:30–37

Barbosa AMC, de Solano MLM, de Umbuzeiro GA (2015) Pesticides in drinking water – The Brazilian monitoring program. Front Public Health 3:246. https://doi.org/10.3389/fpubh.2015.00246

Barbosa MO, Moreira NFF, Ribeiro AR, Pereira MFR, Silva AMT (2016) Occurrence and removal of organic micropollutants: an overview of the watch list of EU Decision 2015/495. Water Res 94:257–279. https://doi.org/10.1016/j.watres.2016.02.047

Barbosa MO, Ribeiro AR, Ratola N, Hain E, Homem V, Pereira MFR, Blaney L, Silva AMT (2018) Spatial and seasonal occurrence of micropollutants in four Portuguese rivers and a case study for fluorescence excitation-emission matrices. Sci Total Environ 644:1128–1140. https://doi.org/10.1016/j.scitotenv.2018.06.355

Barboza LGA, Dick Vethaak A, Lavorante BRBO, Lundebye AK, Guilhermo L (2018) Marine microplastic debris: an emerging issue for food security, food safety and human health. Marine Pollut Bull 133:336–348. https://doi.org/10.1016/j.marpolbul.2018.05.047

Barceló D, Petrovic M (2008) Emerging contaminants from industrial and municipal waste. Removal technologies. Springer, Berlin. ISBN: 978-3-540-79209-3

Barnes DKA, Galgani F, Thompson RC, Barlaz M (2009) Accumulation and fragmentation of plastic debris in global environments. Philos T R Soc B 364:1985–1998. https://doi.org/10.1098/rstb.2008.0205
Barry K, Winquist V, Steenland A (2013) Perfluorooctanoic acid (PFOA) exposures and incident cancers among adults living near a chemical plant. Environ Health Perspect 121:1313–1318. https://doi.org/10.1289/ehp.1306615
Basile T, Petrella A, Petrella M, Boghetich G, Petruzzelli V, Colasuonno S, Petruzzelli D (2011) Review of endocrine-disrupting-compound removal technologies in water and wastewater treatment plants: an EU perspective. Ind Eng Chem Res 50:8389–8401. https://doi.org/10.1021/ie101919v
Bedoya-Ríos DF, Lara-Borrero JA (2018) Occurrence of endocrine disruptor chemicals in the urban cycle of Colombia. In Ahmed RG (ed) Endocrine disruptors. IntechOpen. https://doi.org/10.5772/intechopen.78325
Bedoya-Ríos DF, Lara-Borrero JA, Duke-Pardo V, Wood-Parra CA, Jimenez EM, Toro AF (2018) Study of the occurrence and ecosystem danger of selected endocrine disruptors in the urban water cycle of the city of Bogotá, Colombia. J Env Sci Health A Tox Hazard Subst Environ Eng 53:317–325. https://doi.org/10.1080/10934529.2017.1401372
Benitez FJ, Acero JL, Real FJ, Roldan G, Rodriguez E (2015) Ozonation of benzotriazole and methylindole: kinetic modeling, identification of intermediates and reaction mechanisms. J Hazard Mater 282:224–232. https://doi.org/10.1016/j.jhazmat.2014.05.085
Bhattacharyya S, Bennett J, Short LC, Theisen TS, Wichman MD, White JC, Wright S (2017a) Nanotechnology in the water industry, part 1: occurrence and risks. J AWWA 109:30–37. https://doi.org/10.5942/jawwa.2017.109.0142
Bhattacharyya S, Bennett J, Short LC, Theisen TS, Wichman MD, White JC, Wright S (2017b) Nanotechnology in the water industry, part 2: toxicology and analysis. J AWWA 109:45–53. https://doi.org/10.5942/jawwa.2017.109.0154
Bijlsma L, Botero-Coy AM, Rincón RJ, Peñuela GA, Hernández F (2016) Estimation of illicit drug use in the main cities of Colombia by means or urban wastewater analysis. Sci Total Environ 565:984–993. https://doi.org/10.1016/j.scitotenv.2016.05.078
Biskos G, Schmidt-Ott A (2012) Airborne engineered nanoparticles: potential risks and monitoring challenges for assessing their impacts on children. Paediat Resp Rev 13:79–83. https://doi.org/10.1016/j.prrv.2011.05.011
Boehm PD, Brown JS, Maxon CL, Newton FC, Galperin Y (2005) Aspects of polycyclic aromatic hydrocarbons in offshore sediments in the Azeri sector of the Caspian Sea. In: Armsworthy SL (ed) Offshore oil and gas znvironmental zffects monitoring: approaches and technologies. Battelle Press, Columbus, pp 565–586
Bokony V, Uveges B, Ujhegyi N, Verebelyi V, Nemeshazi E, Csikvari O, Hettyey A (2018) Endocrine disruptors in breeding ponds and reproductive health of toads in agricultural, urban and natural landscapes. Sci Tot Environ 634:1335–1345. https://doi.org/10.1016/j.scitotenv.2018.03.363
Born PJA, Robbins D, Haubold S, Kuhlbusch T, Fissan H, Donaldson K, Schins R, Stone V, Kreyling W, Lademann J, Krutmann J, Warheit D, Oberdorster E (2006) The potential risks of nanomaterials: a review carried out for ECETOC. Part Fibre Toxicol 3:11. https://doi.org/10.1186/1743-8977-3-11
Botero-Coy AM, Martínez-Pachón D, Boix C, Rincón RJ, Castillo N, Arias-Marín NL, Manrique-Losada L, Torres-Palma R, Moncayo-Lasso A, Hernández F (2018) An investigation into the occurrence and removal of pharmaceuticals in wastewater Colombian. Sci Total Environ 642:842–853. https://doi.org/10.1016/j.scitotenv.2018.06.088
Bradley PM, Battaglin WA, Iwanowicz LR, Clark JM, Journey CA (2016) Aerobic biodegradation potential of endocrine-disrupting chemicals in surface-water sediment at Rocky Mountain National Park, USA. Environ Toxicol Chem 35:1087–1096

Brar SK, Verma M, Tyagi RD, Surampalli RY (2010) Engineered nanoparticles in wastewater and wastewater sludge – evidence and impacts. Waste Manag 30:504–520. https://doi.org/10.1016/j.wasman.2009.10.012

Braun U, Jekel M, Gerdts G, Ivleva N, Reiber J (2018) Microplastics analytics sampling, preparation and detection methods. Discussion paper within the scope of the research focus plastics in the environment sources, sink, solutions. German Federal Ministry of Education and Research, 23 p

Brazil (2011) Ministério do Meio Ambiente. Conselho Nacional do Meio Ambiente. Resolução n° 430. Diário Oficial da União, Brasília (in Portuguese)

Brusseau ML (2019) Subsurface pollution (Chapter 15). In: Environmental and pollution science, pp 237–259. https://doi.org/10.1016/B978-0-12-814719-1.00015-X

Burri NM, Weatherl R, Moeck C, Schirmer M (2019) A review of threats to groundwater quality in the anthropocene. Sci Total Environ 684:136–154. . https://doi.org/10.1016/j.scitotenv.2019.05.236

Cabrera-Peralta J, Peña-Alvarez A (2018) Simple method for the determination of personal care product ingredients in lettuce by ultrasound-assisted extraction combined with solid-phase microextraction followed by GC-MS. J Sep Sci 41:2253–2260. https://doi.org/10.1002/jssc.201701244

Cabrerizo A, Muir DCG, De Silva AO, Wang X, Lamoureux SF, Lafrenière MJ (2018) Legacy and emerging persistent organic pollutants (POPs) in terrestrial compartments in the high Arctic: sorption and secondary sources. Environ Sci Technol 52:14187–14197. https://doi.org/10.1021/acs.est.8b05011

Cahyanurani AB, Chiu KH, Wu TM (2017) Glutathione biosynthesis plays an important role against 4-tert-octylphenol-induced oxidative stress in *Ceratophyllum demersum*. Chemosphere 183:565–573. https://doi.org/10.1016/j.chemosphere.2017.05.150

Calatayud-Vernich P, Calatayud F, Simó E, Picó Y (2016) Efficiency of QuEChERS approach for determining 52 pesticide residues in honey and honey bees. MethodsX 3:452–458. https://doi.org/10.1016/j.mex.2016.05.005

Caldas SS, Rombaldi C, Mello LL, Cerqueira MBR, Martins AF, Primel EG (2019) Occurrence of pesticides and PPCPs in surface and drinking water in Southern Brazil: data on 4-year monitoring. J Brazilian Chem Soc 30:71–80. https://doi.org/10.21577/0103-5053.20180154

Calderon-Moreno GM, Vergara-Sanchez J, Saldarriaga-Norena H, Garcia-Betancourt ML, Dominguez-Patino ML, Moeller-Chavez GE, Ronderos-Lara JG, Arias-Montoya MI, Montoya-Balbas IJ, Murillo-Tovar MA (2019) Occurrence and risk assessment of steroidal hormones and phenolic endocrine disrupting compounds in surface water in Cuautla River, Mexico. Water 11:2628. https://doi.org/10.3390/w11122628

Campanha MB, Awan AT, de Sousa DNR, Grosseli GM, Mozeto AA, Fadini PS (2015) A 3-year study on occurrence of emerging contaminants in an urban stream of São Paulo State of Southeast Brazil. Environ Sci Pollut Res 22:7936–7947. https://doi.org/10.1007/s11356-014-3929-x

Cao YQ, Yu M, Dong GH, Chen B, Zhang BY (2020) Digital PCR as an emerging tool for monitoring of microbial biodegradation. Molecules 25:706. https://doi.org/10.3390/molecules25030706

Careghini A, Mastorgio AF, Saponaro S, Sezenna E (2015) Bisphenol A, nonylphenols, benzophenones, and benzotriazoles in soils, groundwater, surface water, sediments, and food: a review. Environ Sci Pollut Res 22:5711–5741. https://doi.org/10.1007/s11356-014-3974-5

Carmona E, Picó Y (2018) The use of chromatographic methods coupled to mass spectrometry for the study of emerging pollutants in the environment. Critic Rev Anal Chem 48:305–316. https://doi.org/10.1080/10408347.2018.1430555

Carmona E, Andreu V, Picó Y (2014) Occurrence of acidic pharmaceuticals and personal care products in Turia River Basin: from waste to drinking water. Sci Total Environ 484:56–63

Carmona E, Andreu V, Picó Y (2017) Multi-residue determination of 47 organic compounds in water, soil, sediment and fish-Turia River as case study. J Pharma Biomed 146:117–125. https://doi.org/10.1016/j.jpba.2017.08.014

Carnevali O, Notarstefano V, Olivotto I, Graziano M, Gallo P, Di Marco PI, Vaccari L, Mandich A, Giorgini E, Maradonna F (2017) Dietary administration of EDC mixtures: a focus on fish lipid metabolism. Aquat Toxicol 185:95–104. https://doi.org/10.1016/j.aquatox.2017.02.007

Carpenter EJ, Smith KL (1972) Plastics on the Sargasso Sea surface. Science 175:1240–1241. https://doi.org/10.1126/science.175.4027.1240

Carpenter EJ, Anderson SJ, Harvey GR, Miklas HP, Peck BB (1972) Polystyrene spherules in coastal waters. Science 178:749–750. https://doi.org/10.1126/science.178.4062.749

Carr SA, Liu J, Tesoro AG (2016) Transport and fate of microplastic particles in wastewater treatment plants. Water Res 91:174–182. https://doi.org/10.1016/j.watres.2016.01.002

Casatta N, Stefani F, Pozzoni F, Guzzella L, Marziali L, Mascolo G, Vigano L (2016) Endocrine-disrupting chemicals in coastal lagoons of the Po River delta: sediment contamination, bioaccumulation and effects on *Manila clams*. Environ Sci Pollut Res 23:10477–10493. https://doi.org/10.1007/s11356-015-5656-3

Cauwenberghe L, Vanreusel A, Mees J, Janssen CR (2013) Microplastic pollution in deep-sea sediments. Environ Pollut 182:495–499. https://doi.org/10.1016/j.envpol.2013.08.013

Cavalheiro J, Zuloaga O, Prieto A, Preudhomme H, Amouroux D, Monperrus M (2017) Occurrence and fate of organic and organometallic pollutants in municipal wastewater treatment plants and their impact on receiving waters (Adour Estuary, France). Arch Environ Contamin Toxicol 73:619–630. https://doi.org/10.1007/s00244-017-0422-9

Ccanccapa-Cartagena A, Pico Y, Ortiz X, Reiner EJ (2019) Suspect, non-target and target screening of emerging pollutants using data independent acquisition: assessment of a Mediterranean River basin. Sci Total Environ 687:355–368. https://doi.org/10.1016/j.scitotenv.2019.06.057

Celeiro M, Lamas JP, Vila M, Garcia-Jares C, Homem V, Ratola N, Dagnac T, Llompart M (2019) Determination of multiclass personal care products in continental waters by solid-phase microextraction followed by gas chromatography-tandem mass spectrometry. J Chromatogr A 1607:460398. https://doi.org/10.1016/j.chroma.2019.460398

Chapman PM (2007) Determining when contamination is pollution – weight of evidence determinations for sediments and effluents. Environ Int 33:492–501. https://doi.org/10.1016/j.envint.2006.09.001

Chau HTC, Kadokami K, Duong HT, Kong L, Nguyen TT, Nguyen TQ, Ito Y (2018) Occurrence of 1153 organic micropollutants in the aquatic environment of Vietnam. Environ Sci Pollut Res 25:7147–7156. https://doi.org/10.1007/s11356-015-5060-z

Chen KL, Elimelech M (2006) Aggregation and deposition kinetics of fullerene (C60) nanoparticles. Langmuir 22:10994–11001. https://doi.org/10.1021/la062072v

Chen KL, Elimelech M (2007) Influence of humic acid on the aggregation kinetics of fullerene (C60) nanoparticles in monovalent and divalent electrolyte solutions. J Colloid Interf Sci 309:126–134. https://doi.org/10.1016/j.jcis.2007.01.074

Cheng JR, Wang K, Yu J, Yu ZX, Yu XB, Zhang ZZ (2018) Distribution and fate modeling of 4-nonylphenol, 4-t-octylphenol, and bisphenol A in the Yong River of China. Chemosphere 195:594–605. https://doi.org/10.1016/j.chemosphere.2017.12.085

Cheng QL, Zhou QF, Jin Z, Jiang Y, Xu LG, Jiang H, Zhao YH (2019) Bioaccumulation, growth performance, and transcriptomic response of *Dictyosphaerium sp.* after exposure to nonylphenol. Sci Tot Environ 687:416–422. https://doi.org/10.1016/j.scitotenv.2019.06.136

Chokwe TB, Okonkwo JO, Sibali LL (2017) Distribution, exposure pathways, sources and toxicity of nonylphenol and nonylphenol ethoxylates in the environment. Water SA 43:529–541. https://doi.org/10.4314/wsa.v43i4.01

Cole M, Lindeque P, Halsband C, Galloway TS (2011) Microplastics as contaminants in the marine environment: a review. Marine Pollut Bull 62:2588–2597. https://doi.org/10.1016/j.marpolbul.2011.09.025

Couto CF, Lange LC, Amaral MCS (2019) Occurrence, fate and removal of pharmaceutically active compounds (PhACs) in water and wastewater treatment plants – a review. J Water Proc Eng 32:100927. . https://doi.org/10.1016/j.jwpe.2019.100927

Cretescu I, Isopescu DN, Lutic D, Soreanu G (2019) Indoor air pollutants and the future perspectives for living space design. Indoor environment and health. Orhan Korhan, IntechOpen. https://doi.org/10.5772/intechopen.87309

Crini G, Badot PM (2007) Traitement et épuration des eaux industrielles polluées (in French). Presses Universitaires de Franche-Comté, France, Besançon. ISBN: 978-2-84867-197-0

Crini G, Badot PM (2010) Sorption processes and pollution. Presses Universitaires de Franche-Comté, France, Besançon. ISBN: 978-2-84867-304-2

Crini G, Lichtfouse E (2018) Wastewater treatment: an overview (Chapter 1). In Crini G, Lichtfouse E (eds) Green adsorbents for pollutant removal. Springer, Cham, pp 1–21. https://doi.org/10.1007/978-3-319-92111-2_1

Crini G, Euvrard E, Druart C, Morin-Crini N, Lagarrigue C, Gavoille S (2016) Alkylphénols et rejets de la filière traitement de surface (in French). GalvanoOrgano 845:53–55

Crini G, Lichtfouse E, Wilson LD, Morin-Crini N (2019) Conventional and non-conventional adsorbents for wastewater treatment. Environ Chem Lett 17:145–155. https://doi.org/10.1007/s10311-018-0785-9

Cruzeiro C, Pardal MÂ, Rocha E, Rocha MJ (2015a) Occurrence and seasonal loads of pesticides in surface water and suspended particulate matter from a wetland of worldwide interest-the Ria Formosa Lagoon, Portugal. Environ Monit Assess 187:669. https://doi.org/10.1007/s10661-015-4824-8

Cruzeiro C, Rocha E, Pardal MÂ, Rocha MJ (2015b) Uncovering seasonal patterns of 56 pesticides in surface coastal waters of the ria Formosa lagoon (Portugal), using a GC-MS method. Int J Environ Anal Chem 95:1370–1384. https://doi.org/10.1080/03067319.2015.1100724

Cruzeiro C, Rocha E, Pardal MÂ, Rocha MJ (2016) Environmental assessment of pesticides in the Mondego River Estuary (Portugal). Marine Pollut Bull 103:240–246. https://doi.org/10.1016/j.marpolbul.2015.12.013

Cui LL (2018) Polllution characteristics study in the Dongting Lake of PCDD/Fs, PCBs and PCNs. Thesis, Hebei Normal University (in Chinese)

Czech T, Bonilla NB, Gambus F, Gonzalez RR, Marin-Saez J, Vidal JLM, Frenich AG (2016) Fast analysis of 4-tertoctylphenol, pentachlorophenol and 4-nonylphenol in river sediments by QuEChERS extraction procedure combined with GC-QqQ-MS/MS. Sci Tot Environ 557-558:681–687. https://doi.org/10.1016/j.scitotenv.2016.03.135

da Costa JP, Reis V, Paço A, Costa M, Duarte AC, Rocha-Santos T (2019) Micro(nano)plastics – analytical challenges towards risk evaluation. Trends Anal Chem 111:173–184. https://doi.org/10.1016/j.trac.2018.12.013

Daesslé LW, Andrade-Tafoya PD, Lafarga-Moreno J, Mahlknecht J, van Geldern R, Beramendi-Orosco LE, Barth JAC (2020) Groundwater recharge sites and pollution sources in the wine-producing Guadalupe Valley (Mexico): restrictions and mixing prior to transfer of reclaimed water from the US-México border. Sci Total Environ 713:136715. https://doi.org/10.1016/j.scitotenv.2020.136715

Dale JG, Cox SS, Vance ME, Marr LC, Hochella MF (2017) Transformation of cerium oxide nanoparticles from a diesel fuel additive during combustion in a diesel engine. Environ Sci Technol 51:1973–1980. https://doi.org/10.1021/acs.est.6b03173

Damkjaer K, Weisser JJ, Msigala SC, Mdegela R, Ztyrishave B (2018) Occurrence, removal and risk assessment of steroid hormones in two wastewater stabilization pond systems in Morogoro, Tanzania. Chemosphere 212:1142–1154. https://doi.org/10.1016/j.chemosphere.2018.08.053

DANE (2011) Departamento Administrativo Nacional, Bogotá, Colombia

Darbre PD (2015) What are endocrine disruptors and where are they found? (chapter 1) In: Darbre PD (eds) Endocrine disruption and human health. Academic, pp 3–26. https://doi.org/10.1016/B978-0-12-801139-3.00001-6

Datta AR, Kang Q, Chen B, Ye X (2018) Fate and transport modelling of emerging pollutants from watersheds to oceans: a review. Adv Marine Biol 81:97–128. https://doi.org/10.1016/bs.amb.2018.09.002

Daughton CG (2004) Non-regulated water contaminants: emerging research. Environ Impact Assess Rev 24:711–732

de Jesus Gaffney V, Almeida CMM, Rodrigues A, Ferreira E, Benoliel MJ, Cardoso VV (2015) Occurrence of pharmaceuticals in a water supply system and related human health risk assessment. Water Res 72:199–208. https://doi.org/10.1016/j.watres.2014.10.027

De Marchi L, Coppola F, Soares AMVM, Pretti C, Monserrat JM, della Torre C, Freitas R (2019) Engineered nanomaterials: from their properties and applications, to their toxicity towards marine bivalves in a changing environment. Environ Res 178:108683. https://doi.org/10.1016/j.envres.2019.108683

de Mora S, Villeneuve JP, Sheikholeslami MR, Cattini C, Tolosa I (2004) Organochlorinated compounds in Caspian Sea sediments. Marine Pollut Bull 48:30–43

de Oliveira M, Frihling BEF, Velasques J, Magalhaes FJG, Cavalheri PS, Migliolo L (2020) Pharmaceuticals residues and xenobiotics contaminants: occurrence, analytical techniques and sustainable alternatives for wastewater treatment. Sci Total Environ 705:135568. https://doi.org/10.1016/j.scitotenv.2019.135568

Deshayes S, Eudes V, Bigourie M, Droguet C, Moilleron R (2017) Alkyphenol and phthalate contamination of all sources of greywater from French households. Sci Total Environ 599–600:883–890. https://doi.org/10.1016/j.scitotenv.2017.05.038

Dettmer-Wilde K, Engewald W (2014) Practical gas chromatography: a comprehensive reference. Springer, Berlin/Heidelberg. https://doi.org/10.1007/978-3-642-54640-2

Dévier MH, Le Menach K, Viglino L, Di Gioia L, Lachassagne P, Budzinski H (2013) Ultra-trace analysis of hormones, pharmaceutical substances, alkylphenols and phthalates in two French natural mineral waters. Sci Total Environ 443:621–632. https://doi.org/10.1016/j.scitotenv.2012.10.015

Deviller G, Lundy L, Fatta-Kassinos D (2020) Recommendations to derive quality standards for chemical pollutants in reclaimed water intended for reuse in agricultural irrigation. Chemosphere 240:124911. https://doi.org/10.1016/j.chemosphere.2019.124911

DeWitt JC (ed) (2015) Toxicological effects of perfluoroalkyl and polyfluoroalkyl substances. Molecular and integrative toxicology. Humana Press. ISBN: 978-3-319-15517-3. https://doi.org/10.1007/978-3-319-15518-0

Dhangar K, Kumar M (2020) Tricks and tracks in removal of emerging contaminants from the wastewater through hybrid treatment systems: a review. Sci Total Environ 738:140320. https://doi.org/10.1016/j.scitotenv.2020.140320

Diatloff E, Smith FW, Asher CJ (1995) Rare earth elements and plant growth-third responses of corn and mungbean to low concentrations of cerium in dilution, continuously flowing nutrient solutions. J Plant Nutr 18:1991–2003

Díaz A, Peña-Alvarez A (2017) A simple method for the simultaneous determination of pharmaceuticals and personal care products in river sediment by ultrasound-assisted extraction followed by solid-phase microextraction coupled with gas chromatography-mass spectrometry. J Chromatogr Sci 55:946–949. https://doi.org/10.1093/chromsci/bmx058

Díaz-Casallas DM, Castro-Fernández MF, Bocos E, Montenegro-Marin CE, Cresp RG (2019) 2008–2017 Bogota river quality assessment based on the water quality index. Sustainability 11:1668. https://doi.org/10.3390/su11061668

Dierkes G, Lauschke T, Becher S, Schumacher H, Földi C, Ternes T (2019) Quantification of microplastics in environmental samples via pressurized liquid extraction and pyrolysis-gas chromatography. Anal Bioanal Chem 411:6959–6968. https://doi.org/10.1007/s00216-019-02066-9

Dimpe KM, Nomngongo PN (2016) Current sample preparation methodologies for analysis of emerging pollutants in different environmental matrices. Trends Anal Chem 82:199–207. https://doi.org/10.1016/j.trac.2016.05.023

Dong WH, Xie W, Su XS, Wen CL, Cao ZP, Wan YY (2018) Review: micro-organic contaminants in groundwater in China. Hydrogeol J 26:1351–1369. https://doi.org/10.1007/s10040-018-1760-z

Dos Santos DM, Buruaem L, Goncalves RM, Williams M, Abessa DMS, Kookana R, De Marchi MRR (2018) Multiresidue determination and predicted risk assessment of contaminants of

emerging concern in marine sediments from the vicinities of submarine sewage outfalls. Marine Pollut Bull 129:299–307. https://doi.org/10.1016/j.marpolbul.2018.02.048
Dougherty J, Swarzenski PW, Dinicola RS, Reinhard M (2010) Occurrence of herbicides and pharmaceutical care personnel and products in surface water and groundwater around Liberty Bay, Puget Sound, Washington. J Environ Qual 39:1173–1180. https://doi.org/10.2134/jeq2009.0189
Dris R, Gasperi J, Rocher V, Saad M, Renault N, Tassin B (2015) Microplastic contamination in an urban area: a case study in greater Paris. J Environ Chem 12:592–599. https://doi.org/10.1071/EN14167
Dris R, Gasperi J, Mirande C, Mandin C, Guerrouache M, Langlois V, Tassin B (2017) A first overview of textile fibers, including microplastics, in indoor and outdoor environments. Environ Pollut 221:453–458. https://doi.org/10.1016/j.envpol.2016.12.013
Duedahl-Olesen L (2013) Polycyclic aromatic hydrocarbons (PAHs) in foods (chapter 13). In Rose M, Fernandes A (eds) Persistent organic pollutants and toxic metals in foods. Woodhead Publishing, pp 308–333. https://doi.org/10.1533/9780857098917.2.308
Duggal V, Rani A, Mehra R, Balaram V (2017) Risk assessment of metals from groundwater in Northeast Rajasthan. J Geolog Soc India 90:77–84
Dulio V, van Bavel B, Brorström-Lundén E, Harmsen J, Hollender J, Schlabach M, Slobodnik J, Thomas K, Koschorreck J (2018) Emerging pollutants in the EU: 10 years of NORMAN in support of environmental policies and regulations. Environ Sci Eur 30:5. https://doi.org/10.1186/s12302-018-0135-3
Dümichen E, Eisentraut P, Bannick CG, Barthel AK, Senz R, Braun U (2017) Fast identification of microplastics in complex environmental samples by a thermal degradation method. Chemosphere 174:572–584. https://doi.org/10.1016/j.chemosphere.2017.02.010
Dykstra CR, Meyer MW, Rasmussen PW, Warnke DK (2005) Contaminant concentrations and reproductive rate of Lake Superior bald eagles, 1989-2001. J Great Lakes Res 31:227–235
Ebele AJ, Abdallah MAE, Harrad S (2017) Pharmaceuticals and personal care products (PPCPs) in the freswater aquatic environment. Emerg Contamin 3:1–16. https://doi.org/10.1016/j.emcon.2016.12.004
Eggen T, Vogelsang C (2015) Occurrence and fate of pharmaceuticals and personal care products in wastewater. In Zeng EY (ed) Persistent organic pollutants (POPs): analytical techniques, environmental fate and biological effects. Comprehensive Anal Chem 67:245–294. https://doi.org/10.1016/B978-0-444-63299-9.00007-7
Eriksen M, Mason S, Wilson S, Box C, Zellers A, Edwards W, Farley H, Amato S (2013) Microplastic pollution in the surface waters of the Laurentian Great Lakes. Marine Pollut Bull 77:177–182. https://doi.org/10.1016/j.marpolbul.2013.10.007
EU (2013) EU-Directive_39/2013 2013/39/EU of the European Parliament and of the Council of 12 August 2013 amending Directives 2000/60/EC and 2008/105/EC as regards priority substances in the field of water policy. Off J Eur Union:1–17
EU (2015) EU-Decision_495/2015 Commission implementing decision (EU) 2015/495 of 20 March 2015 establishing a watch list of substances for Union-wide monitoring in the field of water policy pursuant to Directive 2008/105/EC of the European Parliament and of the Council. Off J Eur Union 78:40–42
EU (2018) EU-Decision_840/2018 Commission implementing decision (EU) 2018/840 of 5 June 2018 establishing a watch list of substances for Union-wide monitoring in the field of water policy pursuant to Directive 2008/105/EC of the European Parliament and of the Council and repealing Commission Implementing Decision (EU) 2015/495 Off J Eur Union 141:9–11
Fairbairn DJ, Arnold WA, Barber BL, Kaufenberg EF, Koskinen WC, Novak PJ, Rice PJ, Swackhamer DL (2016) Contaminants of emerging concern: mass balance and comparison of wastewater effluent and upstream sources in a mixed-use watershed. Environ Sci Technol 50:36–45. https://doi.org/10.1021/acs.est.5b03109

Fekadu S, Alemayehu E, Dewil R, Van der Bruggen B (2019) Pharmaceuticals in freshwater aquatic environments: a comparison of the African and European challenge. Sci Total Environ 654:324–337. https://doi.org/10.1016/j.scitotenv.2018.11.072

Félix-Cañedo TE, Durán-Álvarez JC, Jiménez-Cisneros B (2013) The occurrence and distribution of a group of organic micropollutants in Mexico City's water sources. Sci Total Environ 454:109–118

Fendall LS, Sewell MA (2009) Contributing to marine pollution by washing your face: microplastics in facial cleansers. Marine Pollut Bull 58:1225–1228. https://doi.org/10.1016/j.marpolbul.2009.04.025

Fernandes MJ, Paíga P, Silva A, Llaguno CP, Carvalho M, Vázquez FM, Delerue-Matos C (2020) Antibiotics and antidepressants occurrence in surface waters and sediments collected in the north of Portugal. Chemosphere:239. https://doi.org/10.1016/j.chemosphere.2019.124729

Fernandez P, Grimalt JO, Vilanova RM (2002) Atmospheric gas-particle partitioning of polycyclic aromatic hydrocarbons in high mountain regions of Europe. Environ Sci Technol 36:1162–1168

Fernie K, Letcher RJ (2010) Historical contaminants, flame retardants, and halogenated phenolic compounds in Peregrine Falcon (*Falco peregrinus*) Nestlings in the Canadian Great Lakes Basin. Environ Sci Technol 44:3520–3526

Free CM, Jensen OP, Mason SA, Eriksen M, Williamson NJ, Boldgiv B (2014) High-levels of microplastic pollution in a large, remote, mountain lake. Marine Pollut Bull 85:156–163. https://doi.org/10.1016/j.marpolbul.2014.06.001

Frias J, Pagter E, Nash R, O'Connor I, Carretero O, Figueiras A, Vinas L, Gago J, Antunes JC, Bessa F, Sobral P, Goruppi A, Tirelli V, Pedrotti ML, Suaria G, Aliani S, Lopes C, Raimundo J, Caetano M, Palazzo L, de Lucia GA, Lorenzo SM, Grueiro G, Fernandez-Gonzalez V, Andrade J, Dris R, Laforsch C, Scholz-Böttcher B, Gerdts G (2018) Standardised protocol for monitoring microplastics in sediments. Report Number WP4. https://doi.org/10.13140/RG.2.2.36256.89601/1

Fu F, Akagi T, Yabuki S, Iwaki M (2001) The variation of REE (rare earth elements) patterns in soil-grown plants: a new proxy for the source of rare earth elements and silicon in plants. Plant Soil 235:53–64

Furdui VI, Stock NL, Ellis DA, Butt CM, Whittle DM, Crozier PW, Reiner EJ, Muir DCG, Ferreira RC, Couto Junior OM, De Carvalho KQ, Arroyo PA, Barros MASD (2015) Effect of solution pH on the removal of paracetamol by activated carbon of Dende coconut mesocarp. Chem Biochem Eng Q 29:47–53

Gagné F, Bouchard B, André C, Farcy E, Fournier M (2011) Evidence of feminization in wild *Elliptio complanata* mussels in the receiving waters downstream of a municipal effluent outfall. Comp Biochem Physiol C Toxicol Pharmacol 153:99–106. https://doi.org/10.1016/j.cbpc.2010.09.002

Gago-Ferrero SEL, Bletsou AA, Aalizadeh R, Hollender J, Thomaidis NS (2015) Extended suspect and non-target strategies to characterize emerging polar organic contaminants in raw wastewater with LC-HRMS/MS. Environ Sci Technol 49:12333–12341. https://doi.org/10.1021/acs.est.5b03454

Galgani F, Hanke G, Werner S, De Vrees L (2013) Marine litter within the European marine strategy framework directive. ICES J Marine Sci 70:1055–1064. https://doi.org/10.1093/icesjms/fst122

Gallego-Schmid A, Tarpani RRZ (2019) Life cycle assessment of wastewater in developing countries. Water Res 153:63–79. https://doi.org/10.1016/j.watres.2019.01.010

Gallo V, Leonardi G, Genser B, Lopez-Espinosa MJ, Frisbee SJ, Karlsson L, Ducatman AM, Fletcher T (2012) Serum perfluooctanoate (PFOA) and perfluorooctane sulfonate (PFOS) concentrations and liver function biomarkers in a population with elevated PFOA exposure. Environ Health Perspect 120:655–660. https://doi.org/10.1289/ehp.1104436

Gao LR, Zheng MH, Lv YB, Fu Q, Tan L, Zhu QQ (2016) Persistent organic chemicals in the environment: status and trends in the Pacific Basin Countries. In Loganathan BG, Khim JS, Kodavanti PRS, Masunaga S (eds) Temporal trends. American Chemical Society, Washington, DC, pp 73–102

Gao ZC, Lin YL, Xu B, Xia Y, Hu CY, Zhang TY, Cao TC, Chu WH, Gao NY (2019) Effect of UV wavelength on humic acid degradation and disinfection byproduct formation during the UV/chlorine process. Water Res 154:199–209. https://doi.org/10.1016/j.watres.2019.02.004

García J, García-Galán MJ, Day JW, Boopathy R, White JR, Wallace S, Hunter RG (2020) A review of emerging organic contaminants (EOCs), antibiotic resistant bacteria (ARB), and antibiotic resistance genes (ARGs) in the environment: increasing removal with wetlands and reducing environmental impacts. Bioresour Technol 307:123228. https://doi.org/10.1016/j.biortech.2020.123228

García-Córcoles MT, Rodríguez-Gómez R, de Alarcón-Gómez B, Çipa M, Martín-Pozo L, Kauffmann JM, Zafra-Gómez A (2019) Chromatographic methods for the determination of emerging contaminants in natural water and wastewater samples: a review. Critical Rev Anal Chem 49:160–186. https://doi.org/10.1080/10408347.2018.1496010

García-de la Parra LM, Cervantes-Mojica LJ, González-Valdivia C, Martínez-Cordero FJ, Aguilar-Zárate G, Bastidas-Bastidas P, Betancourt-Lozano M (2012) Distribution of pesticides and PCBs in sediments of agricultural drains in the Culiacan Valley, Sinaloa, Mexico. Arch Environ Contam Toxicol 63:323–336. https://doi.org/10.1007/s00244-012-9780-5

García-Nieto E, Juárez-Santacruz L, Ortiz-Ortiz E, Luna-Zendejas HS, Frías-Márquez DM, Muñoz-Nava H, Romo-Gómez C (2019) Ecotoxicological assessment of sediment from Texcalac River and agricultural soil of riverside area, in Tlaxcala, Mexico. Chem Ecol 35:300–318. https://doi.org/10.1080/02757540.2018.1546297

Gasperi J, Wright SL, Dris R, Collard F, Mandin C, Guerrouache M, Langlois V, Kelly FJ, Tassin B (2018) Microplastics in air: are we breathing it in? Curr Opin Environ Sci Health 1:1–5. https://doi.org/10.1016/j.coesh.2017.10.002

Gee RH, Rockett LS, Rumsby PC (2015) Considerations of endocrine disruptors in drinking water (chapter 18). In Darbre PD (ed) Endocrine disruption and human health. Academic, pp 319–341. https://doi.org/10.1016/B978-0-12-801139-3.00018-1

Gibson JC (2020) Emerging persistent chemicals in human biomonitoring for populations in the Arctic: a Canadian perspective. Sci Total Environ 708:134538. https://doi.org/10.1016/j.scitotenv.2019.134538

Giese E (2018) Rare earth elements: therapeutic and diagnostic applications in modern medicine. Clinical Med Reports 2. https://doi.org/10.15761/CMR.1000139

Gil MJ, Soto AM, Usma JI, Gutiérrez OD (2012) Emerging contaminants in waters: effects and possible treatments (in Spanish). Cleaner Production 7:52–73. http://www.scielo.org.co/pdf/pml/v7n2/v7n2a05.pdf

Gilabert-Alarcón C, Salgado-Méndez SO, Walter Daesslé L, Mendoza-Espinosa LG, Villada-Canela M (2018) Regulatory challenges for the use of reclaimed water in Mexico: a case study in Baja California. Water 10:1432. https://doi.org/10.3390/w10101432

Gioia R, Dachs J, Nizzetto L, Berrojalbiz N, Galbán C, Del Vento S, Méjanelle L, Jones KC (2011) Sources, transport and fate of organic pollutants in the oceanic environment (Chapter 8). In Quante M, Ebinghaus R and Flöser G (eds) Persistent pollution – past, present and future. Springer, Berlin, pp 111–139. https://doi.org/10.1007/978-3-642-17419-3_8

Giovanni P (2016) Rare earth elements in human and environmental health. At the crossroads between toxicity and safety. CRC Press, Singapore, 292p. ISBN: 9789814745000

Gniadek M, Dąbrowska A (2019) The marine nano-and microplastics characterisation by SEM-EDX: the potential of the method in comparison with various physical and chemical approaches. Marine Pollut Bull 148:210. https://doi.org/10.1016/j.marpolbul.2019.07.067

Godage NH, Gionfriddo E (2020) Use of natural sorbents as alternative and green extractive materials: a critical review. Anal Chim Acta 1125:187–200. https://doi.org/10.1016/j.aca.2020.05.045

Gogoi A, Mazumder P, Tyagi VK, Tushara Chaminda GGT, An AK, Kumar M (2018) Occurrence and fate of emerging contaminants in water environment: a review. Groundw Sustain Develop 6:169–180. https://doi.org/10.1016/j.gsd.2017.12.009

Gong J, Duan DD, Yang Y, Ran Y, Chen DY (2016) Seasonal variation and partitioning of endocrine disrupting chemicals in waters and sediments of the Pearl River system, South China. Environ Pollut 219:735–741. https://doi.org/10.1016/j.envpol.2016.07.015

Gonzalez S, Lopez-Roldan R, Cortina JL (2012) Presence and biological effects of emerging contaminants in Llobregat River basin: a review. Environ Poll 161:83–92. https://doi.org/10.1016/j.envpol.2011.10.002

Gonzalez-Rey M, Tapie N, Le Menach K, Dévier MH, Budzinski H, Bebianno MJ (2015) Occurrence of pharmaceutical compounds and pesticides in aquatic systems. Marine Pollut Bull 96:384–400. https://doi.org/10.1016/j.marpolbul.2015.04.029

Good KD, Bergman LE, Klara SS, Leitch ME, VanBriesen JM (2016) Implications of engineered nanomaterials in drinking water sources. J AWWA 108:E1–E17. https://doi.org/10.5942/jawwa.2016.108.0013

Gottschalk F, Nowack B (2011) The release of engineered nanomaterials to the environment. J Environ Monit 13:1145–1155. https://doi.org/10.1039/C0EM00547A

Gottschalk F, Sonderer T, Scholz RW, Nowack B (2009) Modeled environmental concentrations of engineered nanomaterials (TiO_2, ZnO, Ag, CNT, Fullerenes) for different regions. Environ Sci Technol 43:9216–9222. https://doi.org/10.1021/es9015553

Gottschalk F, Sun T, Nowack B (2013) Environmental concentrations of engineered nanomaterials: review of modeling and analytical studies. Environ Pollut 181:287–300. https://doi.org/10.1016/j.envpol.2013.06.003

Grieger KD, Fjordbøge A, Hartmann NB, Eriksson E, Bjerg PL, Baun A (2010) Environmental benefits and risks of zero-valent iron nanoparticles (nZVI) for *in situ* remediation: risk mitigation or trade-off? J Contam Hydrol 118:165–183. https://doi.org/10.1016/j.jconhyd.2010.07.011

Guo X, Wang J (2019) The chemical behaviors of microplastics in marine environment: a review. Marine Pollut Bull 142:1–14. https://doi.org/10.1016/j.marpolbul.2019.03.019

Guo BS, Zhu WM, Xiong PK, Ji YJ, Liu Z, Wu ZM (1988) Rare earths in agriculture. Agricultural Scientific Technological Press, Beijing. pp 23e208

Guo H, Chen LL, Li GL, Deng SP, Zhu CH (2019) Accumulation and depuration of nonylphenol and its effect on the expressions of vitellogenin and vitellogenin receptor in freshwater prawn *Macrobrachium rosenbergii*. Bull Environ Contam Toxicol 103:729–733. https://doi.org/10.1007/s00128-019-02714-x

Guo JJ, Huang XP, Xiang L, Wang YZ, Li YW, Li H, Cai QY, Mo CH, Wong MH (2020) Source, migration and toxicology of microplastics in soil. Environ Int 137:105263. https://doi.org/10.1016/j.envint.2019.105263

Gwenzi W, Mangori L, Danha C, Chaukura N, Dunjana N, Sanganyado E (2018) Sources, behaviour, and environmental and human health risks of high-technology rare earth elements as emerging contaminants. Sci Total Environ 636:299–313. https://doi.org/10.1016/j.scitotenv.2018.04.235

Hammes J, Gallego-Urrea JA, Hassellöv M (2013) Geographically distributed classification of surface water chemical parameters influencing fate and behavior of nanoparticles and colloid facilitated contaminant transport. Water Res 47:5350–5361. https://doi.org/10.1016/j.watres.2013.06.015

Hamza RA, Iorhemen OT, Tay JH (2016) Occurrence, impacts and removal of emerging substances of concern from wastewater. Environ Technol Innovat 5:161–175. https://doi.org/10.1016/j.eti.2016.02.003

Hanus MJ, Harris AT (2013) Nanotechnology innovations for the construction industry. Prog Mater Sci 58:1056–1102. https://doi.org/10.1016/j.pmatsci.2013.04.001

Hao X, Cao Y, Zhang L, Zhang Y, Liu J (2015) Fluoroquinolones in the Wenyu River catchment, China: occurrence simulation and risk assessment: exposure simulation and risk assessment of fluoroquinolones. Environ Toxicol Chem 34:2764–2770

Hardin IR, Kim Y (2016) Nanotechnology for antimicrobial textiles (chapter 6). In Sun G (ed) Antimicrobial textiles. Woodhead Publishing, Cambridge, pp 87–97. https://doi.org/10.1016/B978-0-08-100576-7.00006-7

Harmon SM (2015) The toxicity of persistent organic pollutants to aquatic organisms (chapter 18). In Zeng EY (ed) Persistent organic pollutants (POPs): analytical techniques, environmental fate and biological effects. Comprehensive analytical chemistry, vol 67, pp 587–613. https://doi.org/10.1016/B978-0-444-63299-9.00018-1

Hart CE, Lauth MJ, Hunter CS, Krasny BR, Hardy KM (2016) Effect of 4-nonylphenol on the immune response of the Pacific oyster *Crassostrea gigas* following bacterial infection with *Vibrio campbellii*. Fish Shellfish Immunol 58:449–461. https://doi.org/10.1016/j.fsi.2016.09.054

Hartmann NB, Hüffer T, Thompson RC, Hassellöv M, Verschoor A, Daugaard AE, Rist S, Karlsson T, Brennholt N, Cole M, Herrling MP, Hess MC, Ivleva NP, Lusher AL, Wagner M (2019) Are we speaking the same language? Recommendations for a definition and categorization framework for plastic debris. Environ Sci Technol 53:1039–1047. https://doi.org/10.1021/acs.est.8b05297

Hasegawa H, Begum ZE, Murase K, Sawai H, Mashio AS, Maki T, Rahman IMM (2018) Chelator-induced recovery of rare earths from end-of-life fluorescent lamps with the aid of mechanochemical energy. Waste Manag 80:17–25. https://doi.org/10.1016/j.wasman.2018.08.049

Haxel G et al (2002) Rare earth elements – critical resources for high technology. USGS Fact, Sheet 087-02

He N, Sun X, Zhong Y, Sun KF, Liu WJ, Duan SS (2016) Removal and biodegradation of nonylphenol by four freshwater microalgae. Int J Env Res Public Health 13:1239. https://doi.org/10.3390/ijerph13121239

Helm PA, Jantunen LM, Ridal J, Bidleman TF (2003) Spatial distribution of polychlorinated naphthalenes in air over the great lakes and air-water gas exchange in Lake Ontario. Environ Toxicol Chem 22:1937–1944

Hernández F, Pozo OJ, Sancho JV, López FJ, Marín JM, Ibáñez M (2005) Strategies for quantification and confirmation of multi-class polar products pesticides and transformation in water by LC-MS2 using triple quadrupole and hybrid quadrupole time-of-flight analyzers. TrAC Trends Anal Chem 24:596–612. https://doi.org/10.1016/j.trac.2005.04.007

Hernández F, Ibáñez M, Gracia-Lor E, Sancho JV (2011a) Retrospective LC-QTOF-MS analysis searching for pharmaceutical metabolites in urban wastewater. J Sep Sci 34:3517–3526. https://doi.org/10.1002/jssc.201100540

Hernández F, Portolés T, Pitarch E, López FJ (2011b) Gas chromatography coupled to high-resolution time-of-flight mass spectrometry to analyze trace-level organic compounds in the environment, food safety and toxicology. TrAC Trends Anal Chem 30:388–400. https://doi.org/10.1016/j.trac.2010.11.007

Hernández F, Portolés T, Ibáñez M, Bustos-López MC, Díaz R, Botero-Coy AM, Fuentes CL, Peñuela G (2012) Use of time-of-flight mass spectrometry for large screening of organic pollutants in surface waters and soils from a rice production area in Colombia. Sci Total Environ 439:249–259. https://doi.org/10.1016/j.scitotenv.2012.09.036

Hernández F, Ibáñez M, Botero-Coy AM, Bade R, Bustos-López MC, Rincón J, Moncayo A, Bijlsma L (2015) LC-QTOF MS screening of more than 1,000 licit and illicit drugs and their metabolites in wastewater and surface waters from the area of Bogotá, Colombia Bioanal. Anal Chem 407:6405–6416. https://doi.org/10.1007/s00216-015-8796-x

Hernández-Padilla F, Margni M, Noyola A, Guereca-Hernandez L, Bulle C (2017) Assessing wastewater treatment in Latin America and the Caribbean: enhancing life cycle assessment interpretation by regionalization and impact assessment sensibility. J Clean Prod 142:2140–2215. https://doi.org/10.1016/j.jclepro.2016.11.068

Herrera-Melian J, Guedes-Alonso R, Borreguero-Fabelo A, Santana-Rodriguez J, Sosa-Ferrera Z (2018) Study on the removal of hormones from domestic wastewaters with lab-scale constructed wetlands with different substrates and flow directions. Environ Sci Pollut Res 25:20374–20384. https://doi.org/10.1007/s11356-017-9307-8

Hidalgo-Ruz V, Gutow L, Thompson RC, Thiel M (2012) Microplastics in the marine environment: a review of the methods used for identification and quantification. Environ Sci Technol 46:3060–3075. https://doi.org/10.1021/es2031505

Hillery BR, Basu I, Sweet CW, Hites RA (1997) Temporal and spatial trends in a long-term study of gas phase PCB concentrations near the Great Lakes. Environ Sci Technol 31:1811–1816

Hochella MF, Lower SK, Maurice PA, Penn RL, Sahai N, Sparks DL, Twining BS (2008) Nanominerals, mineral nanoparticles, and earth systems. Science 319:1631–1635. https://doi.org/10.1126/science.1141134

Houde M, Wang X, Colson TLL, Gagnon P, Ferguson SH, Ikonomou MG, Dubetz C, Addison RF, Muir DCG (2019) Trends of persistent organic pollutans in ringed seals (*Phoca hispida*) from the Canadian Artic. Sci Total Environ 665:1135–1146. https://doi.org/10.1016/j.scitotenv.2019.02.138

Hu Z, Haneklaus S, Sparovek G, Schnug E (2006) Rare earth elements in soils. Commun Soil Sci Plant Anal 37:1381–1420. https://doi.org/10.1080/00103620600628680

Hu JD, Zevi Y, Kou XM, Xiao J, Wang XJ, Jin Y (2010) Effect of dissolved organic matter on the stability of magnetite nanoparticles under different pH and ionic strength conditions. Sci Total Environ 408:3477–3489. https://doi.org/10.1016/j.scitotenv.2010.03.033

Hu Y, Yan X, Shen Y, Di MX, Wang J (2019) Occurrence, behavior and risk assessment of estrogens in surface water and sediments from Hanjiang River, Central China. Ecotoxicology 28:143–153. https://doi.org/10.1007/s10646-018-2007-4

Hurst C (2010) China's rare earth elements industry: what we can the west learn? Department of Defense, The U.S. Government, pp 1–43

Iavicoli I, Leso V, Beezhold DH, Shvedova AA (2017) Nanotechnology in agriculture: opportunities, toxicological implications, and occupational risks. Toxicol Appl Pharmacol 329:96–111. https://doi.org/10.1016/j.taap.2017.05.025

Ibáñez M, Sancho JV, Hernández F, McMillan D, Rao R (2008) Rapid screening of non-target organic pollutants in water by ultraperformance liquid chromatography coupled to time-of-light mass spectrometry. TrAC Trends Anal Chem 27:481–489. https://doi.org/10.1016/j.trac.2008.03.007

Ibáñez M, Guerrero C, Sancho JV, Hernández F (2009) Screening of antibiotics in surface and wastewater samples by ultra-high-pressure liquid chromatography to hybrid coupled quadrupole time-of-flight mass spectrometry. J Chromatogr A 1216:2529–2539. https://doi.org/10.1016/j.chroma.2009.01.073

Illés E, Tombácz E (2006) The effect of humic acid adsorption on pH-dependent surface charging and aggregation of magnetite nanoparticles. J Colloid Interface Sci 295:115–123. https://doi.org/10.1016/j.jcis.2005.08.003

Imhof HK, Sigl R, Brauer E, Feyl S, Giesemann P, Klink S, Leupolz K, Loder MG, Loschel LA, Missun J, Muszynski S, Ramsperger AF, Schrank I, Speck S, Steibl S, Trotter B, Winter I, Laforsch C (2017) Spatial and temporal variation of macro-, meso- and microplastic abundance on a remote coral island of the Maldives, Indian Ocean. Marine Pollut Bull 116:340–347. https://doi.org/10.1016/j.marpolbul.2017.01.010

Inyinbor AA, Bello OS, Oluyori AP, Inyinbor HE, Fadiji AE (2019) Wastewater conservation and reuse in quality vegetable cultivation: overview, challenges and future prospects. Food Control 98:489–500. https://doi.org/10.1016/j.foodcont.2018.12.008

IOM (2014) Institute of Medicine. Identifying and reducing environmental health risks of chemicals in our society: workshop summary. National Academic Press, Washington, DC, 176 p. https://doi.org/10.17226/18710

Iroegbu AOC, Sadiku RE, Ray SS, Hamam Y (2020) Plastics in municipal drinking water and wastewater treatment plant effluents: challenges and opportunities for South Africa – a review. Environ Sci Pollut Res. https://doi.org/10.1007/s11356-020-08194-5

ISO (2013) ISO 472:2013 plastics – vocabulary. https://www.iso.org/obp/ui/#iso:std:iso:472:ed-4:v1:en

ISO (2020) ISO/CD 24187 principles for the development of standards for investigation procedures of plastics in environmental matrices and related materials

Ivey CD, Wang N, Alvarez D, Hammer EJ, Bauer CR (2018) Chronic toxicity of 4-nonylphenol to two unionid mussels in water-only exposures. Bull Environ Contam Toxicol 101:423–427. https://doi.org/10.1007/s00128-018-2422-5

Janousek RM, Muller J, Knepper TP (2020) Combined study of source, environmental monitoring and fate of branched alkylphenols: the chain length matters. Chemosphere 241:124950. https://doi.org/10.1016/j.chemosphere.2019.124950

Janssens I, Tanghe T, Verstraete W (1997) Micropollutants: a bottleneck in sustainable wastewater treatment. Water Sci Technol 35:12–26. https://doi.org/10.1016/S0273-1223(97)00221-7

Jardak K, Drogui P, Daghrir R (2016) Surfactants in aquatic and terrestrial environment: occurrence, behavior, and treatment processes. Environ Sci Pollut Res 23:3195–3216. https://doi.org/10.1007/s11356-015-5803-x

Javedankherad I, Esmaili-Sari A, Bahramifar N (2013) Levels and distribution of organochlorine pesticides and polychlorinated biphenyls in water and sediment from the International Anzali Wetland, North of Iran. Bull Environ Contamin Toxicol 90:285–290

Jeremiason JD, Hornbuckle KC, Eisenreich SJ (1994) PCBs in lake-superior, 1978–1992 – decreases in water concentrations reflect loss by volatilization. Environ Sci Technol 28:903–914

Jiang Q, Ngo HH, Nghiem LD, Hai FI, Price WE, Zhang J, Guo W (2018) Effect of hydraulic retention time on the performance of a hybrid moving bed biofilm reactor-membrane bioreactor system for micropollutants removal from municipal wastewater. Bioresour Technol 247:1228–1232. https://doi.org/10.1016/j.biortech.2017.09.114

Jin H, Zhu L (2016) Occurrence and partitioning of bisphenol analogues in water and sediment from Liaohe River Basin and Taihu Lake, China. Water Res 103:343–351. https://doi.org/10.1016/j.watres.2016.07.059

Jordao R, Garreta E, Campos B, Lemos MFL, Soares A, Tauler R, Barata C (2016) Compounds altering fat storage in *Daphnia magna*. Sci Tot Environ 545–546:127–136. https://doi.org/10.1016/j.scitotenv.2015.12.097

Jowitt SM, Werner TT, Weng Z, Mudd G (2018) Recycling of the rare earth elements. Curr Opin Green Sustain Chem 13:1–7. https://doi.org/10.1016/j.cogsc.2018.02.008

Juliano C, Magrini GA (2017) Cosmetic ingredients as emerging pollutants of environmental and health concern. A mini-review. Cosmetics 4:11. https://doi.org/10.3390/cosmetics4020011

Jurado A, Walther M, Díaz-Cruz MS (2019) Occurrence, fate and environmental risk assessment of the organic microcontaminants included in the watch lists set by EU decisions 2015/495 and 2018/840 in the groundwater of Spain. Sci Total Environ 663:285–296. https://doi.org/10.1016/j.scitotenv.2019.01.270

Kah M, Hofmann T (2014) Nanopesticide research: current trends and future priorities. Environ Int 63:224–235. https://doi.org/10.1016/j.envint.2013.11.015

Kallenborn R, Hung H, Brorström-Lundén E (2015) Atmospheric long-range transport of persistent organic pollutants (POPs) into Polar Regions. In Zeng EY (ed) Persistent organic pollutants (POPs): analytical techniques, environmental fate and biological effects. Comprehensive Anal Chem 67:411–432. https://doi.org/10.1016/B978-0-444-63299-9.00013-2

Kanhai LDK, Johansson C, Frias JPGL, Gardfeldt K, Thompson RC, O'Connor I (2019) Deep sea sediments of the Arctic Central Basin: a potential sink for microplastics. Deep-Sea Res Pt I 145:137–142. https://doi.org/10.1016/j.dsr.2019.03.003

Kannan K, Yamashita N, Imagawa T, Decoen W, Khim JS, Day RM, Summer CL, Giesy JP (2000) Polychlorinated naphthalenes and polychlorinated biphenyls in fishes from Michigan waters including the Great Lakes. Environ Sci Technol 34:566–572

Käppler A, Fischer M, Scholz-Böttcher BM, Oberbeckmann S, Labrenz M, Fischer D, Eichhorn K-J, Voit B (2018) Comparison of μ-ATR-FTIR spectroscopy and py-GCMS as identification tools for microplastic particles and fibers isolated from river sediments. Anal Bioanal Chem 410:5313–5327. https://doi.org/10.1007/s00216-018-1185-5

Kasprzyk-Hordern B, Dinsdale RM, Guwy AJ (2008) The occurrence of pharmaceuticals, personal care products, endocrine disruptors and illicit drugs in surface water in South Wales, UK. Water Res 42:3498–3518

Katsigiannis A, Noutsopoulos C, Mantziaras J, Gioldasi M (2015) Removal of emerging pollutants through granular activated carbon. Chem Eng J 280:49–57. https://doi.org/10.1016/j.cej.2015.05.109

Khan AM, Yusoff I, Abu Baker NK, Abu Baker AF, Alias Y, Mispan MS (2017) Accumulation, uptake and bioavailability of rare earth elements (REESs) in soil grown plants from ex-mining area in Perak, Malaysia. Appl Ecol Environ Res 15:117–133. https://doi.org/10.15666/aeer/1503_117133

Khan NA, Khan SU, Ahmed S, Farooqi IH, Yousefi M, Mohammadi AA, Changani F (2020) Recent trends in disposal and treatment technologies of emerging-pollutants – a critical review. Trends Anal Chem 122:115744. https://doi.org/10.1016/j.trac.2019.115744

Kim RO, Kim H, Lee YM (2019) Evaluation of 4-nonylphenol and bisphenol A toxicity using multiple molecular biomarkers in the water flea *Daphnia magna*. Ecotoxicology 28:167–174. https://doi.org/10.1007/s10646-018-2009-2

Klaine SJ, Alvarez PJJ, Batley GE, Fernandes TF, Handy RD, Lyon DY, Mahendra S, McLaughlin MJ, Lead JR (2008) Nanomaterials in the environment: behavior, fate, bioavailability, and effects. Environ Toxicol Chem 27:1825–1851. https://doi.org/10.1897/08-090.1

Klamerth N, Miranda N, Malato S, Agüera A, Fernández-Alba AR, Maldonado MI, Coronado JM (2009) Degradation of emerging contaminants at low concentrations in MWTPs effluents with mild solar photo-Fenton and TiO_2. Catal Today 144:124–130. https://doi.org/10.1016/j.cattod.2009.01.024

Klemes MJ, Ling Y, Ching C, Wu C, Xiao L, Helbling DWR (2019) Reduction of a tetrafluoroterephthalonitrile-β-cyclodextrin polymer to remove anionic micropollutants and perfluorinated alkyl substances from water. Angew Chem Int Ed 58:12049–12053. https://doi.org/10.1002/anie.201905142

Knepper TP, Sacher F, Lange FT, Brauch HJ, Karrenbrock F, Roeden O, Linder K (1999) Detection of polar organic substances relevant for drinking water. Waste Manag 19:77–99. https://doi.org/10.1016/S0956-053X(99)00003-3

Koba O, Grabicova K, Cerveny D, Turek J, Kolarova J, Randak T, Zlabek V, Grabic R (2018) Transport of pharmaceuticals and their metabolites between water and sediments as a further potential exposure for aquatic organisms. J Hazard Mater 342:401–407. https://doi.org/10.1016/j.jhazmat.2017.08.039

Kolpin DW, Furlong ET, Meyer MT, Thurman EM, Zaugg SD, Barber LB, Buxton HT (2002) Pharmaceuticals, hormones, and other organic wastewater contaminants in US streams, 1999-2000: a national reconnaissance. Environ Sci Technol 36:1202–1211. https://doi.org/10.1021/es011055j

Kosma CI, Lambropoulou DA, Albanis TA (2014) Investigation of PPCPs in wastewater treatment plants in Greece: occurrence, removal and environmental risk assessment. Sci Total Environ 466-467:421–438. https://doi.org/10.1016/j.scitotenv.2013.07.044

Koumaki E, Mamais D, Noutsopoulos C, Nika MC, Bletsou AA, Thomaidis NS, Eftaxias A, Stratogianni G (2015) Degradation of emerging contaminants from water under natural sunlight: the effect of season, pH, humic acids and nitrate and identification of photodegradation by-products. Chemosphere 138:675–681

Koumaki E, Mamais D, Noutsopoulos C (2017) Environmental fate of non-steroidal anti-inflammatory drugs in river water/sediment systems. J Hazard Mater 323:233–241

Koumaki E, Mamais D, Noutsopoulos C (2018) Assessment of the environmental fate of endocrine disrupting chemicals in rivers. Sci Total Environ 628-629:947–958. https://doi.org/10.1016/j.scitotenv.2018.02.110

Kovalakova P, Cizmas L, McDonald TJ, Marsalek B, Feng B, Sharma VK (2020) Occurrence and toxicity of antibiotics in the aquatic environment: a review. Chemosphere 251:126351. https://doi.org/10.1016/j.chemosphere.2020.126351

Krauss M, Singer H, Hollender J (2010) LC-high resolution MS in environmental analysis: from target screening to the identification of unknowns. Anal Bioanal Chem 397:943–951. https://doi.org/10.1007/s00216-010-3608-9

Kroon FJ, Berry KLE, Brinkman DL, Kookana R, Leusch FDL, Melvin SD, Neale PA, Negri AP, Puotinen M, Tsang JJ, van de Merwe JP, Williams M (2020) Sources, presence and potential effects of contaminants of emerging concern in the marine environments of the Great Barrier Reef and Torres Strait, Australia. Sci Total Environ 719:135140. https://doi.org/10.1016/j.scitotenv.2019.135140

Kuhar N, Sil S, Verma T, Umapathy S (2018) Challenges in application of Raman spectroscopy to biology and materials. RSC Adv 8:25888–25908. https://doi.org/10.1039/C8RA04491K

Kumar M, Jaiswal S, Sodhi KK, Shree P, Singh DK, Agrawal PK, Shukla P (2019) Antibiotics bioremediation: perspectives on its ecotoxicity and resistance. Environ Int 124:448–461. https://doi.org/10.1016/j.envint.2018.12.065

Kumar M, Xiong X, He M, DCW T, Gupta J, Khan E, Harrad S, Hou D, Ok YS, Bolan NS (2020) Microplastics as pollutants in agricultural soils. Environ Pollut 265:114980. https://doi.org/10.1016/j.envpol.2020.114980

Kung TA, Lee SH, Yang TC, Wang WH (2018) Survey of selected personal care products in surfacewater of coral reefs in Kenting National Park, Taiwan. Sci Tot Environ 635:1302–1307. https://doi.org/10.1016/j.scitotenv.2018.04.115

Kurwadkar S (2019) Occurrence and distribution of organic and inorganic pollutants in groundwater. Water Environ Res 91:1001–1008. https://doi.org/10.1002/wer.1166

Kuster M, de Alda MJL, Hernando MD, Petrovic M, Martín-Alonso J, Barceló D (2008) Analysis and occurrence of pharmaceuticals, estrogens, progestogens and polar pesticides in sewage treatment plant effluents, river water and drinking water in the Llobregat river basin (Barcelona, Spain). J Hydrol 358:112–123. https://doi.org/10.1016/j.jhydrol.2008.05.030

Lambert S, Wagner M (2018) Microplastics are contaminants of emerging concern in freshwater environments: an overview. In: Wagner M, Lambert S (eds) Freshwater microplastics: emerging environmental contaminants? Springer, Cham, pp 1–23. https://doi.org/10.1007/978-3-319-61615-5_1

Lamprea K, Bressy A, Mirande-Bret C, Caupos E, Gromaire MC (2018) Alkylphenol and bisphenol A contamination of urban runoff: an evaluation of the emission potentials of various construction materials and automotive supplies. Environ Sci Pollut Res 25:21887–21900. https://doi.org/10.1007/s11356-018-2272-z

Lapworth DJ, Lopez B, Laabs V, Kozel R, Wolter R, Ward R, Amelin EV, Besien T, Claessens J, Delloye F, Ferretti E, Grath J (2019) Developing a groundwater watch list for substance of emerging concern: a European perspective. Environ Res Lett 14:035004. https://doi.org/10.1088/1748-9326/aaf4d7

Laranjeiro CSM, da Silva LJG, Pereira AMPT, Pena A, Lino CM (2018) The mycoestrogen zearalenone in Portuguese flowing waters and its potential environmental impact. Mycotoxin Res 34:77–83. https://doi.org/10.1007/s12550-017-0301-2

Lebedev AT, Polyakova OV, Mazur DM, Artaev VB, Canet I, Lallement A, Vaïtilingom M, Deguillaume L, Delort AM (2018) Detection of semi-volatile compounds in cloud waters by GCxGC-TOF-MS. Evidence of phenols and phthalates as priority pollutants. Environ Pollut 241:616–625. https://doi.org/10.1016/j.envpol.2018.05.089

Lesser LE, Mora A, Moreau C, Mahlknecht J, Hernandez-Antonio A, Ramírez AI, Barrios-Piña H (2018) Survey of 218 organic contaminants in groundwater derived from the world's largest untreated wastewater irrigation system: Mezquital Valley, Mexico. Chemosphere 198:510–521. https://doi.org/10.1016/j.chemosphere.2018.01.154

Leyva-Morales JB, Valdez-Torres JB, Bastidas-Bastidas PJ, Angulo-Escalante MA, Sarmiento-Sánchez JI, Barraza-Lobo AL, Olmeda-Rubio C, Chaidez-Quiroz C (2017) Monitoring of pesticides residues in northwestern Mexico rivers. Acta Universitaria 27:45–54. https://doi.org/10.15174/au.2017.1203

Li X, Lenhart JJ (2012) Aggregation and dissolution of silver nanoparticles in natural surface water. Environ Sci Technol 46:5378–5386. https://doi.org/10.1021/es204531y
Li Y, Chen Y, Song B, Olson D, Yu N, Chen L (2009) Ecosystem structure and functioning of Lake Taihu (China) and the impacts of fishing. Fisheries Res 95:309–324. https://doi.org/10.1016/j.fishres.2008.09.039
Li X, Chu Z, Yang J, Li M, Du M, Zhao X, Zhu ZJ, Li Y (2018a) Synthetic musks: a class of commercial fragrance additives in personal care products (PCPs) causing concern as emerging contaminants. Adv Marine Biol 81:213–280. https://doi.org/10.1016/bs.amb.2018.09.008
Li J, Liu H, Paul Chen J (2018b) Microplastics in freshwater systems: a review on occurrence, environmental effects, and methods for microplastics detection. Water Res 137:362–374. https://doi.org/10.1016/j.watres.2017.12.056
Li C, Yang L, Shi M, Liu G (2019a) Persistent organic pollutants in typical lake ecosystems. Ecotoxicol Environ Saf 180:668–678. https://doi.org/10.1016/j.ecoenv.2019.05.060
Li Y, Zhang L, Liu X, Ding J (2019b) Ranking and prioritizing pharmaceuticals in the aquatic environment in China. Sci Total Environ 658:333–342. https://doi.org/10.1016/j.scitotenv.2018.12.048
Lin W, Li X, Yang M, Lee K, Chen B, Zhang BH (2018a) Brominated flame retardants, microplastics, and biocides in the marine environment: recent updates of occurrence, analysis, and impacts. Adv Marine Biol 81:167–211. https://doi.org/10.1016/bs.amb.2018.09.007
Lin L, Zuo LZ, Peng JP, Cai LQ, Fok L, Yan Y, Li HX, Xu XR (2018b) Occurrence and distribution of microplastics in an urban river: a case study in the Pearl River along Guangzhou City, China. Sci Total Environ 644:375–381. https://doi.org/10.1016/j.scitotenv.2018.06.327
Liu Z (1988) The effects of rare earth elements on growth of crops V. In Pais I (ed) Proceedings of the international symposium. New results in the research of hardly known trace elements and their role in food chain. University of Horticulture and Food Industry, Budapest, p 23
Liu D, Liu JN, Guo M, Xu HZ, Zhang SH, Shi LL, Yao C (2016) Occurrence, distribution, and risk assessment of alkylphenols, bisphenol A, and tetrabromobisphenol A in surface water, suspended particulate matter, and sediment in Taihu Lake and its tributaries. Mar Pollut Bull 112:142–150. https://doi.org/10.1016/j.marpolbul.2016.08.026
Liu DH, Tao YQ, Zhou WZ (2018) Research progress on the distribution and enrichment of persistent organic pollutants in lake organisms in China. Lake Sci 30:581–596. (in Chinese)
Liu B, Zhang SG, Chang CC (2019) Emerging pollutants – part II: treatment. Water Environ Res 91:1390–1401. https://doi.org/10.1002/wer.1233
Llompart M, Celeiro M, Dagnac T (2019) Microwave-assisted extraction of pharmaceuticals, personal care products and industrial contaminants in the environment. Trends Anal Chem 116:136–150. https://doi.org/10.1016/j.trac.2019.04.029
Löder MGJ, Gerdts G (2015) Methodology used for the detection and identification of microplastics – a critical appraisal. Springer, pp 201–227. https://doi.org/10.1007/978-3-319-16510-3_8
Lofthus S, Almas IK, Evans P, Pelz O, Brakstad OG (2016) Biotransformation of potentially persistent alkylphenols in natural seawater. Chemosphere 156:191–194. https://doi.org/10.1016/j.chemosphere.2016.04.132
López de Alda M, Diaz-Cruz S, Petrovic M, Barceló D (2003) Liquid chromatography-(tandem) mass spectrometry of selected emerging pollutants (steroid sex hormones, drugs and alkylphenolic surfactants) in the aquatic environment. J Chromatogr A 1000:503–526. https://doi.org/10.1016/S0021-9673(03)00509-0
Lorenzo M, Campo J, Pico Y (2018) Analytical challenges to determine emerging persistent organic pollutants in aquatic ecosystems. Trends Anal Chem 103:137–155. https://doi.org/10.1016/j.trac.2018.04.003
Lu Z, Gan J (2014a) Analysis, toxicity, occurrence and biodegradation of nonylphenol isomers: a review. Environ Int 73:334–345. https://doi.org/10.1016/j.envint.2014.08.017
Lu Z, Gan J (2014b) Isomer-specific biodegradation of nonylphenol in river sediments and structure-biodegradability relationship. Environ Sci Technol 48:1008–1014

Lu PJ, Huang SC, Chen YP, Chiueh LC, Shih DYC (2015) Analysis of titanium dioxide and zinc oxide nanoparticles in cosmetics. J Food Drug Anal 23:587–594. https://doi.org/10.1016/j.jfda.2015.02.009

Lu VM, McDonald KL, Townley HE (2017) Realizing the therapeutic potential of rare earth elements in designing nanoparticles to target and treat glioblastoma. Nanomedicine 12:19. https://doi.org/10.2217/nnm-2017-0193

Lu Z, Smyth SA, De Silva AO (2019) Distribution and fate of synthetic phenolic antioxidants in various wastewater treatment processes in Canada. Chemosphere 219:826–835. https://doi.org/10.1016/j.chemosphere.2018.12.068

Luo Y, Guo W, Ngo HH, Nghiem LD, Hai FI, Zhang J, Liang S, Wang XC (2014) A review on the occurrence of micropollutants in the aquatic environment and their fate and removal during wastewater treatment. Sci Total Environ 473:619–641

Luo ZF, Tu Y, Li HP, Qiu B, Liu Y, Yang ZG (2019) Endocrine-disrupting compounds in the Xiangjiang River of China: spatio-temporal distribution, source apportionment, and risk assessment. Ecotoxicol Environ Saf 167:476–484. https://doi.org/10.1016/j.ecoenv.2018.10.053

Luross JM, Alaee M, Sergeant DB, Cannon CM, Whittle DM, Solomon KR, DCG M (2002) Spatial distribution of polybrominated diphenyl ethers and polybrominated biphenyls in lake trout from the Laurentian Great Lakes. Chemosphere 46:665–672

Ma W, Yan Y, Ma M, Zhang Y, Nie C, Lun X (2016) Effect of biochar on migration and biodegradation of 4-n-nonylphenol (NP) during river-based groundwater recharge with reclaimed water. Des Water Treat 57:29316–29327. https://doi.org/10.1080/19443994.2016.1167631

Ma JX, Dai RB, Chen M, Khan SJ, Wang ZW (2018) Applications of membrane bioreactors for water reclamation: micropollutant removal, mechanisms and perspectives. Bioresour Technol 269:532–543. https://doi.org/10.1016/j.biortech.2018.08.121

Machado KC, Grassi MT, Vidal C, Pescara IC, Jardim WF, Fernandes AN, Sodré FF, Almeida FV, Santana JS, Canela CC, Nunes CRO, Bichinho KM, Severo FJR (2016) A preliminary nationwide survey of the presence of emerging contaminants in drinking and source waters in Brazil. Sci Total Environ 572:138–146. https://doi.org/10.1016/j.scitotenv.2016.07.210

Mahugija JAM, Nambela L, Mmochi AJ (2018) Determination of dichlorodiphenyltrichloroethane (DDT) and metabolites residues in fish species from Eastern Lake Tanganyika. South Afr J Chem 71:86–93

Manirakiza P, Covaci A, Nizigiymana L, Ntakimazi G, Schepens P (2002) Persistent chlorinated pesticides and polychlorinated biphenyls in selected fish species from Lake Tanganyika, Burundi, Africa. Environ Pollut 117:447–455

Marin P, Bergamasco R, Módenes AN, Paraiso PR, Hamoudi S (2019) Synthesis and characterization of graphene oxide functionalized with $MnFe_2O_4$ and supported on activated carbon for glyphosate adsorption in fixed bed. Process Saf Environ 123:59–71. https://doi.org/10.1016/j.psep.2018.12.027

Martin J, Zafra-Gomez A, Hidalgo F, Ibanez-Yuste AJ, Alonso E, Vilchez JL (2017) Multi-residue analysis of 36 priority and emerging pollutants in marine echinoderms (*Holothuria tubulosa*) and marine sediments by solid-liquid extraction followed by dispersive solid phase extraction and liquid chromatography tandem mass spectrometry analysis. Talanta 166:336–348. https://doi.org/10.1016/j.talanta.2017.01.062

Martínez Vidal JL, Arrebola Liébanas FJ, González MJ, Garrido Frenich A, Fernández Moreno JL (2006) Validation of a gas chromatography/triple quadrupole mass spectrometry based method for the quantification of pesticides in food commodities. Rapid Commun Mass Spectrom 20:365–375. https://doi.org/10.1002/rcm.2315

Martinez-Gomez DA, Baca S, Walsh EJ (2015) Lethal and sublethal effects of selected PPCPs on the freshwater rotifer, *Plationus patulus*. Environ Toxicol Chem 34:913–922. https://doi.org/10.1002/etc.2873

Mashroofeh A, Bakhtiari AR, Pourkazemi M (2015) Distribution and composition pattern of polycyclic aromatic hydrocarbons in different tissues of sturgeons collected from Iranian coastline of the Caspian Sea. Chemosphere 120:575–583

Matich EK, Chavez Soria NG, Aga DS, Atilla-Gokcumen GE (2019) Applications of metabolomics in assessing ecological effects of emerging contaminants and pollutants on plants. J Hazard Mater 373:527–535. https://doi.org/10.1016/j.jhazmat.2019.02.084

Mauricio R, Dias R, Ribeiro V, Fernandes S, Vicente AC, Pinto MI, Mano AP (2018) 17α-ethinylestradiol and 17β-estradiol removal from a secondary urban wastewater using an RBC treatment system. Environ Monitor Assess 190:1

Mauter MS, Elimelech M (2008) Environmental applications of carbon-based nanomaterials. Environ Sci Technol 42:5843–5859. https://doi.org/10.1021/es8006904

Mautner A (2020) Nanocellulose water treatment membranes and filters: a review. Polym Int. https://doi.org/10.1002/pi.5993

McDonald RJ, McDonald JS, Kallmes DF, Jentoft ME, Murray DL, Thielen KR, Williamson EE, Eckel LJ (2015) Intracranial gadolinium deposition after contrast-enhanced MR imaging. Radiology 275:772–782. https://doi.org/10.1148/radiol.15150025

McNesby KL, Pesce-Rodriguez RA (2006) Handbook of vibrational spectroscopy. https://doi.org/10.1002/0470027320

Mearns AJ, Reish DJ, Bissell M, Morrison AM, Rempel-Hester MA, Arthur C, Rutherford N, Pryor R (2018) Effects of pollution on marine organisms. Water Environ Res 90:1206–1300. https://doi.org/10.2175/106143018X15289915807218

Mearns AJ, Bissell M, Morrison AM, Rempel-Hester MA, Arthur C, Rutherford N (2019) Effects of pollution on marine organisms. Water Environ Res 91:1229–1252. https://doi.org/10.1002/wer.1218

Melzer D, Rice N, Depledge MH, Henley WE, Galloway TS (2010) Association between serum perfluorooctanoic acid (PFOA) and thyroid disease in the U.S. national health and nutrition examination survey. Environ Health Perspect 118:686–692. https://doi.org/10.1289/ehp.0901584

Mezzelani M, Gorbi S, Regoli F (2018) Pharmaceuticals in the aquatic environments: evidence of emerged threat and future challenges for marine organisms. Marine Environ Res 140:41–60. https://doi.org/10.1016/j.marenvres.2018.05.001

Minh TLT, Phuoc DN, Quoc TD, Ngo HH, Lan CDH (2016) Presence of e-EDCs in surface water and effluents of pollution sources in Sai Gon and Dong Nai river basin. Sustain Environ Res 26:20–27. https://doi.org/10.1016/j.serj.2015.09.001

Mishra S, Rath CC, Das AP (2019) Marine microfiber pollution: a review on present status and future challenges. Marine Pollut Bull 140:188–197. https://doi.org/10.1016/j.marpolbul.2019.01.039

Mohapatra DP, Kirpalani DM (2019) Advancement in treatment of wastewater: fate of emerging contaminants. Can J Chem Eng 97:2621–2631. https://doi.org/10.1002/cjce.23533

Möller JN, Löder MGJ, Laforsch C (2020) Finding microplastics in soils: a review of analytical methods. Environ Sci Technol 54:2078–2090. https://doi.org/10.1021/acs.est.9b04618

Monneret C (2017) What is an endocrine disruptor? C R Chim 340:403–405. https://doi.org/10.1016/j.crvi.2017.07.004

Montagner CC, Jardim WF (2011) Spatial and seasonal variations of pharmaceuticals and endocrine disruptors in the Atibaia River, Sao Paulo State (Brazil). J Braz Chem Soc 22:1452–1462. https://doi.org/10.1590/S0103-50532011000800008

Montagner CC, Sodré FF, Acayaba RD, Vidal C, Campestrini I, Locatelli MAF, Pescara IC, Albuquerque AF, Umbuzeiro GA, Jardim WF (2019) Ten years-snapshot of the occurrence of emerging contaminants in drinking, surface and ground waters and wastewaters from São Paulo State, Brazil. J Braz Chem Soc 30:614–632. https://doi.org/10.21577/0103-5053.20180232

Moreau M, Hadfield J, Hughey J, Sanders F, Lapworth DJ, White D, Civil W (2019) A baseline assessment of emerging organic contaminants in New Zealand groundwater. Sci Total Environ 686:425–439. https://doi.org/10.1016/j.scitotenv.2019.05.210

Morin-Crini N, Crini G (eds) (2017) Eaux industrielles contaminées (in French). PUFC, Besançon, 513 p

Mortimer DN (2013) Persistent organic pollutants in foods: science, policy and regulation (chapter 1). In Rose M, Fernandes A (eds) Persistent organic pollutants and toxic metals in foods. Woodhead Publishing, pp 3–19. https://doi.org/10.1533/9780857098917.1.3

Mousavi SZ, Nafisi S, Maibach HI (2017) Fullerene nanoparticle in dermatological and cosmetic applications. Nanomed Nanotechnol Biol Med 13:1071–1087. https://doi.org/10.1016/j.nano.2016.10.002

MSFD (2013) MSFD technical subgroup on marine litter. Guidance on monitoring of marine litter in European seas – a guidance document within the common implementation strategy for the Marine Strategy Framework Directive. JRC Scientific and Policy Reports

Mubashir M (2018) Rare earth elements – a review. J Ecol Nat Resour. https://doi.org/10.23880/JENR-16000128

Mueller NC, Nowack B (2008) Exposure modeling of engineered nanoparticles in the nnvironment. Environ Sci Technol 42:4447–4453. https://doi.org/10.1021/es7029637

Murdoch FK, Sanin FD (2016) Biotransformation of nonylphenol diethoxylate in anaerobic digesters: accumulation of metabolites and their effects on digester performance. Int Biodeterior Biodegrad. https://doi.org/10.1016/j.ibiod.2016.02.017

Murga MN, Gutiérrez R, Vega S, Pérez JJ, Ortiz R, Schettino B, Yamasaki A, Ruíz JL (2016) Organochlorine pesticide distribution in an organic production system for cow's milk in Chiapas, Mexico. J Environ Sci Health B 51:589–593. https://doi.org/10.1080/03601234.2016.1181901

Nemirovskaya IA, Brekhovskikh VF (2008) Origin of hydrocarbons in the particulate matter and bottom sediments of the northern shelf of the Caspian Sea. Oceanology 48:43–53

Ng A, Weerakoon D, Lim E, Padhye LP (2019) Fate of environmental pollutants. Water Environ Res 91:1294–1325. https://doi.org/10.1002/wer.1225

Nika MC, Bletsou A, Koumaki E, Noutsopoulos C, Mamais D, Stasinakis AS, Thomaidis NS (2016) Chlorination of benzothiazoles and benzotriazoles and transformation products identification by LC-HR-MS/MS. J Hazard Mater 323:400–413. https://doi.org/10.1016/j.jhazmat.2016.03.035

Norén F (2007) Small plastic particles in Coastal Swedish waters. Report by N-research on behalf of KIMO Sweden, pp 1–11

Norman Network (2016) Network of reference laboratories, research centres and related organisations for monitoring of emerging environmental substances. List of emerging substances. www.norman-network.com. Accessed on 18 Mar 2020

Noutsopoulos C, Koumaki E, Mamais D, Nika MC, Bletsou A, Thomaidis N (2015) Removal of endocrine disruptors and non-steroidal anti-inflammatory drugs through wastewater chlorination: the effect of pH, total suspended solids and humic acids and identification of degradation by-products. Chemosphere 119:S109–S114. https://doi.org/10.1016/j.chemosphere.2014.04.107

Noutsopoulos C, Koumaki E, Sarantopoulos V, Mamais D (2019) Analytical and mathematical assessment of emerging pollutants fate in a river system. J Hazard Mater 364:48–58. https://doi.org/10.1016/j.jhazmat.2018.10.033

Nowack B, Bucheli TD (2007) Occurrence, behavior and effects of nanoparticles in the environment. Environ Pollut 150:5–22. https://doi.org/10.1016/j.envpol.2007.06.006

Oberbeckmann S, Löder MGJ, Labrenz M (2015) Marine microplastic-associated biofilms – a review. Environ Chem 12:551. https://doi.org/10.1071/EN15069

Olaniyan LWB, Okoh OO, Mkwetshana NT, Okoh AI (2020) Environmental water pollution, endocrine interference and ecotoxicity of 4-tert-octylphenol: a review. Rev Environ Contamin Toxicol 248:81–109. https://doi.org/10.1007/398_2018_20

Onyekwere O, Okonkwo CJ, Okoroafor AB, Okonkwo CJ (2019) Occurrence and risk assessment of phenolic endocrine disrupting chemicals in shallow groundwater resource from selected Nigerian rural settlements. Ovidius Univ Annals Chem 30:101–107. https://doi.org/10.2478/auoc-2019-0018

Ort C, Lawrence MG, Rieckermann J, Joss A (2010) Sampling for pharmaceuticals and personal care products (PPCPs) and illicit drugs in wastewater systems: are your conclusions valid? A critical review. Environ Sci Technol 44:6024–6035

Osorio V, Larrañaga A, Aceña J, Pérez S, Barceló D (2016) Concentration and risk of pharmaceuticals in freshwater systems are related to the population density and the livestock units in Iberian rivers. Sci Total Environ. https://doi.org/10.1016/j.scitotenv.2015.06.143

Osuna-Flores I, Riva MC (2002) Organochlorine pesticide residue concentrations in shrimps, sediments, and surface water from Bay of Ohuira, Topolobampo, Sinaloa, Mexico. Bull Environ Contam Toxicol 68:532–539. https://doi.org/10.1007/s001280287

Pagano G, Guida M, Tommasi F (2015) Health effects and toxicity mechanisms of rare earth elements-knowledge gaps and research prospects. Ecotoxicol Environ Saf 115:40–48. https://doi.org/10.1016/j.ecoenv.2015.01.030

Paíga P, Ramos S, Jorge S, Silva JG, Delerue-Matos C (2019) Monitoring survey of caffeine in surface waters (Lis River) and wastewaters located at Leiria Town in Portugal. Environ Sci Pollut Res 26:33440–33450. https://doi.org/10.1007/s11356-019-06168-w

Pan YT, Wang JH, Yeung LWY, Wei S, Dai JY (2020) Analysis of emerging per- and polyfluoroalkyl substances: progress and current issues. Trends Anal Chem 124:115481. https://doi.org/10.1016/j.trac.2019.04.013

Panichev AM (2015) Rare earth elements: review of medical and biological properties and their abundance in the rock materials and mineralized spring waters in the context of animal and human geophagia reasons evaluation. Achievements Life Sci 9:95–103

Park CM, Chu KH, Her N, Jang M, Baalousha M, Heo J, Yoon Y (2017) Occurrence and removal of engineered nanoparticles in drinking water treatment and wastewater treatment processes. Sep Purif Rev 46:255–272. https://doi.org/10.1080/15422119.2016.1260588

Parvez MK, Parveen S (2017) Evolution and emergence of pathogenic viruses: past, present, and future. Intervirology 60:1–7. https://doi.org/10.1159/000478729

Patel M, Kumar R, Kishor K, Mlsna T, Pittman CU Jr, Mohan D (2019) Pharmaceuticals of emerging concern in aquatic systems: chemistry, occurrence, effects, and removal methods. Chem Rev 119:3510–3673. https://doi.org/10.1021/acs.chemrev.8b00299

Patel N, Khan MZA, Shahane S, Rai D, Chauhan D, Kant C, Chaudhary VK (2020) Emerging pollutants in aquatic environment: source, effect, and challenges in biomonitoring and bioremediation – a review. Pollution 6:99–113. https://doi.org/10.22059/poll.2019.285116.646

Pearson RF, Swackhamer DL, Eisenreich SJ, Long DT (1998) Atmospheric inputs of polychlorinated dibenzo-p-dioxins and dibenzofurans to the Great Lakes: compositional comparison of PCDD and PCDF in sediments. J Great Lakes Res 24:65–82. https://doi.org/10.1016/S0380-1330(98)70800-7

Pedrazzani R, Bertanza G, Brnardić I, Cetecioglu Z, Dries J, Dvarionienė J, García-Fernández AJ, Langehoff A, Libralato G, Lofrano G, Škrbić B, Martínez-López E, Meriç S, Pavlović DM, Papa M, Schröder P, Tsagarakis KP, Vogelsang C (2019) Opinion paper about organic trace pollutants in wastewater: toxicity assessment in a European perspective. Sci Total Environ 651:3202–3221. https://doi.org/10.1016/j.scitotenv.2018.10.027

Peeken I, Beyer SPB, Gütermann J, Katlein C, Krumpen T, Bergmann M, Hehemann L, Gerdts G (2018) Arctic Sea ice is an important temporal sink and means of transport for microplastic. Nat Commun 9:1505. https://doi.org/10.1038/s41467-018-03825-5

Peez N, Becker J, Ehlers S, Fritz M, Fischer C, Koop J, Winkelmann C, Imhof W (2019) Quantitative analysis of PET microplastics in environmental model samples using quantitative 1H-NMR spectroscopy: validation of an optimized and consistent sample clean-up method. Anal Bioanal Chem 411:7409–7418. https://doi.org/10.1007/s00216-019-02089-2

Peña-Guzmán C, Ulloa-Sánchez S, Mora K, Helena-Bustos R, Lopez-Barrera E, Alvarez J, Rodriguez-Pinźon M (2019) Emerging pollutants in the urban cycle in Latin America: a review of the current literature. J Environ Manag 237:408–423. https://doi.org/10.1016/j.jenvman.2019.02.100

Peng FJ, Pan CG, Zhang M, Zhang NS, Windfeld R, Salvito D, Selck H, Van Den Brink PJ, Ying GG (2017) Occurrence and ecological risk assessment of emerging organic chemicals in urban rivers: Guangzhou as a case study in China. Sci Tot Environ 589:46–55. https://doi.org/10.1016/j.scitotenv.2017.02.200

Peng L, Dai X, Liu Y, Sun J, Song S, Ni B (2018) Model-based assessment of estrogen removal by nitrifying activated sludge. Chemosphere 197:430–437. https://doi.org/10.1016/j.chemosphere.2018.01.035

Pereira AMPT, Silva LJG, Laranjeiro CSM, Meisel LM, Lino CM, Pena A (2017a) Human pharmaceuticals in Portuguese rivers: the impact of water scarcity in the environmental risk. Sci Total Environ 609:1182–1191. https://doi.org/10.1016/j.scitotenv.2017.07.200

Pereira AS, Daam MA, Cerejeira MJ (2017b) Evaluation of FOCUS surface water pesticide concentration predictions and risk assessment of field-measured pesticide mixtures-a crop-based approach under Mediterranean conditions. Environ Sci Pollut Res 24:17394–17406. https://doi.org/10.1007/s11356-017-9393-7

Pérez-Alvarez I, Islas-Flores H, Gómez-Oliván LM, Baceló D, López-de Alda M, Pérez-Solsona S, Sánchez-Aceves L, SanJuan-Reyes N, Galar-Martínez M (2018) Determination of metals and pharmaceutical compounds released in hospital wastewater from Toluca, Mexico, and evaluation of their toxic impact. Environ Pollut 240:330–341. https://doi.org/10.1016/j.envpol.2018.04.116

Pérez-Coyotl I, Galar-Martínez M, García-Medina S, Gómez-Oliván LM, Gasca-Pérez E, Martínez-Galero E, Islas-Flores H, Pérez-Pastén BR, Barceló D, López-de Alda M, Pérez-Solsona S, Serra-Roig MP, Montemurro N, Peña-Herrera JM, Sánchez-Aceves LM (2019) Polluted water from an urban reservoir (Madín dam, México) induces toxicity and oxidative stress in *Cyprinus carpio* embryos. Environ Pollut 251:510–521. https://doi.org/10.1016/j.envpol.2019.04.095

Pesqueira JFJR, Pereira MFR, Silva AMT (2020) Environmental impact assessment of advanced urban wastewater treatment technologies for the removal of priority substances and contaminants of emerging concern: a review. J Clean Prod 261:121078. https://doi.org/10.1016/j.jclepro.2020.121078

Peters RJB, van Bemmel G, Milani NBL, den Hertog GCT, Undas AK, van der Lee M, Bouwmeester H (2018) Detection of nanoparticles in Dutch surface waters. Sci Total Environ 621:210–218. https://doi.org/10.1016/j.scitotenv.2017.11.238

Petrie B, Barden R, Kasprzk-Hordern B (2015) A review on emerging contaminants in wastewaters and the environment: current knowledge, understudied areas and recommendations for future monitoring. Water Res 72:3–27. https://doi.org/10.1016/j.watres.2014.08.053

Piccoli E, Mazzilis D, Gava E, De Martin S (2018) Study of nanoparticles in a few rivers in North East Italy using SP-ICP-MS. Nano Res Appl 4:1. https://doi.org/10.21767/2471-9838.100030

Picos-Corrales LA, Sarmiento-Sánchez JI, Ruelas-Leyva JP, Crini G, Hermosillo-Ochoa E, Gutierrez-Montes JA (2020) Environment-friendly approach toward the treatment of raw agricultural wastewater and river water via flocculation using chitosan and bean straw flour as bioflocculants. ACS Omega 5:3943–3951. https://doi.org/10.1021/acsomega.9b03419

Plastics Europe (2017) Plastics – the facts 2017: an analysis of European plastics production, demand and waste data. https://www.issuelab.org/resource/plastics-the-facts-2017-an-analysis-of-european-plastics-production-demand-and-waste-data.html

Plastics Europe (2019) Plastics – the facts 2019: an analysis of European plastics production, demand and waste data. https://www.plasticseurope.org/fr/resources/publications/1804-plastics-facts-2019

Polanco-Rodríguez AG, Riba-López MA, DelValls-Casillas A, Araujo-León JA, Datta-Banik S (2018) Impact of pesticides in karst groundwater. Review of recent trends in Yucatan, Mexico. Groundw Sustain Develop 7:20–29. https://doi.org/10.1016/j.gsd.2018.02.003

Polder A, Muller MB, Lyche JL, Mdegela RH, Nonga HE, Mabiki FP, Mbise TJ, Skaare JU, Sandvik M, Skjerve E, Lie E (2014) Levels and patterns of persistent organic pollutants (POPs) in tilapia (*Oreochromis sp.*) from four different lakes in Tanzania: geographical differences and implications for human health. Sci Total Environ 488:252–260

Portilla A, Torres D, Machado-Duque ME, Machado-Alba JE (2017) Intervention to rationalize use of losartan. Revista Colombiana Cardiologiá 24:10–14. https://doi.org/10.1016/j.rccar.2016.04.004

Prata JC, da Costa JP, Duarte AC, Rocha-Santos T (2019) Methods for sampling and detection of microplastics in water and sediment: a critical review. Trends Anal Chem 110:150–159. https://doi.org/10.1016/j.trac.2018.10.029

Priac A, Morin-Crini N, Druart C, Gavoille S, Bradu C, Lagarrigue C, Torri G, Winterton P, Crini G (2017) Alkylphenol and alkylphenol polyethoxylates in water and wastewater: a review of options for their elimination. Arab J Chem 10:S3749–S3773. https://doi.org/10.1016/j.arabjc.2014.05.011

Puzyn T, Mostrag-Szlichtyng A (2012) Organic pollutants ten years after the Stockholm convention. Environmental and analytical update. InTech, Rijeka. ISBN: 978-953-307-917-2

Qiao M, Wang CX, Huang SB, Wang DH, Wang ZJ (2006) Composition, sources, and potential toxicological significance of PAHs in the surface sediments of the Meiliang Bay, Taihu Lake, China. Environ Int 32:28–33

Qiu XH, Zhu T, Jing L, Pan HS, Li QL, Miao GF, Gong JC (2004) Organochlorine pesticides in the air around the Taihu Lake, China. Environ Sci Technol 38:1368–1374

Rajaei F, Sari AE, Bahramifar N, Savabieasfahani M, Ghasempouri M (2011) Persistent organic pollutants in muscle and feather of ten Avian species from Mazandaran Province of Iran, on the Coast of the Caspian Sea. Bull Environl Contamin Toxicol 87:678–683

Raju CSK, Cossmer A, Scharf H, Panne U, Lück D (2010) Speciation of gadolinium-based MRI contrast agents in environmental water samples using hydrophilic interaction chromatography hyphenated with inductively coupled plasma mass spectrometry. J Anal Atomic Spectrom 25:55–61. https://doi.org/10.1039/B919959D

Rana VK, Kissner R, Gaspard S, Levalois-Grützmacher J (2016) Cyclodextrin as a complexation agent in the removal of chlordecone from water. Chem Eng J 293:82–89. https://doi.org/10.1016/j.cej.2016.02.032

Rasheed T, Hassan AA, Bilal M, Hussain T, Rizwan K (2020) Metal-organic frameworks based adsorbents: a review from removal perspective of various environmental contaminants from wastewater. Chemosphere 259:127369. https://doi.org/10.1016/j.chemosphere.2020.127369

Reddy VM, Babu KS, Balaram V, Satyanarayanan M (2012) Assessment of the effects of municipal sewage, immersed idols and boating on the heavy metal and other elemental pollution of surface water of the eutrophic Hussainsagar Lake (Hyderabad, India). Environ Monitor Assess 184:1991–2000. https://doi.org/10.1007/s10661-011-2094-7

Redling K (2006) Rare earth elements in agriculture with emphasis on animal husbandry. PhD thesis. University of Munich, Germany, 360p

Reed RB, Martin DP, Bednar AJ, Montaño MD, Westerhoff P, Ranville JF (2017) Multi-day diurnal measurements of Ti-containing nanoparticle and organic sunscreen chemical release during recreational use of a natural surface water. Environ Sci Nano 4:69–77. https://doi.org/10.1039/C6EN00283H

Reichert G, Hilgert S, Fuchs S, Azevedo JCR (2019) Emerging contaminants and antibiotic resistance in the different environmental matrices of Latin America. Environ Pollut 255:113140. https://doi.org/10.1016/j.envpol.2019.113140

Reis-Santos P, Pais M, Duarte B, Caçador I, Freitas A, Vila Pouca AS, Barbosa J, Leston S, Rosa J, Ramos F, Cabral HN, Gillanders BM, Fonseca VF (2018) Screening of human and veterinary pharmaceuticals in estuarine waters: a baseline assessment for the Tejo estuary. Marine Pollut Bull 135:1079–1084. https://doi.org/10.1016/j.marpolbul.2018.08.036

Rennie RA (2016) Dictionary of chemistry. Oxford University Press, Oxford. https://doi.org/10.1093/acref/9780198722823.001.0001

Ribeiro AR, Maia A, Santos M, Tiritan ME, Ribeiro CMR (2016a) Occurrence of natural contaminants of emerging concern in the Douro River Estuary, Portugal. Arch Environ Contam Toxicol 70:361–371. https://doi.org/10.1007/s00244-015-0212-1

Ribeiro C, Ribeiro AR, Tiritan ME (2016b) Priority substances and emerging organic pollutants in Portuguese aquatic environment: a review. In De Voogt P (ed) Reviews of environmental contamination tand toxicology , volume 238, pp 1–44. https://doi.org/10.1007/398_2015_5006

Ribeiro CMR, Maia AS, Ribeiro AR, Couto C, Almeida AA, Santos M, Tiritan ME (2016c) Anthropogenic pressure in a Portuguese river: endocrine-disrupting compounds, trace elements and nutrients. J Environ Sci Health A Toxic/Hazard Subst Environ Eng 51:1043–1052. https://doi.org/10.1080/10934529.2016.1198622

Rigueto CVT, Nazari MT, De Souza CF, Cadore JS, Briao VB, Piccin JS (2020) Alternative techniques for caffeine removal from wastewater: an overview of opportunities and challenges. J Water Proc Eng 35:101231. https://doi.org/10.1016/10.1016/j.jwpe.2020.101231

Rim KT (2016) Effects of rare earth elements on the environment and human health: a literature review. Toxicol Environ Health Sci 8:189e200. https://doi.org/10.1007/s13530-016-0276-y

Rivera-Jaimes JA, Postigo C, Melgoza-Alemán RM, Aceña J, Barceló D, López-de Alda M (2018) Study of pharmaceuticals in surface and wastewater from Cuernavaca, Morelos, Mexico: occurrence and environmental risk assessment. Sci Total Environ 613–614:1263–1274. https://doi.org/10.1016/j.scitotenv.2017.09.134

Roberts J, Kumar A, Du J, Hepplewhite C, Ellis DJ, Christy AG, Beavis SG (2016) Pharmaceuticals and personal care products (PPCPs) in Australia's largest inland sewage treatment plant, and its contribution to a major Australian river during high and low flow. Sci Total Environ. https://doi.org/10.1016/j.scitotenv.2015.03.145

Rocha MJ, Cruzeiro C, Reis M, Pardal MÂ, Rocha E (2016) Pollution by oestrogenic endocrine disruptors and β-sitosterol in a south-western European river (Mira, Portugal). Environ Monit Assess 188. https://doi.org/10.1007/s10661-016-5236-0

Rodrigues ET, Alpendurada MF, Ramos F, Pardal MÂ (2018a) Environmental and human health risk indicators for agricultural pesticides in estuaries. Ecotoxicol Environ Saf 150:224–231. https://doi.org/10.1016/j.ecoenv.2017.12.047

Rodrigues MO, Abrantes N, Gonçalves FJM, Nogueira H, Marques JC, Gonçalves AMM (2018b) Spatial and temporal distribution of microplastics in water and sediments of a freshwater system (Antuã River, Portugal). Sci Total Environ 633:1549–1559. https://doi.org/10.1016/j.scitotenv.2018.03.233

Rodrigues ET, Alpendurada MF, Guimarães A, Avó R, Ferreira B, Pardal MA (2019a) The environmental condition of an estuarine ecosystem disturbed by pesticides. Environ Sci Pollut Res 26:24075–24087. https://doi.org/10.1007/s11356-019-05751-5

Rodrigues SM, Almeida CMR, Silva D, Cunha J, Antunes C, Freitas V, Ramos S (2019b) Microplastic contamination in an urban estuary: abundance and distribution of microplastics and fish larvae in the Douro estuary. Sci Total Environ 659:1071–1081. https://doi.org/10.1016/j.scitotenv.2018.12.273

Routti H, Atwood TC, Bechshoft T, Boltunov A, Ciesielski TM, Desforges JP, Dietz R, Gabrielsen GW, Jenssen BM, Letcher RJ, McKinney MA, Morris AD, Rigét FF, Sonne C, Styrishave B, Tartu S (2019) State of knowledge on current exposure, fate and potential health effects on contaminants in polar bears from the circumpolar Artic. Sci Total Environ 664:1063–1083. https://doi.org/10.1016/j.scitotenv.2019.02.030

Różalska S, Soboń A, Pawłowska J, Wrzosek M, Długoński J (2015) Biodegradation of nonylphenol by a novel entomopathogenic *Metarhizium robertsii* strain. Bioresour Technol. https://doi.org/10.1016/j.biortech.2015.05.011

Rudnick RL, Gao S (2003) Composition of the continental crust. Treatise on geochemistry, pp 1–64. https://doi.org/10.1016/b0-08-043751-6/03016-4

Ruge Z, Muir D, Helm P, Lohmann R (2018) Concentrations, trends, and air-water exchange of PCBs and organochlorine pesticides derived from passive samplers in Lake Superior in 2011. Environ Sci Technol 52:14061–14069

Ryan PG, Moore CJ, van Franeker JA, Moloney CL (2009) Monitoring the abundance of plastic debris in the marine environment. Philos T R Soc B 364:1999–2012. https://doi.org/10.1098/rstb.2008.0207

Sackaria M, Elango L (2020) Organic micropollutants in groundwater of India – a review. Water Environ Res 92:504–523. https://doi.org/10.1016/10.1002/wer.1243

Sadutto D, Álvarez-Ruiz R, Picó Y (2020) Systematic assessment of extraction of pharmaceuticals and personal care products in water and sediment followed by liquid chromatography-tandem mass spectrometry. Anal Bioanal Chem 412:113–127. https://doi.org/10.1007/s00216-019-02207-0

Saeed T, Al-Jandal N, Abusam A, Taqi H, Al-Khabbaz A, Zafar J (2017) Sources and levels of endocrine disrupting compounds (EDCs) in Kuwait's coastal areas. Mar Pollut Bull 118:407–412. https://doi.org/10.1016/j.marpolbul.2017.03.010

Salgueiro-Gonzalez N, Muniategui-Lorenzo S, Lopez-Mahia P, Prada-Rodriguez D (2017) Trends in analytical methodologies for the determination of alkylphenols and bisphenol A in water samples. Anal Chim Acta 962:1–14. https://doi.org/10.1016/j.aca.2017.01.035

Salgueiro-Gonzalez N, Castiglioni S, Zuccato E, Turnes-Carou I, Lopez-Mahia P, Muniategui-Lorenzo S (2018) Recent advances in analytical methods for the determination of 4-alkylphenols and bisphenol A in solid environmental matrices: a critical review. Anal Chim Acta 1024:39–51. https://doi.org/10.1016/j.aca.2018.02.081

Salthammer T (2014) Very volatile organic compounds: an understudied class of indoor air pollutants. Indoor Air 26:25–38. https://doi.org/10.1111/ina.12173

Salthammer T (2020) Emerging indoor pollutants. Int J Hyg Environ Health 224:113423. https://doi.org/10.1016/j.ijheh.2019.113423

Santos MSF, Franquet-Griell H, Alves A, Lacorte S (2018) Development of an analytical methodology for the analysis of priority cytostatics in water. Sci Total Environ 645:1264–1272. https://doi.org/10.1016/j.scitotenv.2018.07.232

Saravanan M, Nam SE, Eom HJ, Lee DH, Rhee JS (2019) Long-term exposure to waterborne nonylphenol alters reproductive physiological parameters in economically important marine fish. Comparative Biochem Physiol C Toxicol Pharmacol 216:10–18. https://doi.org/10.1016/j.cbpc.2018.11.009

Sarria-Villa R, Ocampo-Duque W, Páez M, Schuhmacher M (2016) Presence of PAHs in water and sediments of the Colombian Cauca river during heavy rain episodes, and implications for risk assessment. Sci Total Environ 540:455–465. https://doi.org/10.1016/j.scitotenv.2015.07.020

Sauvé S, Desrosiers M (2014) A review of what is an emerging contaminant. Chem Cent J 8:15. https://doi.org/10.1186/1752-153X-8-15

Sayed AEH, Kataoka C, Oda S, Kashiwada S, Mitani H (2018) Sensitivity of medaka (*Oryzias latipes*) to 4-nonylphenol subacute exposure; erythrocyte alterations and apoptosis. Environ Toxicol Pharmacol 58:98–104. https://doi.org/10.1016/j.etap.2017.12.023

Sayed AEH, Abd-Elkareem M, Abou Khalil NS (2019) Immunotoxic effects of 4-nonylphenol on *Clarias gariepinus*: cytopathological changes in hepatic melanomacrophages. Aquat Toxicol 207:83–90. https://doi.org/10.1016/j.aquatox.2018.12.002

Schrenk D, Chopra M (2013) Dioxins and polychlorinated biphenyls (PCBs) in foods (chapter 9). In Rose M, Fernandes A (eds) Persistent organic pollutants and toxic metals in foods. Woodhead Publishing, pp 191–214. https://doi.org/10.1533/9780857098917.2.191

Schulze T, Ahel M, Ahlheim J, Aït-Aïssa S, Brion F, Di Paolo C, Froment J, Hidasi AO, Hollender J, Hollert H, Meng H, Kloß A, Koprivica S, Krauss M, Muz OP, Petre M, Schollée J, Brack W (2017) Assessment of a novel device for onsite integrative large-volume solid phase extraction of water samples to enable a comprehensive chemical and effect-based analysis. Sci Total Environ 581-582:350–358. https://doi.org/10.1016/j.scitotenv.2016.12.140

Serna-Galvis EA, Botero-Coy AM, Martínez-Pachón D, Moncayo-Lasso A, Ibáñez M, Hernández F, Torres-Palma RA (2019) Degradation of seventeen contaminants of emerging concern in municipal wastewater effluents by sonochemical advanced oxidation processes. Water Res 154:349–360. https://doi.org/10.1016/j.watres.2019.01.045

Serrão Sousa V, Corniciuc C, Ribau Teixeira M (2017) The effect of TiO_2 nanoparticles removal on drinking water quality produced by conventional treatment C/F/S. Water Res:109. https://doi.org/10.1016/j.watres.2016.11.030

Shahbazi A, Bahramifa N, Smolders E (2012) Elevated concentrations of pesticides and PCBs in soils at the Southern Caspian Sea (Iran) are related to land use. Soil Sediment Contamin 21:160–175

Shajib MTI, Hansen HCB, Liang T, Holm PE (2020) Rare earth elements in surface specific urban runoff in Northern Beijing. Sci Total Environ 136969. https://doi.org/10.1016/j.scitotenv.2020.136969

Sharma M, Chadha P (2017a) 4-Nonylphenol induced DNA damage and repair in fish, Channa punctatus after subchronic exposure. Drug Chem Toxicol 320–325. https://doi.org/10.1080/01480545.2016.1223096

Sharma M, Chadha P (2017b) Widely used non-ionic surfactant 4-nonylphenol: showing genotoxic effects in various tissues of Channa punctatus. Environ Sci Pollut Res Int 24:11331–11339. https://doi.org/10.1007/s11356-017-8759-1

Shen L, Gewurtz SB, Reiner EJ, MacPherson KA, Kolic TM, Khurana V, Helm PA, Howell ET, Burniston DA, Brindle ID, Marvin CH (2009) Occurrence and sources of polychlorinated dibenzo-p-dioxins, dibenzofurans and dioxin-like polychlorinated biphenyls in surficial sediments of Lakes Superior and Huron. Environ Pollut 157:1210–1218. https://doi.org/10.1016/j.envpol.2008.12.012

Shirdel I, Kalbassi MR (2016) Effects of nonylphenol on key hormonal balances and histopathology of the endangered Caspian brown trout (*Salmo trutta caspius*). Comparative Biochem Physiol C Toxicol Pharmacol 183–184:28–35. https://doi.org/10.1016/j.cbpc.2016.01.003

Shiu RF, Jiang JJ, Kao HY, Fang MD, Liang YJ, Tang CC, Lee CL (2019) Alkylphenol ethoxylate metabolites in coastal sediments off southwestern Taiwan: spatiotemporal variations, possible sources, and ecological risk. Chemosphere 225:9–18. https://doi.org/10.1016/j.chemosphere.2019.02.136

Siemens J, Huschek G, Siebe C, Kaupenjohann M (2008) Concentrations and mobility of human pharmaceuticals in the world's largest wastewater irrigation system, Mexico City-Mezquital Valley. Water Res 42:2124–2134. https://doi.org/10.1016/j.watres.2007.11.019

Silva E, Daam MA, Cerejeira MJ (2015a) Aquatic risk assessment of priority and other river basin specific pesticides in surface waters of Mediterranean river basins. Chemosphere 135:394–402. https://doi.org/10.1016/j.chemosphere.2015.05.013

Silva E, Daam MA, Cerejeira MJ (2015b) Predicting the aquatic risk of realistic pesticide mixtures to species assemblages in Portuguese river basins. J Environ Sci 31:12–20. https://doi.org/10.1016/j.jes.2014.11.006

Skrbic BD, Kadokami K, Antic I (2018) Survey on the micro-pollutants presence in surface water system of northern Serbia and environmental and health risk assessment. Environ Res 166:130–140. https://doi.org/10.1016/j.envres.2018.05.034

Slezakova K, Morais S, do Carmo Pereira M (2012) Indoor air pollutants: relevant aspects and health impacts. Environmental healty – emerging issues and practice. Intech, pp 125–146. https://doi.org/10.5772/30364

Sneller FEC, Kalf DF, Weltje L, Van Wezel AP (2000) Maximum permissible concentrations and negligible concentrations for rare earth elements (REES). RIVM Report 601501011. National Institute of Public Health and the Environment, Bilthoven, The Netherlands, p 66

Sodré FF, Locatelli MAF, Jardim WF (2010) Occurrence of emerging contaminants in Brazilian drinking waters: a sewage-to-tap issue. Water Air Soil Pollut 206:57–67. https://doi.org/10.1007/s11270-009-0086-9

Sodré FF, Santana JS, Sampaio TR, Brandão CCS (2018) Seasonal and spatial distribution of caffeine, atrazine, atenolol and DEET in surface and drinking waters from the Brazilian Federal District. J Brazilian Chem Soc 29:1854–1865. https://doi.org/10.21577/0103-5053.20180061

Song J, Nagae M, Takao Y, Soyano K (2020) Field survey of environmental estrogen pollution in the coastal area of Tokyo Bay and Nagasaki City using the Japanese common goby *Acanthogobius flavimanus*. Environ Pollut 258:113673. https://doi.org/10.1016/j.envpol.2019.113673

Sousa VS, Teixeira MR (2013) Aggregation kinetics and surface charge of CuO nanoparticles: the influence of pH, ionic strength and humic acids. Environ Chem 10:313–322. https://doi.org/10.1071/EN13001
Sousa VS, Teixeira MR (2020) Metal-based engineered nanoparticles in the drinking water treatment systems: a critical review. Sci Total Environ 707:136077. https://doi.org/10.1016/j.scitotenv.2019.136077
Sousa JCG, Ribeiro AR, Barbosa MO, Pereira MFR, Silva AMT (2018) A review on environmental monitoring of water organic pollutants identified by EU guidelines. J Hazard Mater 344:146–162. https://doi.org/10.1016/j.jhazmat.2017.09.058
Sousa JCG, Ribeiro AR, Barbosa MO, Ribeiro C, Tiritan ME, Pereira MFR, Silva AMT (2019) Monitoring of the 17 EU Watch List contaminants of emerging concern in the Ave and the Sousa Rivers. Sci Total Environ 649:1083–1095. https://doi.org/10.1016/j.scitotenv.2018.08.309
Sousa JCG, Barbosa MO, Ribeiro ARL, Ratola N, Pereira MFR, Silva AMT (2020) Distribution of micropollutants in estuarine and sea water along the Portuguese coast. Marine Pollut Bull 154:111120. https://doi.org/10.1016/j.marpolbul.2020.111120
Sparks DL (2005) Toxic metals in the environment: the role of surfaces. Elements 1:193–197. https://doi.org/10.2113/gselements.1.4.193
Stackelberg PE, Gibs J, Furlong ET, Meyer MT, Zaugg SD, Lippincott RL (2007) Efficiency of conventional drinking-water-treatment processes in removal of pharmaceuticals and other organic compounds. Sci Total Environ 377:255–272. https://doi.org/10.1016/j.scitotenv.2007.01.095
Starling MCVM, Amorim CC, Leão MMD (2019) Occurrence, control and fate of contaminants of emerging concern in environmental compartments in Brazil. J Hazard Mater 372:17–36. https://doi.org/10.1016/j.jhazmat.2018.04.043
Stock F, Kochleus C, Bänsch-Baltruschat B, Brennholt N, Reifferscheid G (2019) Sampling techniques and preparation methods for microplastic analyses in the aquatic environment – a review. Trends Anal Chem 113:84–92. https://doi.org/10.1016/j.trac.2019.01.014
Strandberg B, Dodder NG, Basu I, Hites RA (2001) Concentrations and spatial variations of polybrominated diphenyl ethers and other organohalogen compounds in Great Lakes air. Environ Sci Technol 35:1078–1083
Sun TY, Gottschalk F, Hungerbühler K, Nowack B (2014) Comprehensive probabilistic modelling of environmental emissions of engineered nanomaterials. Environ Pollut 185:69–76. https://doi.org/10.1016/j.envpol.2013.10.004
Sun GY, Li ZG, Liu T, Chen J, Wu TT, Feng XB (2017) Rare earth elements in street dust and associated health risk. Environ Geochem Health 39:1469–1486. https://doi.org/10.1007/s10653-017-9982-x
Sun CY, Ma RS, Xu L, Chen LB, Xu M, Cao HN, Zhang ZX (2019) Spatial distribution, risk assessment, and seasonal variations of 4-nonylphenol in China's Yinma River basin. Polish J Environ Studies 28:845–852. https://doi.org/10.15244/pjoes/85264
Syafiuddin A, Salmiati S, Hadibarata T, Kueh ABH, Salim MR, Zaini MAA (2018) Silver nanoparticles in the water environment in Malaysia: inspection, characterization, removal, modeling, and future perspective. Sci Rep 8:986. https://doi.org/10.1038/s41598-018-19375-1
Székács S, Darvas B (2018) Re-registration challenges of glyphosate in the European Union. Frontiers Environ Sci 6:78. https://doi.org/10.3389/fenvs.2018.00078
Tahar A, Tiedeken EJ, Clifford E, Cummins E, Rowan N (2017) Development of a semi-quantitative risk assessment model for evaluating environmental threat posed by the three first EU watch-list pharmaceuticals to urban wastewater treatment plants: an Irish case study. Sci Total Environ 603-604:627–638. https://doi.org/10.1016/j.scitotenv.2017.05.227
Tang Y, Yin M, Yang W, Li H, Mo L, Liang Y, Ma X, Sun X (2019) Emerging pollutants in water environment: occurrence, monitoring, fate, and risk assessment. Water Environ Res 91:984–991. https://doi.org/10.1002/wer.1163
Teijón G, Candela L, Tamoh K, Molina-Diáz A, Fernández-Alba AR (2010) Occurrence of emerging contaminants, priority substances (2008/105/EC) and heavy metals in treated wastewater

and groundwater at Depurbaix facility (Barcelona, Spain). Sci Total Environ 408:3584–3595. https://doi.org/10.1016/j.scitotenv.2010.04.041

Teo RD, Termini J, Gray HB (2016) Lanthanides: applications in cancer diagnosis and therapy. J Med Chem 59:6012–6024. https://doi.org/10.1021/acs.jmedchem.5b01975

Tetreault GR, Bennett CJ, Shires K, Knight B, Servos MR, McMaster ME (2011) Intersex and reproductive impairment of wild fish exposed to multiple municipal wastewater discharges. Aquat Toxicol 104:278–290. https://doi.org/10.1016/j.aquatox.2011.05.008

Thomaidi VS, Stasinakis AS, Borova VL, Thomaidis NS (2015) Is there a risk for the aquatic environment due to the existence of emerging organic contaminants in treated domestic wastewater? Greece as a case-study. J Hazard Mater 283:740–747. https://doi.org/10.1016/j.jhazmat.2014.10.023

Thomas PJ, Carpenter D, Boutin C, Allison JE (2014) Rare earth elements (REE): effects on germination and growth of selected crop and native plant species. Chemosphere 96:57–66. https://doi.org/10.1016/j.chemosphere.2013.07.020

Thompson RC (2015) Microplastics in the marine environment: sources, consequences and solutions. In: Bergmann M, Gutow L, Klages M (eds) Marine anthropogenic litter. Springer, Cham, pp 185–200. https://doi.org/10.1007/978-3-319-16510-3_7

Tijani JO, Fatoba OO, Babajide OO, Petrik LF (2016) Pharmaceuticals, endocrine disruptors, personal care products, nanomaterials and perfluorinated pollutants: a review. Environ Chem Lett 14:27–49

Tiwari N, Santhiya D, Sharma JG (2020) Microbial remediation of micro-nano plastics: current knowledge and future trends. Environ Pollut 265:115044. https://doi.org/10.1016/j.envpol.2020.115044

Tobón-Marulanda FA, López-Giraldo LA, Paniagua-Suárez RE (2010) Water pollution caused by pesticides in an área of Antioquia (in Spanish). Rev Salud Publica 12:300–307. https://doi.org/10.1590/S0124-00642010000200013

Tolboom SN, Carrillo-Nieves D, de Jesús R-AM, de la Cruz QR, Barceló D, Iqbal HMN, Parra-Saldivar R (2019) Algal-based removal strategies for hazardous contaminants from the environment – a review. Sci Total Environ 665:358–366. https://doi.org/10.1016/j.scitotenv.2019.02.129

Tongue ADW, Reynolds SJ, Fernie KJ, Harrad S (2019) Flame retardant concentrations and profiles in wild birds associated with landfill: a critical review. Environ Pollut 248:646–658. https://doi.org/10.1016/j.envpol.2019.01.103

Tousova Z, Oswald P, Slobodnik J, Blaha L, Muz M, Hu M, Brack W, Krauss M, Di Paolo C, Tarcai Z, Seiler TB, Hollert H, Koprivica S, Ahel M, Schollee JE, Hollender J, Suter MJF, Hidasi AO, Schirmer K, Sonavane M, Ait-Aissa S, Creusot N, Brion F, Froment J, Almeida AC, Thomas K, Tollefsen KE, Tufi S, Ouyang XY, Leonards P, Lamoree M, Torrens VO, Kolkman A, Schriks M, Spirhanzlova P, Tindall A, Schulze T (2017) European demonstration program on the effect-based and chemical identification and monitoring of organic pollutants in European surface waters. Sci Tot Environ 601-602:1849–1868. https://doi.org/10.1016/j.scitotenv.2017.06.032

Troester M, Brauch HJ, Hofmann T (2016) Vulnerability of drinking water supplies to engineered nanoparticles. Water Res 96:255–279. https://doi.org/10.1016/j.watres.2016.03.038

Tuan Omar TF, Ahmad A, Zaharin Aris A, Md Yusoff F (2016) Endocrine disrupting compounds (EDCs) in environmental matrices: review of analytical strategies for pharmaceuticals, estrogenic hormones, and alkylphenol compounds. Trends Anal Chem 85:241–259. https://doi.org/10.1016/j.trac.2016.08.004

Turra A, Manzano AB, DiasRJ MMM, Barbosa L, Balthazar-Silva D, Moreira FT (2014) Three-dimensional distribution of plastic pellets in sandy beaches: shifting paradigms. Sci Rep 4:4435. https://doi.org/10.1038/srep04435

Turner S, Horton AA, Rose NL, Hall C (2019) A temporal sediment record of microplastics in an urban lake, London, UK. J Paleolimnology 61:449–462. https://doi.org/10.1007/s10933-019-00071-7

UNEP (2009) Marine litter: a global challenge. United Nations Environment Programme, 232p. http://hdl.handle.net/20.500.11822/7787
USGS (2020). https://prd-wret.s3-us-west-2.amazonaws.com/assets/palladium/production/atoms/files/mcs-2019-raREEs.pdf
Vajda AM, Barber LB, Gray JL, Lopez EM, Woodling JD, Norris DO (2008) Reproductive disruption in fish downstream from an estrogenic wastewater effluent. Environ Sci Technol 42:3407–3414. https://doi.org/10.1021/es0720661
Vara S, Karnena MK (2020) Fungal enzymatic degradation of industrial effluents – a review. Current Res Environ Appl Mycol 10:417–442. https://doi.org/10.5943/cream/10/1/33
Vargas-Berrones K, de León-Martínez LD, Bernal-Jácome L, Rodriguez-Aguilar M, Ávila-Galarza A, Flores-Ramirez R (2020a) Rapid analysis of 4-nonylphenol by solid phase microextraction in water samples. Talanta 209:120546. https://doi.org/10.1016/j.talanta.2019.120546
Vargas-Berrones K, Bernal-Jácome L, de León-Martínez LD, Flores-Ramírez R (2020b) Emerging pollutants (Eps) in Latin América: a critical review of under-studied Eps, case of study – nonylphenol (2020b) Sci Total Environ 726:138493. https://doi.org/10.1016/j.scitotenv.2020.138493
Varnosfaderany MN, Bakhtiari AR, Gu ZY, Chu GQ (2015) Distribution and characteristic of PAHs in sediments from the Southwest Caspian Sea, Guilan Province, Iran. Water Sci Technol 71:1587–1596
Vazquez GR, Meijide FJ, Lo Nostro FL (2016) Recovery of the reproductive capability following exposure to 4-tert-octylphenol in the neotropical cichlid fish *Cichlasoma dimerus*. Bull Environ Contam Toxicol 96:585–590. https://doi.org/10.1007/s00128-016-1766-y
Venkatesan AK, Reed RB, Lee S, Bi X, Hanigan D, Yang Y, Ranville JF, Herckes P, Westerhoff P (2018) Detection and sizing of Ti-containing particles in recreational waters using single particle ICP-MS. Bull Environ Contam Toxicol 100:120–126. https://doi.org/10.1007/s00128-017-2216-1
Vestergren R, Cousins IT (2013) Human dietary exposure to per- and poly-fluoroalkyl substances (PFASs) (chapter 12). In Rose M, Fernandes A (eds) Persistent organic pollutants and toxic metals in foods. Woodhead Publishing, pp 279–307. https://doi.org/10.1533/9780857098917.2.279
von der Kammer F, Ferguson PL, Holden PA, Masion A, Rogers KR, Klaine SJ, Koelmans AA, Horne N, Unrine JM (2012) Analysis of engineered nanomaterials in complex matrices (environment and biota): general considerations and conceptual case studies. Environ Toxicol Chem 31:32–49. https://doi.org/10.1002/etc.723
Wagner M, Lambert S (2018) Freshwater microplastics: emerging environmental contaminants? Springer. 303p, Cham. https://doi.org/10.1007/978-3-319-61615-5
Wagner M, Scherer C, Alvarez-Muñoz D, Brennholt N, Bourrain X, Buchinger S, Fries E, Grosbois C, Klasmeier J, Marti T, Rodriguez-Mozaz S, Urbatzka R, Vethaak AD, Winther-Nielsen M, Reifferscheid G (2014) Microplastics in freshwater ecosystems: what we know and what we need to know. Environ Sci Eur 26:1–9. https://doi.org/10.1186/s12302-014-0012-7
Wagner S, Hüffer T, Kloeckner P, Wehrhahn M, Hofmann T, Reemtsma T (2018) Tire wear particles in the aquatic environment – a review on generation, analysis, occurrence, fate and effects. Water Res 139:83–100. https://doi.org/10.1016/j.watres.2018.03.051
Wakabayashi T, Ymamoto A, Kazaana A, Nakano Y, Nojiri Y, Kashiwazaki M (2016) Antibacterial, antifungal and nematicidal activities of rare earths ions. Biol Trace Elem Res 174:464–470
Wang JH, Zhuan R (2020) Degradation of antibiotics by advanced oxidation processes: an overview. Sci Total Environ 701:135023. https://doi.org/10.1016/j.scitotenv.2019.135023
Wang F, Jiang X, Bian YR, Yao FX, Gao HJ, Yu GF, Munch JC, Schroll R (2007) Organochlorine pesticides in soils under different land usage in the Taihu Lake region, China. J Environ Sci 19:584–590
Wang L, Cai Y, Fang L (2009) Pollution in Taihu Lake, China: causal chain and policy options analyses. Front Earth Sci China 3:437–444. https://doi.org/10.1007/s11707-009-0043-3
Wang X, Xi B, Huo S, Sun W, Pan H, Zhang J, Ren Y, Liu H (2013) Characterization, treatment and releases of PBDEs and PAHs in a typical municipal sewage treatment plant situ-

ated beside an urban river, East China. J Environ Sci 25:1281–1290. https://doi.org/10.1016/S1001-0742(12)60201-0
Wang CL, Zou XQ, Zhao YF, Li BJ, Song QC, Li YL, Yu WW (2016a) Distribution, sources, and ecological risk assessment of polycyclic aromatic hydrocarbons in the water and suspended sediments from the middle and lower reaches of the Yangtze River, China. Environ Sci Pollut Res 23:17158–17170
Wang JZ, Jia XW, Gao ST, Zeng XY, Li HR, Zhou Z, Sheng GY, Yu ZQ (2016b) Levels and distributions of polybrominated diphenyl ethers, hexabromocyclododecane, and tetrabromobisphenol A in sediments from Taihu Lake, China. Environ Sci Pollut Res 23:10361–10370
Wang WF, Ndungu AW, Wang J (2016c) Monitoring of endocrine-disrupting compounds in surface water and sediments of the three gorges reservoir region, China. Arch Environ Contam Toxicol 71:509–517. https://doi.org/10.1007/s00244-016-0319-z
Wang L, Kang Y, Liang SY, Chen DY, Zhang QY, Zeng LX, Luo JW, Jiang F (2018a) Synergistic effect of co-exposure to cadmium (II) and 4-n-nonylphenol on growth inhibition and oxidative stress of *Chlorella sorokiniana*. Ecotoxicol Environ Saf 154:145–153. https://doi.org/10.1016/j.ecoenv.2018.02.039
Wang S, Zhu ZL, He JF, Yue XY, Pan JX, Wang ZZ (2018b) Steroidal and phenolic endocrine disrupting chemicals (EDCs) in surface water of Bahe River, China: distribution, bioaccumulation, risk assessment and estrogenic effect on *Hemiculter leucisculus*. Environ Pollut 243:103–114. https://doi.org/10.1016/j.envpol.2018.08.063
Wang X, Zhu Q, Yan X, Wang Y, Liao C, Jiang G (2020) A review of organophosphate flame retardants and plasticizers in the environment: analysis, occurrence and risk assessment. Sci Total Environ 731:139071. https://doi.org/10.1016/j.scitotenv.2020.139071
Watanabe H, Horie Y, Takanobu H, Koshio M, Flynn K, Iguchi T, Tatarazako N (2017) Medaka extended one-generation reproduction test evaluating 4-nonylphenol. Environ Toxicol Chem 36:3254–3266. https://doi.org/10.1002/etc.3895
Watkinson AJ, Murby EJ, Kolpin DW, Costanzo SD (2009) The occurrence of antibiotics in an urban watershed: from wastewater to drinking water. Sci Total Environ 407:2711–2723
Wei LF, Tadesse AW, Wang J (2019) Organohalogenated Contaminants (OHCs) in surface sediments and water of East Dongting Lake and Hong Lake, China. Archiv Environ Contamin Toxicol 76:157–170
Weir A, Westerhoff P, Fabricius L, Hristovski K, von Goetz N (2012) Titanium dioxide nanoparticles in food and personal care products. Environ Sci Technol 46:2242–2250. https://doi.org/10.1021/es204168d
Wen S, Hui Y, Yang FX, Liu ZT, Xu Y (2008) Polychlorinated dibenzo-p-dioxins (PCDDs) and dibenzofurans (PCDFs) in surface sediment and bivalve from the Changjiang Estuary, China. Chin J Oceanol Limnol 26:35–44
Westerhoff P, Atkinson A, Fortner J, Wong MS, Zimmerman J, Gardea-Torresdey J, Ranville J, Herckes P (2018) Low risk posed by engineered and incidental nanoparticles in drinking water. Nat Nanotechnol 13:661–669. https://doi.org/10.1038/s41565-018-0217-9
Wille K, De Brabander HF, Vanhaecke L, De Wulf E, Van Caeter P, Janssen CR (2012) Coupled chromatographic and mass-spectrometric techniques for the analysis of emerging pollutants in the aquatic environment. Trends Anal Chem 35:87–108. https://doi.org/10.1016/j.trac.2011.12.003
Wimmer A, Markus AA, Schuster M (2019) Silver nanoparticle levels in river water: real environmental measurements and modeling approaches – a comparative study. Environ Sci Technol Lett 6:353–358. https://doi.org/10.1021/acs.estlett.9b00211
Wong JKH, Lee KK, Tang KHD, Yap PS (2020) Microplastics in the freshwater and terrestrial environments: prevalence, fates, impacts and sustainable solutions. Sci Total Environ 719:137512. https://doi.org/10.1016/j.scitotenv.2020.137512
Wu S, Zhang S, Gong Y, Shi L, Zhou B (2020a) Identification and quantification of titanium nanoparticles in surface water: a case study in Lake Taihu, China. J Hazard Mater 382:121045. https://doi.org/10.1016/j.jhazmat.2019.121045

Wu Z, He C, Han W, Song J, Li H, Zhang Y, Jing X, Wu W (2020b) Exposure pathways, levels and toxicity of polybrominated diphenyl ethers in humans: a review. Environ Res 187:109531. https://doi.org/10.1016/j.envres.2020.109531

Xiang M, Bian ZH, Xie JQ (2019) Contamination characteristics and ecological risk assessment of polybrominated diphenyl ethers in the North Pacific Ocean. Environ Chem (in press) (in Chinese)

Xiao F (2017) Emerging poly- and perfluoroalkyl substances in the aquatic environment: a review of current literature. Water Res 124:482–495. https://doi.org/10.1016/j.watres.2017.07.024

Xiao F, Simcik MF, Gulliver JS (2013) Mechanisms for removal of perfluorooctane sulfonate (PFOS) and perfluorooctanoate (PFOA) from drinking water by conventional and enhanced coagulation. Water Res 47:49–56. https://doi.org/10.1016/j.watres.2012.09.024

Xiao CY, Wang LH, Zhou Q, Huang XH (2020) Hazards of bisphenol A (BPA) exposure: a systematic review of plant toxicology studies. J Hazard Mater 384:121488. https://doi.org/10.1016/j.jhazmat.2019.121488

Xie WP, Zhao JG, Zhang QY, Ye CP, Zheng GM, Shan Q, Li LC, Shao XF (2020) Occurrence, distribution and bioaccumulation of alkylphenols in the Pearl River networks, South China. Ecol Indic 110:105847. https://doi.org/10.1016/j.ecolind.2019.105847

Xiong Y, Yusheng Z (1991) The mobility of rare-earth elements during hydrothermal activity: a review. Chin J Geochem 10:295–306. https://doi.org/10.1007/BF02841090

Xu J, Zhang Y, Guo CS, He Y, Li L, Meng W (2013) Levels and distribution of tetrabromobisphenol A and hexabromocyclododecane in Taihu Lake, China. Environ Toxicol Chem 32:2249–2255

Xu Y, Wei S, Qin QD, Lam MHW, Giesy JP (2015) AhR-mediated activities and compounds in sediments of Meiliang Bay, Taihu Lake, China determined by *in vitro* bioassay and instrumental analysis. RSC Adv 5:55746–55755

Xu G, Ma SH, Tang L, Sun R, Xiang JJ, Xu BT, Bao YY, Wu MH (2016) Occurrence, fate, and risk assessment of selected endocrine disrupting chemicals in wastewater treatment plants and receiving river of Shanghai, China. Environ Sci Pollut Res 23:25442–25450. https://doi.org/10.1007/s11356-016-7669-y

Xu EG, Chan SN, Choi KW, Lee JHW, Leung KMY (2018) Tracking major endocrine disruptors in coastal waters using an integrative approach coupling field-based study and hydrodynamic modeling. Environ Pollut 233:387–394. https://doi.org/10.1016/j.envpol.2017.10.086

Xu M, Huang H, Li N, Li F, Wang D, Luo Q (2019) Occurrence and ecological risk of pharmaceuticals and personal care products (PPCPs) and pesticides in typical surface watersheds, China. Ecotoxicol Environ Safety 175:289–298. https://doi.org/10.1016/j.ecoenv.2019.01.131

Xu S, Ma J, Ji R, Pan K, Miao AJ (2020) Microplastics in aquatic environments: occurrence, accumulation, and biological effects. Sci Total Environ 703:134699. https://doi.org/10.1016/j.scitotenv.2019.134699

Yadav B, Sellamuthu B, Tyagi RD (2020) Degradation of organic micro-pollutants by ultraviolet (Chapter 17). In Varjani S, Pandey A, Tyagi RD, Ngo HH, Larroche C (eds) Current developments in biotechnology and bioengineering. Emerging organic micro-pollutants. Elsevier, Amsterdam, pp 427–456

Yang LP, Zhu LY, Liu ZT (2011) Occurrence and partition of perfluorinated compounds in water and sediment from Liao River and Taihu Lake, China. Chemosphere 83:806–814

Yang Y, Wang Z, He T, Dai Y, Xie S (2015) Sediment bacterial communities associated with anaerobic biodegradation of bisphenol A. Microb Ecol 70:97–104. https://doi.org/10.1007/s00248-014-0551-x

Yang LH, Cheng Q, Tam NFY, Lin L, Su WQ, Luan TG (2016) Contributions of abiotic and biotic processes to the aerobic removal of phenolic endocrine-disrupting chemicals in a simulated estuarine aquatic environment. Environ Sci Technol 50:4324–4334. https://doi.org/10.1021/acs.est.5b06196

Yang ZL, Shi YQ, Zhan Y, Cheng QZ, Li XG, Zhao CH, Zhang DH (2018) Different pathways for 4-n-nonylphenol biodegradation by two Aspergillus strains derived from estuary sediment: evidence from metabolites determination and key-gene identification. J Hazard Mater 359:203–212. https://doi.org/10.1016/j.jhazmat.2018.07.058

Ying GG, Williams B, Kookana R (2002) Environmental fate of alkylphenols and alkylphenol ethoxylates_a review. Environ Int 28:215–226. https://doi.org/10.1016/S0160-4120(02)00017-X

Yoon HK, Moon HS, Park SH, Song JS, Lim Y, Kohyama N (2005) Dendriform pulmonary ossification in a patient with rare earth pneumoconiosis. Thorax 60:701–703

Yoon SJ, Hong S, Kim S, Lee J, Kim T, Kim B, Kwon BO, Zhou YQ, Shi B, Liu P, Hu WY, Huang B, Wang TY, Khim JS (2020) Large-scale monitoring and ecological risk assessment of persistent toxic substances in riverine, estuarine, and coastal sediments of the Yellow and Bohai seas. Environ Int 137:105517. https://doi.org/10.1016/j.envint.2020.105517

Yu YX, Zhang SH, Huang NB, Li JL, Pang YP, Zhang XY, Yu ZQ, Xu Z (2012) Polybrominated diphenyl ethers and polychlorinated biphenyls in freshwater fish from Taihu Lake, China: their levels and the factors that influence biomagnification. Environ Toxicol Chem 31:542–549

Yu J, Zhou J, Luo Y, Yang XS, Yang J, Yang Y, Yang JQ, Xu J (2017) Pollution by nonylphenol in river, tap water, and aquatic in an acid rain-plagued city in Southwest China. Int J Environ Health Res 27:179–190. https://doi.org/10.1080/09603123.2017.1332345

Yu WW, Du BH, Yang L, Zhang Z, Yang C, Yuan SC, Zhang MN (2019) Occurrence, sorption, and transformation of free and conjugated natural steroid estrogens in the environment. Environ Sci Pollut Res 26:9443–9468. https://doi.org/10.1007/s11356-019-04402-z

Zad TJ, Astuti MP, Padhye LP (2018) Fate of environmental pollutants. Water Environ Res 90:1104–1170. https://doi.org/10.2175/106143018X15289915807191

Zaied BK, Rashid M, Nasrullah M, Zularisam AW, Pant D, Singh L (2020) A comprehensive review on contaminants removal from pharmaceutical wastewater by electrocoagulation process. Sci Total Environ 726:138095. https://doi.org/10.1016/j.scitotenv.2020.138095

Zaytseva TB, Medvedeva NG (2019) Sorption and biodegradation of octyl- and nonylphenols by the cyanobacterium Planktothrix agardhii (Gomont) Anagn. & Komarek. Inland Water Biol 12:337–345. https://doi.org/10.1134/S1995082919030192

Zeng EY (2015) Persistent organic pollutants (POPs): analytical techniques, environmental fate and biological effects. Comprehensive Anal Chem 67:1–660. ISBN: 978-0-444-63299-9

Zhang QH, Jiang GB (2005) Polychlorinated dibenzo-p-dioxins/furans and polychlorinated biphenyls in sediments and aquatic organisms from the Taihu Lake, China. Chemosphere 61:314–322

Zhang J, Cheng H, Gao Q, Zhang ZL, Liu QD (2000) Effect of Lanthanum on growth and biochemical property of Sclerotinia sclerotiorum. Chin J Appl Ecol 11:382e384

Zhang Y, Chen Y, Westerhoff P, Hristovski K, Crittenden JC (2008) Stability of commercial metal oxide nanoparticles in water. Water Res 42:2204–2212. https://doi.org/10.1016/j.watres.2007.11.036

Zhang DP, Zhang XY, Yu YX, Li JL, Yu ZQ, Wang DQ, Wu MH, Sheng GY, Fu JM (2012) Intakes of omega-3 polyunsaturated fatty acids, polybrominated diphenyl ethers and polychlorinated biphenyls via consumption of fish from Taihu Lake, China: a risk-benefit assessment. Food Chem 132:975–981

Zhang W, Rattanaudompol U, Li H, Bouchard D (2013) Effects of humic and fulvic acids on aggregation of aqu/nC60 nanoparticles. Water Res 47:1793–1802. https://doi.org/10.1016/j.watres.2012.12.037

Zhang Y, Liu Y, Dong H, Li XG, Zhang DH (2016a) The nonylphenol biodegradation study by estuary sediment-derived fungus *Penicillium simplicissimum*. Environ Sci Pollut Res 23:15122–15132. https://doi.org/10.1007/s11356-016-6656-7

Zhang M, Mao Q, Feng J, Yuan S, Wang Q, Huang D, Zhang J (2016b) Validation and application of an analytical method for the determination of selected acidic pharmaceuticals and estrogenic hormones in wastewater and sludge. J Environ Sci Health A Tox Hazard Subst Environ Eng 51:914–920. https://doi.org/10.1080/10934529.2016.1191304

Zhang LM, Wei CD, Zhang H, Song MW (2017) Criteria for assessing the ecological risk of nonylphenol for aquatic life in Chinese surface fresh water. Chemosphere 184:569–574. https://doi.org/10.1016/j.chemosphere.2017.06.035

Zhang X, Gu P, Liu (2019) Decontamination of radioactive wastewater: state of the art and challenges forward. Chemopshere 215:543–553. https://doi.org/10.1016/j.chemosphere.2018.10.029

Zhao Y, Ye L, Zhang XX (2018) Emerging pollutants – part I: occurrence, fate and transport. Water Environ Res 90:1301–1322. . https://doi.org/10.2175/106143018X15289915807236

Zhong MQ, Yin PH, Zhao L (2017) Toxic effect of nonylphenol on the marine macroalgae *Gracilaria lemaneiformis* (Gracilariales, Rhodophyta): antioxidant system and antitumor activity. Environ Sci Pollut Res 24:10519–10527. https://doi.org/10.1007/s11356-017-8669-2

Zhou L, Li HR, Yu ZQ, Ren M, Zeng XY, Peng PA, Sheng GY, Fu JM (2012a) Chlorinated and brominated dibenzo-p-dioxins and dibenzofurans in surface sediment from Taihu Lake, China. J Environ Monitor 14:1935–1942

Zhou Z, Shi YL, Li WH, Xu L, Cai YQ (2012b) Perfluorinated compounds in surface water and organisms from Baiyangdian Lake in North China: source profiles, bioaccumulation and potential risk. Bull Environ Contamin Toxicol 89:519–524

Zhou H, Yin N, Faiola F (2020) Tetrabromobisphenol A (TBBPA): a controversial environmental pollutant. J Environ Sci 97:54–66. https://doi.org/10.1016/j.jes.2020.04.039

Zhuang MQ, Zhao JS, Li SY, Liu DR, Wang K, Xiao PR, Yu LL, Jiang Y, Song J, Zhou JG, Wang LS, Chu ZH (2017) Concentrations and health risk assessment of rare earth elements in vegetables from mining area in Shandong, China. Chemosphere 168:578–582. https://doi.org/10.1016/j.chemosphere.2016.11.023

Zuiderveen EAR, Slootweg JC, de Boer J (2020) Novel brominated flame retardants – a review of their occurrence in indoor air, dust, consumer goods and food. Chemosphere 255:126816. https://doi.org/10.1016/j.chemosphere.2020.126816

Chapter 2
Occurrence, Fate and Associated Risks of Organic Micropollutants from the Watch List of European Groundwaters

Anna Jurado, Marc Walther, and M. Silvia Díaz-Cruz

Abstract Groundwater is the main source of water supply in the European Union and therefore, the conservation and protection of this freshwater resource against pollution are key issues. Organic micropollutants have raised concern as wastewater treatment plants partly remove them and thus, they continuously enter the aquifers, threatening groundwater quality.

This chapter summarizes the occurrence, fate and human and environmental risk assessment of organic micropollutants defined in the Watch List of the Decision 2018/840/EU in European groundwater bodies. This Watch List includes 15 substances: 7 pesticides namely acetamiprid, clothianidin, imidacloprid, metaflumizone, methiocarb, thiacloprid and thiamethoxam, 5 antibiotics that are amoxicillin, azithromycin, clarithromycin, ciprofloxacin and erythromycin and 3 steroid hormones, which are the natural hormones estrone and 17-beta-estradiol and the synthetic hormone 17-alpha-ethinylestradiol. These substances are likely to enter the aquifers because of the close interaction between surface water and groundwater bodies through discharges of wastewater treatment plants, which are not designed to

A. Jurado (✉)
Institute for Groundwater Management, Technische Universität Dresden, Dresden, Germany

GHS, Institute of Environmental Assessment and Water Research (IDAEA), Severo Ochoa Excellence Center of the Spanish Council for Scientific Research (CSIC), Barcelona, Spain
e-mail: anna.jurado@idaea.csic.es

M. Walther
Chair of Forest Biometrics and Forest Systems Analysis, Faculty of Environmental Sciences, Department of Forest Sciences, Technische Universität Dresden, Dresden, Germany
e-mail: marc.walther@tu-dresden.de

M. S. Díaz-Cruz
Department of Environmental Chemistry, Institute of Environmental Assessment and Water Research (IDAEA), Severo Ochoa Excellence Center of the Spanish Council for Scientific Research (CSIC), Barcelona, Spain
e-mail: silvia.diaz@idaea.csic.es

N. Morin-Crini et al. (eds.), *Emerging Contaminants Vol. 1*, Environmental Chemistry for a Sustainable World 65,
https://doi.org/10.1007/978-3-030-69079-3_2

efficiently remove the organic micropollutants. Most of these substances are usually detected at low ng/L concentration range or not detected in European groundwater bodies; however, there are an increasing number of studies that reported concentrations above 100 ng/L for some of these organic micropollutants, such as azithromycin, ciprofloxacin, erythromycin, imidacloprid and methiocarb. Therefore, there is an urgent need to define drinking water standards and/or threshold quality values for groundwater because it is the largest body of freshwater in the European Union. Generally, the effluents of wastewater treatment plants and surface waters are more contaminated than groundwater, suggesting that during the soil passage sediments act as a natural filter reducing the concentration of the organic micropollutants due to the occurrence of chemical processes such as sorption, transformation and biodegradation. Currently, most research on these processes has been carried out at a laboratory scale, which does not consider the potential synergistic effects associated with the heterogeneous and complex hydrochemical conditions that are inherent in the subsurface.

The human health and environmental risk assessment posed by the organic micropollutants has been estimated using risk quotients and hazard quotients, respectively. According to the estimated risk quotients for pesticides, pharmaceuticals and steroid hormones at different life stages, these substances do not pose risk for human health if groundwater is a source of water for human consumption. Infants present higher risk than adolescents and adults. For aquatic organisms, the hazard quotients have been evaluated for fish, plants, algae and invertebrates. Overall, most organic micropollutants do not represent a risk for the aquatic organisms tested but there are some exceptions. For instance, *Americamysis bahia*, *Mysidopsis bahia* and *Daphnia magna* are impacted by some pesticides because hazard quotients far above 1 are estimated for imidacloprid, thiacloprid and methiocarb, which hazard quotients were equal to 49, 333 and 38, respectively. Concerning pharmaceuticals, ciprofloxacin is toxic against the plant *Lemna minor* and ciprofloxacin as well as erythromycin for the alga *Microcystis aeruginosa,* which hazard quotients were equal to 2.2, 26 and 81, respectively.

Keywords Groundwater · Organic micropollutants · Monitoring · Risk assessment · Watch List · European Union

Abbreviations

ACE	acetamiprid
ADI	acceptable daily intake
AMX	amoxicillin
AZY	azithromycin
BW	body weight
CIP	ciprofloxacin
CLA	clarithromycin

CLO	clothianidin
C_{max}	maximum concentration
D_{ow}	pH-dependent *n*-octanol-water distribution ratio
DT_{50}	soil half-life
DWELs	drinking water equivalent levels
DWI	drinking water intake
E1	estrone
E2	17-beta-estradiol
EC	European Commission
EC_{50}	half-maximal effective concentration
EE2	17-alpha-ethinylestradiol
ERY	erythromycin
FD	frequency of detection
GUS	groundwater ubiquity score
HQ	hazard quotient
IMD	imidacloprid
ISPRA	Italian national institute for environmental protection and research
K_{oc}	partition coefficient between soil organic carbon and water
K_{ow}	n-octanol-water partition coefficient
LC_{50}	lethal concentration that kills 50% of the test animals during an observation period
LOD	limit of detection
LOQ	limit of quantification
MEC	measured environmental concentrations
MET	methiocarb
MFM	metaflumizone
n.d.	not detected
pK_a	the negative base-10 logarithm of the acid dissociation constant (K_a) of a solution
PNEC	predicted no-effect concentration
RQ	risk quotient
RR	removal rate
S_w	water solubility
THI	thiacloprid
TMX	thiamethoxam
WWTP	wastewater treatment plant

2.1 Introduction

The consumption of groundwater is increasing rapidly driven by the high global population growth coupled with agricultural and industrial activities (Ross 2016). Groundwater is the main source of water supply for human consumption for 75% of the European Union (Tanwar et al. 2014). For example, 75% of the drinking water is produced from groundwater in Belgium (SPW 2020), 67% in France (Corso et al. 2018) and 60% in the Netherlands (van den Brink and Wuijts 2016). This percentage is lower in semi-arid countries such as Spain, where groundwater represents only 26% of the total water used for domestic purposes (De Stefano et al. 2015). According to the European Directive 2006/118/EC (European Commission, EC 2006) groundwater is the largest body of freshwater in the European Union, but it is also the most sensitive since it may suffer contamination from many sources (Fig. 2.1). The pollution sources include loss from sewage systems and the water supply networks, leakage from septic systems, hospital effluents, among others in urban areas (Vázquez-Suñé et al. 2010) and the runoff from agricultural and livestock activities in rural areas (Boy-Roura et al. 2018). Another important aquifer contamination source is the infiltrated water used in managed aquifer recharge, such as reclaimed wastewater and river water (Petrie et al. 2015). As a result, a wide array of pollutants may enter the aquifers posing a risk to groundwater quality.

Among the different pollutants, research on organic micropollutants in the aquatic environment has been extended in the last years, largely due to the advances in analytical techniques. The organic micropollutants are also called "emerging organic contaminants", "trace organic contaminants" and more recently "contaminants of emerging concern" and include pharmaceuticals, personal care products, pesticides, industrial additives, illicit drugs, steroid hormones, food additives, among others (Lapworth et al. 2012). These substances are of particular concern for several reasons. First, wastewater treatment plants and the regulations to build them are not specifically designed to remove the organic micropollutants and thus they end up in the aquatic environment due to the continuous interaction of the

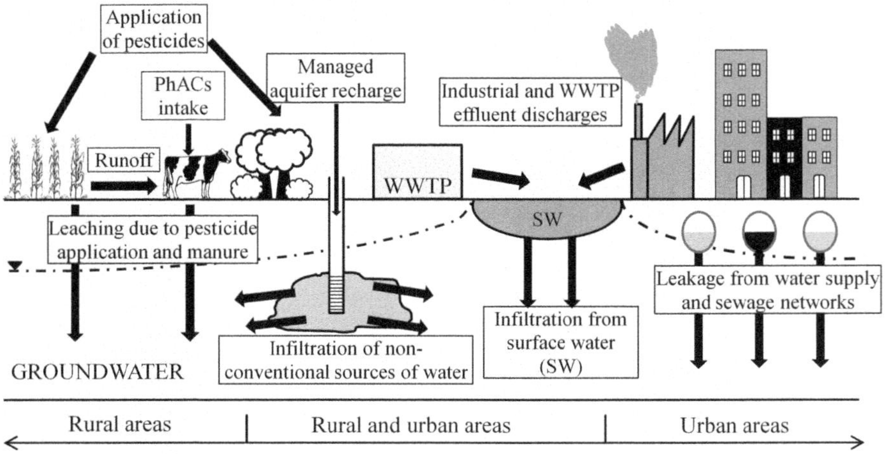

Fig. 2.1 Most common sources and potential pathways of groundwater contamination by organic micropollutants in urban and rural areas (abbreviations: *PhACs* pharmaceuticals, *SW* surface water, *WWTPs* wastewater treatment plants)

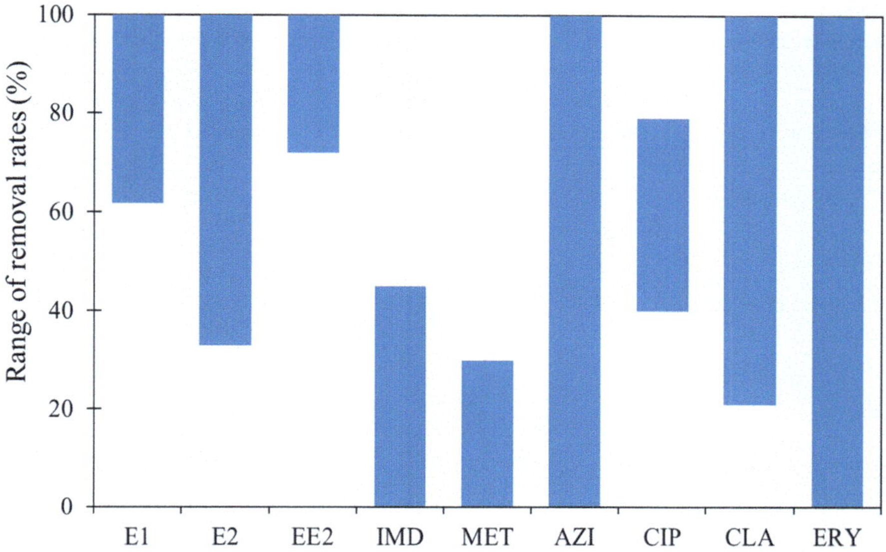

Fig. 2.2 Range of removal rates (%) of the organic micropollutants in wastewater treatment plants based on the bibliographic references of the Table 2.1 (abbreviations: *AZI* azithromycin, *CIP* ciprofloxacin, *CLA* clarithromycin, *E1* estrone, *E2* 17-beta-estradiol, *EE2* 17-alpha-ethinylestradiol, *ERY* erythromycin, *IMD* imidacloprid, *MET* methiocarb)

hydrological compartments (Luo et al. 2014) (Fig. 2.2 and Table 2.1). Hence, these substances are also widely detected in groundwater at concentrations ranging from ng/L to µg/L (Jurado et al. 2012). Second, due to their design and mode of action, they might pose an ecological risk even at trace levels because many of them interfere with the endocrine system of high organisms, can promote microbiological resistance, accumulate in soils, plants and organisms (Arnold et al. 2014) and their long-term effects on human health are largely unknown (Fahrenfeld et al. 2013). Moreover, the effects of organic micropollutants' mixtures are subject to uncertainty and are assumed to have unforeseen consequences on ecosystems and human health. Finally, most of them are not included in the Watch Lists of priority pollutants due to the existing regulatory gap.

To tackle the aforementioned problems, it is required to establish groundwater quality standards for these unregulated substances. So far, regulation on groundwater contamination has been in place only for pesticides (EC 2006), the allowed concentrations are 0.1 µg/L for the active substances and relevant metabolites and 0.5 µg/L for the sum of all them. The occurrence of an increasing number of organic micropollutants in European groundwater bodies have been compiled in some reviews (Jurado et al. 2012, 2019; Lapworth et al. 2012; Stuart et al. 2012; Meffe and de Bustamante 2014, Postigo and Barceló 2015; Sui et al. 2015), demonstrating the urgent need to define quality standards for groundwater. According to the European Directive 2014/80/EC (EC 2014), a Watch List for groundwater

Table 2.1 Average and range *(in parentheses)* concentrations (Ci and Ce in ng/L) for organic micropollutants in influent and effluent of wastewater treatment plants (WWTPs), and removal rate (%) after treatment *(in parentheses)*

Substance	WWTP concentration (ng/L)		Removal rate (%) (Treatment)	References
	Influent (Ci)	Effluent (Ce)		
Estrone	<32.3	<32.3		Camacho-Muñoz et al. (2012)
17-beta-estradiol	2780/930	220/110/<5	33 (CAS), 46 (HF-MBR), 74 (FS-MBR), 100 (HF/RO-MBR)	
17-alpha-ethinylestradiol	2870/2970	1470/590/510/<45	79 (CAS), 72 (HF-MBR), 73 (FS-MBR), 100 (HF/RO-MBR)	
Estrone		20.2	61.8 (UV + TiO_2), 100 (O_3), 100 (O_3 + UV + TiO2)	Moreira et al. (2015)
17-beta-estradiol		110.4	100 (UV + TiO_2), 100 (O_3), 100 (O_3 + UV + TiO2)	
17-alpha-ethinylestradiol		391.4	72.2 (UV + TiO_2), 100 (O_3), 100 (O_3 + UV + TiO2)	
Estrone	19.22 (2.3–37)	0.62 (0.3–0.9)	96.8[a, b]	Zhang and Fent (2018)
17-beta-estradiol	2.4 (n.d.–5.7)	0	100[a, b]	
Imidacloprid			(0–45)[c, d]	Campo et al. (2013)
Methiocarb			(0–30)[c, d]	
Acetamiprid	(0.97–2.4)	1.95	23.2[b]	Iancu and Radu (2018)
Clothianidin	n.d.	n.d.		
Imidacloprid	(60.8–80.2)	53.3	22.4[b]	
Thiamethoxam	(16.4–23.6)	14.6	20.3[b]	
Azithromycin	129/153	143/115	0 (BIO), 15 (UV)	Collado et al. (2014)
Ciprofloxacin	392/178	176/137	40 (BIO), 55 (UV)	
Clarithromycin	100/87	99/77	21 (BIO), 42 (UV)	
Erythromycin	15/17	18/15	0 (BIO, UV)	
Azithromycin	118		75 (MBR), >99 (MBR + RO)	Dolar et al. (2012)
Clarithromycin	2020		87 (MBR), >99 (MBR + RO)	
Erythromycin	49		80 (MBR), >99 (MBR + RO)	

(continued)

Table 2.1 (continued)

Substance	WWTP concentration (ng/L)		Removal rate (%) (Treatment)	References
	Influent (Ci)	Effluent (Ce)		
Azithromycin		140	91.4 (UV + TiO_2), 92.2 (O_3), 100 (O_3 + UV + TiO_2)	Moreira et al. (2015)
Clarithromycin		116.4	97 (UV + TiO_2), 98.2 (O_3), 100 (O_3 + UV + TiO_2)	
Erythromycin		27	76.6 (UV + TiO_2), 100 (O_3), 100 (O_3 + UV + TiO_2)	
Azithromycin	42.2 (n.d.–719)	6.7 (n.d.–200)	94.6 (0–100)[c, d]	Pereira et al. (2015)
Ciprofloxacin	4374 (n.d.–17500)	1225 (n.d.–1224.7)	79.1 (0–100)[c, d]	
Azithromycin	5837 (269–22730)	5453 (316–20178)	0 (BIO)	Senta et al. (2019)
Clarithromycin	828 (11–10491)	266 (25–2517)	61 (BIO)	
H_2O-Ery	26 (10–129)	19 (4.2–75)	34 (BIO)	

Negative removal rates are considered zero
n.d. not detected
Abbreviations: *BIO* biological treatment, *CAS* conventional activated sludge, *FS-MBR* microfiltration membrane bioreactor, *HF-MBR* ultrafiltration membrane bioreactor, *O_3* ozonation; *O_3 + UV + TiO_2* photocatalytic ozonation, *RO* reverse osmosis, *UV + TiO_2* photocatalysis, *H_2O-Ery* anhydroerythromycin
[a][(Ci–Ce)/Ci] × 100
[b]Treatment not specified
[c]Range of removal rates
[d]CAS, CAS with conventional and extended aeration, CAS with medium load aeration, biofiltration, tricking filters, microfiltration and lagoons with extended aeration

pollutants should be established to increase the availability of monitoring data on substances posing a risk or potential risk to groundwater and thereby facilitate the identification of substances, including the organic micropollutants, for which groundwater quality standards or threshold values should be set. Recently, a voluntary groundwater Watch List was proposed by the European groundwater-working group that included per- and polyfluoroalkyl substances and pharmaceuticals (Lapworth et al. 2018a), but it has not been yet defined in the European groundwater directive. In contrast to groundwater, surface water Watch Lists (EC 2015, 2018) have already been defined as established by the 2013 amendment (EC 2013) to the European Environmental Quality Standards Directive 2008/105/EC (EC 2008). The second Watch List of substances for Union-wide monitoring in the field of water policy was proposed by the European Union in the Decision 2018/840/EU (EC

2018), which amends the first Watch List of substances set in the Decision 2015/495/ EU (EC 2015). This list includes 15 organic micropollutants:

- Three estrogens: the natural hormones estrone (E1) and 17-beta-estradiol (E2) and the synthetic hormone 17-alpha-ethinylestradiol (EE2);
- Seven pesticides: the carbamate methiocarb, the semicarbazone insecticide metaflumizone and the neonicotinoids imidacloprid, thiacloprid, thiamethoxam, clothianidin and acetamiprid;
- Five pharmaceuticals: the β-lactam antibiotic amoxicillin, the fluoroquinolone antibiotic ciprofloxacin and 3 macrolide antibiotics namely azithromycin, clarithromycin and erythromycin.

These substances might reach the aquifers because of the close interaction of groundwater with surface waters, such as rivers and the contamination of groundwater-dependent ecosystems, i.e., springs, lakes and wetlands (Zuehlke et al. 2004; Heberer et al. 2008; Kløve et al. 2011; Eamus et al. 2016). Surface waters are usually heavily contaminated because they are impacted by discharges of secondary and tertiary effluents from wastewater treatment plants. Figure 2.2 shows the range of removal rates (%) of the organic micropollutants in wastewater treatment plants and Table 2.1 lists the average concentrations (ng/L) of wastewater treatment plants influent and effluent, as well as the removal rates for some of the organic micropollutants included in the surface water Watch List (EC 2018).

The removal efficiency varies significantly with the wastewater treatment applied and the substance (Ratola et al. 2012; Barbosa et al. 2016; Garcia-Becerra and Ortiz 2018). Overall, steroid hormones have high removal rates, whereas for pesticides they are low (generally lower than 30%) or even negative, suggesting that the pollutant concentration in the effluent is higher than that in the influent (Table 2.1). The removal rates of antibiotics are variable (Fig. 2.2). For instance, several authors reported that azithromycin and erythromycin are barely or not removed when using conventional activated sludge, biological and ultraviolet radiation as treatments (Collado et al. 2014; Pereira et al. 2015; Senta et al. 2019) (Fig. 2.2). In contrast, removal rates higher than 75% for macrolide antibiotics were reported by Moreira et al. (2015) when applying advanced processes such as ozonation, photolysis and photolytic ozonation. However, sometimes the pollutants' load of wastewater effluents is high even achieving removal rates of up to 90% as reported by Pereira et al. (2015) for ciprofloxacin (1200 ng/L). Thus, the discharge of wastewater treatment plant effluents to surface waters potentially leads to the contamination of groundwater bodies with the organic micropollutants included in the surface water Watch List, especially in aquifers where rivers are the main source of groundwater recharge.

The present chapter aims to compile data on the occurrence of the 15 organic micropollutants listed in the Watch List for the European Union monitoring defined in the Decision 2018/840/EU (5 June 2018, EC 2018) in the groundwater bodies of Europe. The occurrence data and fate of these substances in European groundwater bodies are presented and discussed. Besides, a preliminary estimation of the potential risk posed by the substances reported in the groundwater of European countries is carried out for human health and aquatic organisms.

2.2 Occurrence and Spatial Distribution of the Organic Micropollutants in the Groundwater of Europe

The occurrence and the spatial distribution of the 15 organic micropollutants listed in the Decision 2018/840/EU (EC 2018) in the groundwater bodies of Europe are summarized in this section. A total of 40 studies were revised from 1996 to 2019. These studies include a wide variety of topics such as the occurrence and monitoring of the organic micropollutants in groundwater, the development of new methodologies and their application in the analysis of the micropollutants in groundwater samples and managed aquifer recharge studies (Tables 2.2, 2.3, and 2.4). The number of studies (%) and the reported maximum concentrations of each substance are shown in Fig. 2.3.

2.2.1 Steroid Hormones

Hormones are not unique to humans. They are present in nature, both in animal and plant species. In the group of vertebrate organisms, there is a lot of similarity among the hormones existing in the different species, both in their structural characteristics and in the function they perform. In vegetables, although there are differences in structure and function, the mechanism of action is similar. This explains why any substance that interferes with the hormonal mechanism of action can alter the development, reproduction, and other functions of living beings of different species (Damstra et al. 2002).

Steroid hormones are natural and synthetic substances that are classified as endocrine-disrupting compounds since they might interfere with the function of the endocrine system in animals and humans' metabolism. Therefore, any change in their normal concentration affects the homeostasis, being compensated by positive or negative feedback mechanisms, corresponding to diverse direct or indirect effects (Diamanti-Kandarakis et al. 2009). These substances can disrupt the functioning of the endocrine system different strategies, i.e., by mimicking natural hormones, blocking receptors in a cell, activating the synthesis and secretion of natural hormones, inactivating enzymes responsible for hormone secretion and stimulating the ability of hormones to interact with hormones cell receivers. This endogenous behaviour is responsible for the adverse effects when released to the environment. For example, the natural hormone 17-beta-estradiol and the synthetic 17-alpha-ethinylestradiol can cause sex reversal of amphibians and fish (Palace et al. 2002; Hayes et al. 2002). Moreover, these substances can bioaccumulate in aquatic organisms and biomagnificate through the food web, ultimately reaching human beings through the diet, including food and tap water containing them (Santos et al. 2010).

The most studied substances were the natural hormones estrone and 17-beta-estradiol ($n = 10$) followed by the synthetic hormone

Table 2.2 Minimum, maximum and mean concentrations (ng/L) of the steroid hormones estrone (E1), 17-beta-estradiol (E2) and 17-alpha-ethinylestradiol (EE2) in European groundwater bodies

	Concentration (ng/L)			LOD	LOQ	FD (%)			
Substance	Min	Max	Mean	(ng/L)	(ng/L)	(np/nt)	Country	Purpose and type of study/Scale	References
E1	n.d.	1.6			<0.1	18 (20/109)	Austria	O&M, R/N	Hohenblum et al. (2004)
(n = 10)	n.d.				0.07	0 (4)	Belgium	MD&A, R/L	Glineur et al. (2018)
	n.d.			0.04–2		0 (0/16)	Denmark/Sweden	MAR, R/L	Kuster et al. (2010)
	n.d.	3.5			0.02	67 (2/3)	France	MD&A, R/L	Vulliet et al. (2008)
	n.d.	9			5	0.1 (1/954)	France	O&M, R/N	Lopez et al. (2015)
	n.d.	0.1			0.1	13 (1/8)	Germany	MAR, R/L	Zuehlke et al. (2004)
	n.d.			0.16	0.52	0 (0/17)	Italy	O&M, R/L	Pignotti et al. (2017)
	n.d.			7.24	21.7	- (13)	Portugal	O&M, R/L	Carvalho et al. (2015)
	n.d.			2.5		0 (0/7)	Spain	MD&A, R/L	Rodríguez-Mozaz et al. (2004)
	n.d.			<10		0 (0/29)	Spain	O&M, R/L	Corada-Fernández et al. (2017)
E2	n.d.	0.79	0.13		<0.1	52 (58/112)	Austria	O&M, R/N	Hohenblum et al. (2004)
(n = 10)	n.d.	< 0.13			0.13	25 (1/4)	Belgium	MD&A, R/L	Glineur et al. (2018)
	n.d.			0.04–2		0 (0/16)	Denmark/ Sweden	MAR, R/L	Kuster et al. (2010)
	n.d.	1.3			0.01	67 (2/3)	France	MD&A, R/L	Vulliet et al. (2008)
	n.d.				0.1	0 (0/8)	Germany	MAR, R/L	Zuehlke et al. (2004)
	n.d.			0.45	1.5	0 (0/17)	Italy	O&M, R/L	Pignotti et al. (2017)
	n.d.	3		1	3		Luxembourg	O&M, R/L	Banzhaf et al. (2013)
	n.d.			6.86	20.6	0 (13)	Portugal	O&M, R/L	Carvalho et al. (2015)
	n.d.			2.5		0 (0/7)	Spain	MD&A, R/L	Rodríguez-Mozaz et al. (2004)
	n.d.			<10		0 (0/29)	Spain	O&M, R/L	Corada-Fernández et al. (2017)

EE2	n.d.	0.94			<0.1	0.01 (1/112)	Austria	O&M, R/N	Hohenblum et al. (2004)
(n = 9)	n.d.	0.02			0.06	50 (2/4)	Belgium	MD&A, R/L	Glineur et al. (2018)
	n.d.			0.04–2		0 (0/16)	Denmark/Sweden	MAR, R/L	Kuster et al. (2010)
	n.d.	3			0.2	67 (2/3)	France	MD&A, R/L	Vulliet et al. (2008)
	n.d.				0.2	0 (0/8)	Germany	MAR, R/L	Zuehlke et al. (2004)
	n.d.			0.56	1.88	0 (0/17)	Italy	O&M, R/L	Pignotti et al. (2017)
	n.d.			6.83	20.5	0 (13)	Portugal	O&M, R/L	Carvalho et al. (2015)
	n.d.			3.22		0 (0/7)	Spain	MD&A, R/L	Rodríguez-Mozaz et al. (2004)
	n.d.			<10		0 (0/29)	Spain	O&M, R/L	Corada-Fernández et al. (2017)

Abbreviations: *LOD* limit of detection, *LOQ* limit of quantification, *FD* frequency of detection, *n.d.* not detected, *n* number of studies, *np* number of positive samples, *nt* total number of samples collected, *MD&A* method development and application, *O&M* occurrence and monitoring, *MAR* managed aquifer recharge, *R* research paper, *L* local scale, *N* nationwide scale

Table 2.3 Minimum, maximum and mean concentrations (ng/L) of pesticides in European groundwater bodies

Substance	Concentration (ng/L)			LOD	LOQ	FD (%)	Country	Purpose and type of study/Scale	References
	Min	Max	Mean	(ng/L)	(ng/L)	(np/nt)			
Acetamiprid	n.d.			1	10	0 (0/90)	Belgium/The Netherlands	O&M, R/N	Sjerps et al. (2019)
(n = 7)	n.d.	70			20	0.2 (2/1067)	France	O&M, R/N	Lopez et al. (2015)
	n.d.			2.5	5	0 (0/28)	Italy	O&M, R/L	De Liguoro et al. (2014)
	n.d.	290	<10		10	3.8 (59/1562)	Italy	O&M, Rp/N	ISPRA (2018)
	n.d.			15.6	52	0 (0/40)	Portugal	O&M, R/L	Melo et al. (2012)
	n.d.			3.5	11.6	0 (0/6)	Serbia	MD&A, R/L	Dujaković et al. (2010)
	n.d			5	17	0 (0/44)	Serbia	MD&A, R/L	Radović et al. (2015)
Clothianidin	n.d.	120		2	10	1.1 (1/90)	Belgium/The Netherlands	O&M, R/N	Sjerps et al. (2019)
(n = 4)	n.d.			2.5	5	0 (0/28)	Italy	O&M, R/L	De Liguoro et al. (2014)
	n.d.	20	<10		10	2.3 (6/258)	Italy	O&M, Rp/N	ISPRA (2018)
	n.d.	90	3	1		33 (89/267)	United Kingdom	O&M, Rp/N	Lapworth et al. (2018b)
Imidacloprid	n.d.	5		2.5	5	3.6 (1/28)	Italy	O&M, R/L	De Liguoro et al. (2014)
(n = 7)	n.d.	1670	<20		20	7.7 (202/2628)	Italy	O&M, Rp/N	ISPRA (2018)
	n.d.			4.1	13.5	0 (0/6)	Serbia	MD&A, R/L	Dujaković et al. (2010)
	n.d			3	10	0 (0/44)	Serbia	MD&A, R/L	Radović et al. (2015)
	n.d.			300		0 (0/1)	Spain	MD&A, R/L	Martínez Galera et al. (1998)
	n.d.	120			9	35 (8/23)	Spain	O&M, R/L	Pitarch et al. (2016)
	n.d.	20		1		16 (43/267)	United Kingdom	O&M, Rp/N	Lapworth et al. (2018b)
Methiocarb	n.d.	120	<20		20	4.3 (54/1264)	Italy	O&M, Rp/N	ISPRA (2018)
(n = 3)	n.d.	300		10		53 (21/40)	Spain	O&M, B/L	Barceló et al. (1996)
	n.d.				50	0 (0/4)	Spain	MD&A, R/L	Kharbouche et al. (2019)

Thiacloprid	n.d.				<100	0 (0/100)	Denmark	O&M, Rp/N	Brüsch et al. (2016)
(n = 3)	n.d			2.5	5	0 (0/28)	Italy	O&M, R/L	De Liguoro et al. (2014)
	n.d.	200	<10		10	2.9 (47/1594)	Italy	O&M, Rp/N	ISPRA (2018)
Thiamethoxam	n.d.	10		2	10	2.2 (2/90)	Belgium/The Netherlands	O&M, R/N	Sjerps et al. (2019)
(n = 6)	n.d.				<100	0 (0/359)	Denmark	O&M, Rp/N	Brüsch et al. (2016)
	n.d.			2.5	5	0 (0/28)	Italy	O&M, R/L	De Liguoro et al. (2014)
	n.d.	180	<10		10	7 (113/1617)	Italy	O&M, Rp/N	ISPRA (2018)
	n.d.	3		1		7.5 (20/267)	United Kingdom	O&M, Rp/N	Lapworth et al. (2018b)
	n.d.	>100		6.4	21.2	35 (14/40)	Portugal	O&M, R/L	Melo et al. (2012)

Abbreviations: *LOD* limit of detection, *LOQ* limit of quantification, *FD* frequency of detection, *n.d.* not detected, *n* number of studies, *np* number of positive samples, *nt* total number of samples collected, *MD&A* method development and application, *O&M* occurrence and monitoring, *R* research paper, *Rp* report, *B* book chapter, *L* local scale, *N* nationwide scale

Table 2.4 Minimum, maximum and mean concentrations (ng/L) of pharmaceuticals in European groundwater bodies

Substance	Concentration (ng/L) Min	Max	Mean	LOD (ng/L)	LOQ (ng/L)	FD (%) (np/nt)	Country	Purpose and type of study/Scale	References
Amoxicillin	n.d.			4.6	15	<3 (<3/105)	Germany	MD&A, R/L	Sacher et al. (2001)
(n = 8)	n.d.			0.79	2.63	0 (0/10)	Serbia	MD&A, R/L	Grujić et al. (2009)
	n.d			3	10	0 (0/44)	Serbia	MD&A, R/L	Radović et al. (2015)
	n.d.					0	Spain	O&M, R/L	Cabeza et al. (2012)
	n.d.				500	0	Spain	MD&A, R/L	Tong et al. (2014)
	n.d.					0	Spain	O&M, R/L	Candela et al. (2016)
	n.d.			0.12	0.4	0 (0/47)	Spain	O&M, R/L	Boy-Roura et al. (2018)
	n.d.			1.1	3.6	0 (0/6)	Sweden	O&M, MT/L	Rostvall (2017)
Azithromycin	n.d.				10–50	0	Italy	MAR, R/L	Ernst et al. (2012)
(n = 8)	n.d.	140		2.58	8.59	20 (2/10)	Serbia	MD&A, R/L	Grujić et al. (2009)
	n.d	68		3	10	4.5 (2/44)	Serbia	MD&A, R/L	Radović et al. (2015)
	18	83		4.5		3.2 (4/123)	Serbia	MAR, R/L	Kovačević et al. (2017)
	n.d.					0	Spain	O&M, R/L	Cabeza et al. (2012)
	n.d	1620	117	0.16	0.53	97 (30/31)	Spain	O&M, R/L	López-Serna et al. (2013)
	n.d.					0	Spain	O&M, R/L	Candela et al. (2016)
	n.d.	13	0.93	0.04	0.13	15(7/47)	Spain	O&M, R/L	Boy-Roura et al. (2018)
Ciprofloxacin	n.d				<60	0 (0/5)	France	O&M, R/L	Guillon et al. (2015)
(n = 9)	n.d.	43			5	0.6 (6/954)	France	O&M, R/N	Lopez et al. (2015)
	n.d.				2	0 (0/161)	Germany	MAR, R/L	Heberer et al. (2008)
	n.d			0.02	0.08		Portugal	O&M, R/L	de Jesus Gaffney et al. (2015)
	n.d	130.8				50 (3/6)	Romania	O&M, R/L	Szekeres et al. (2018)
			14–323.8				Spain	O&M, R/L	Cabeza et al. (2012)
	n.d.	443	47	2.47	8.23	94 (29/31)	Spain	O&M, R/L	López-Serna et al. (2013)
	2	406	126				Spain	O&M, R/L	Candela et al. (2016)
	n.d	298.3	34	1.04	3.47	51 (24/47)	Spain	O&M, R/L	Boy-Roura et al. (2018)
Clarithromycin	n.d.	42			5	0.3 (3/954)	France	O&M, R/N	Lopez et al. (2015)
(n = 11)	n.d.			3.6	13	<3 (<3/105)	Germany	MD&A, R/L	Sacher et al. (2001)

	<25				25	-	Germany	O&M, R/L	Ternes et al. (2007)
	n.d.	3.4			0.2	<6 (10/161)	Germany	MAR, R/L	Heberer et al. (2008)
	8.7	24.5		7.5	18	3.7 (6/163)	Germany	O&M, R/L	Reh et al. (2013)
	n.d.		n.d.–190		10		Italy	MAR, R/L	Ernst et al. (2012)
	n.d.					0	Spain	O&M, R/L	Cabeza et al. (2012)
	1.6	20.5	6	0.19	0.65	100 (31/31)	Spain	O&M, R/L	López-Serna et al. (2013)
	n.d.					0	Spain	O&M, R/L	Candela et al. (2016)
	n.d.	0.54	0.34	0.33	1.11	62 (29/47)	Spain	O&M, R/L	Boy-Roura et al. (2018)
	n.d.	1		0.004	0.01	67 (4/6)	Sweden	O&M, MT/L	Rostvall (2017)
Erythromycin	n.d.				< 4	0 (0/5)	France	O&M, R/L	Guillon et al. (2015)
(n = 17)	n.d.	1872			5	2.5 (24/954)	France	O&M, R/N	Lopez et al. (2015)
	n.d.	49		3.6	12	10 (10/105)	Germany	MD&A, R/L	Sacher et al. (2001)[a]
	<25				25		Germany	O&M, R/L	Ternes et al. (2007)
	n.d.	10.4			0.3	(-/161)	Germany	MAR, R/L	Heberer et al. (2008)[a]
	5.5	392.4		7.5	29	1.8 (3/163)	Germany	O&M, R/L	Reh et al. (2013)
	n.d.		n.d.–25		10		Italy	MAR, R/L	Ernst et al. (2012)
	n.d	5		0.02	0.07		Portugal	O&M, R/L	de Jesus Gaffney et al. (2015)
	n.d.	568.4				83 (5/6)	Romania	O&M, R/L	Szekeres et al. (2018)
	n.d.			1.65	5.48	0 (0/10)	Serbia	MD&A, R/L	Grujić et al. (2009)
	n.d			2	7	0 (0/44)	Serbia	MD&A, R/L	Radović et al. (2015)
	n.d.			1.65		0 (0/123)	Serbia	MAR, R/L	Kovačević et al. (2017)[a]
			n.d.–154.3				Spain	O&M, R/L	Cabeza et al. (2012)
	n.d.	41.3	12	7.34	24.48	42 (13/31)	Spain	O&M, R/L	López-Serna et al. (2013)
	7	428	63				Spain	O&M, R/L	Candela et al. (2016)
	0.1	2.6		0.04	0.1	100 (6/6)	Sweden	O&M, MT/L	Rostvall (2017)
	n.d	<25			10	7 (1/15)	The Netherlands	MD&A, R/L	Stolker et al. (2004)[a]

Abbreviations: *LOD* limit of detection, *LOQ* limit of quantification, *FD* frequency of detection, *n.d.* not detected, *n* total number of studies, *np* number of positive samples, *nt* total number of samples collected, *MD&A* method development and application, *O&M* occurrence and monitoring, *MAR* managed aquifer recharge, *R* research paper, *MT* master thesis, *L* Local scale, *N* nationwide scale

[a]Anhydroerythromycin

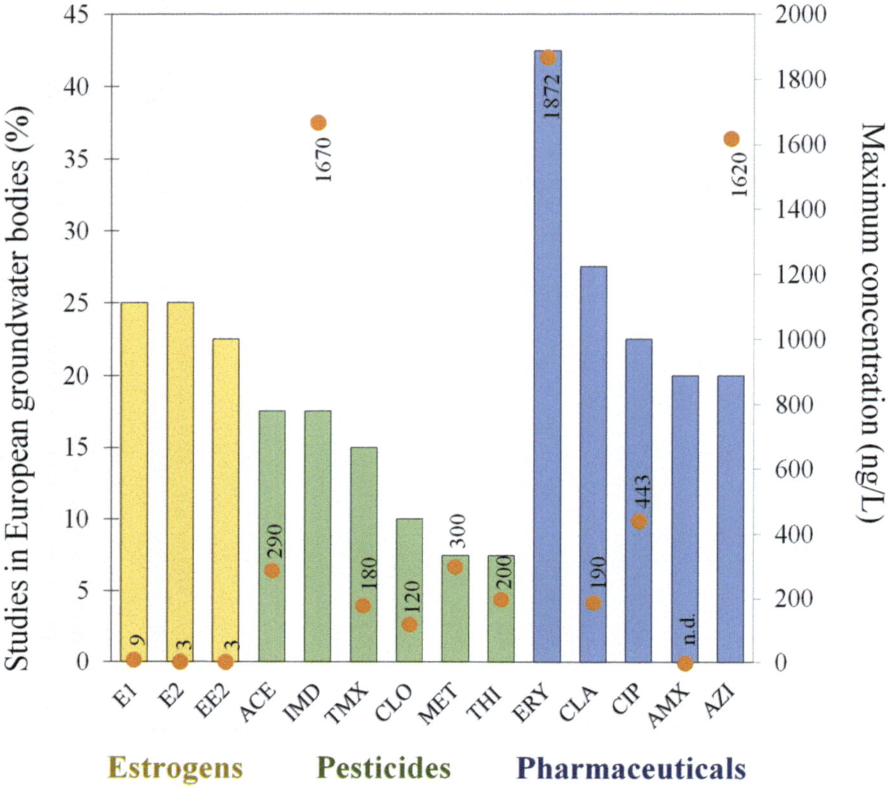

Fig. 2.3 Number of studies (n = 40) investigating the occurrence of the micropollutants listed in the decision 2018/840/EU in groundwater bodies of Europe (%) and maximum groundwater concentrations (ng/L) reported (n.d. not detected; abbreviations: *ACE* acetamiprid, *AMX* amoxicillin, *AZI* azithromycin, *CIP* ciprofloxacin, *CLA* clarithromycin, *CLO* clothianidin, *E1* estrone, *E2* 17-beta-estradiol, *EE2* 17-alpha-ethinylestradiol, *ERY* erythromycin, *IMD* imidacloprid, *MET* methiocarb, *THI* thiacloprid, *TMX* thiamethoxam)

17-alpha-ethinylestradiol (n = 9). Steroid hormones have been widely investigated in groundwater bodies across Europe but they were quantified in only a few samples and at low concentrations (Table 2.2). So far, they have been detected in the groundwater bodies of France (Vulliet et al. 2008; Lopez et al. 2015), Austria (Hohenblum et al. 2004), Germany (Zuehlke et al. 2004) and Luxembourg (Banzhaf et al. 2013). Zuehlke et al. (2004) investigated the behaviour of estrone, 17-beta-estradiol and 17-alpha-ethinylestradiol in a river bank filtration system in the Lake Tegel that supplies drinking water in Berlin (Germany). This lake is used for groundwater recharge and receives effluents of wastewater treatment plants. Estrone was detected in surface water at concentrations around or below 1 ng/L and in one groundwater sample at concentration of 0.1 ng/L. 17-beta-estradiol and 17-alpha-ethinylestradiol were not present at concentrations above the limit of quantification of 0.2 ng/L in surface water and groundwater. Hohenblum et al. (2004) conducted a regional study on surface and groundwater in Austria to assess the risk of endocrine disrupting compounds in humans and aquatic life. Results revealed the presence of low but detectable levels of estrone, 17-beta-estradiol and 17-alpha-ethinylestradiol at maximum concentrations of 1.6 ng/L, 0.79 ng/L and 0.94 ng/L, respectively. The most observed substances were 17-beta-estradiol and estrone, with a frequency of detection of 52% and 18%, respectively.

Vulliet et al. (2008) and Lopez et al. (2015) also observed these substances in France. The maximum concentrations in French groundwater bodies were 9 ng/L for estrone, 1.3 for 17-beta-estradiol and 3 ng/L for 17-alpha-ethinylestradiol. In Luxembourg, Banzhaf et al. (2013) reported concentrations of 3 ng/L for 17-beta-estradiol.

Many other studies have attempted to quantify steroid hormones in other European groundwater bodies (Rodríguez-Mozaz et al. 2004; Kuster et al. 2010; Carvalho et al. 2015; Corada-Fernández et al. 2017; Glineur et al. 2018; Pignotti et al. 2017). Despite the high sensitivity of the analytical methods applied, for example, limit of detection below 0.16 for estrone (Pignotti et al. 2017), these substances were never detected in the studied groundwater bodies of Portugal, Spain, Belgium, Denmark, Sweden and Italy (Table 2.2).

2.2.2 *Pesticides*

Pesticides are bioactive substances intended to eliminate, prevent and mitigate pests, such as pathogens, rodents, weeds, fungi, among others (Imfeld and Vuilleumier 2012). By their nature, pesticides are toxic to specific organisms, including humans and wildlife and have to be used safely (Mostafalou and Abdollahi 2017). Definitely, pesticides usage has improved the production and storage of crops preventing them from pests; however their widespread use has resulted in the pollution of freshwater, including groundwater, predominantly in agricultural lands (Zhao and Pei 2012).

Among the 7 pesticides included in the Watch List, only metaflumizone has not been studied yet in the groundwater bodies of Europe. Methiocarb and the 5 neonicotinoids were detected at concentrations above 100 ng/L, reaching levels up to 1670 ng/L for imidacloprid in Italy (Table 2.3). In addition, the occurrence of neonicotinoids was reported in groundwater bodies of France, Belgium, Italy, the Netherlands, Portugal, Spain and the United Kingdom (Melo et al. 2012; De Liguoro et al. 2014; Lopez et al. 2015; Pitarch et al. 2016; ISPRA 2018; Lapworth et al. 2018b; Sjerps et al. 2019).

In Italy, the Italian National Institute for Environmental Protection and Research (ISPRA) is responsible for the technical management and assessment of the monitoring of pesticides in surface waters and groundwater. The neonicotinoid imidacloprid was the most frequently reported insecticide in groundwater in 2016 at maximum concentration of 1670 ng/L (ISPRA 2018), with a detection frequency of

7.7%. The neonicotinoids acetamiprid, thiacloprid and thiamethoxam were detected at concentrations above the threshold of 100 ng/L in Italian groundwater at maximum concentrations of 290 ng/L, 200 ng/L and 180 ng/L, respectively (Table 2.3). Similarly, concentrations above 100 ng/L of the neonicotinoids clothianidin (120 ng/L) were reported in Dutch/Belgian groundwater bodies used as water sources for human consumption in the Netherlands (Sjerps et al. 2019) and of imidacloprid (120 ng/L) in Spanish groundwater in the surroundings of a solid-waste treatment plant (Pitarch et al. 2016). Melo et al. (2012) carried out a monitoring program of pesticides in water samples from an intensive horticulture area in a vulnerable zone from north of Portugal. In this area, thiamethoxam was reported at concentration higher than 100 ng/L.

Neonicotinoids were also reported at low but measurable levels. For example, clothianidin, imidacloprid and thiamethoxam were measured in groundwater bodies of the United Kingdom at maximum concentrations of 90 ng/L, 20 ng/L and 3 ng/L, respectively (Lapworth et al. 2018b) and acetamiprid was measured at up to 70 ng/L in French groundwater (Lopez et al. 2015). Several studies conducted in other European groundwater bodies such as Spain, Serbia, Portugal, Italy and Denmark investigated the occurrence of neonicotinoids but they were not detected (Martínez-Galera et al. 1998; Dujaković et al. 2010; Melo et al. 2012; De Liguoro et al. 2014; Radović et al. 2015; Brüsch et al. 2016). Overall, the detection frequencies of neonicotinoids in groundwater bodies of Europe were below 20%. Some exceptions were for clothianidin in the United Kingdom, imidacloprid in Spain and thiamethoxam in Portugal with detection frequencies of 33%, 35% and 35%, respectively (Table 2.3).

Few literature data exists for the occurrence of the carbamate methiocarb in European groundwater bodies and no data was found for the semicarbazone metaflumizone (Table 2.3). Metaflumizone was included in the Watch List because both there are a lack of data on occurrence in water and the risk quotient estimated for certain organisms is high (Pietrzak et al. 2019). Methiocarb was reported at high concentrations (300 ng/L) and quite frequently (53%) in an agricultural region of the southeast of the Mediterranean coast in Spain with 36,460 hectares of cultivated fields, mainly greenhouses (Barceló et al. 1996). Methiocarb was also reported at high concentration but low frequency of detection in Italian groundwater, being 120 ng/L and 4.3%, respectively (ISPRA 2018).

2.2.3 Pharmaceuticals

The pharmaceuticals included in the Watch List (EC 2018) are five antibiotics: three belong to the macrolide antibiotics family, i.e., azithromycin, clarithromycin and erythromycin, one is the β-lactam antibiotic amoxicillin and the other is the fluoroquinolone antibiotic ciprofloxacin. Despite antibiotics are used to protect humans and livestock from microbial diseases and improve agricultural yields, they have recently become a matter of great concern because their overuse has boosted the development of antibiotic-resistant bacteria (Roose-Amsaleg and Laverman 2016; Szekeres et al. 2018).

Macrolide antibiotics were frequently studied and detected in the groundwater bodies of Europe (Table 2.4). Particularly, these antibiotics were widely investigated in some Spanish groundwater bodies (Cabeza et al. 2012; López-Serna et al. 2013; Candela et al. 2016; Boy-Roura et al. 2018). The occurrence of macrolide antibiotics was investigated in the low confined aquifer of the Llobregat Delta, located in the northeast of Spain, where tertiary treated wastewater was injected through wells to build up a hydraulic barrier against seawater intrusion from 2007 to 2010 (Cabeza et al. 2012; Candela et al. 2016). Azithromycin and clarithromycin were not detected in this aquifer (Cabeza et al. 2012; Candela et al. 2016) but erythromycin was detected at a maximum concentration of 428 ng/L (Candela et al. 2016). Candela et al. (2016) reported lower average concentrations in the injected treated wastewater than those found in the aquifer, being 31 and 63 ng/L, respectively. The authors concluded that these substances might be released in the aquifer recharge area before the injection of the treated wastewater by different sources and mechanisms such as disposal of treated effluents and sludge to the land or surface streams hydraulically connected to aquifers, the possible leakage from sanitary sewer infrastructure and illegal dumping in abandoned wells.

Similarly, López-Serna et al. (2013) investigated the occurrence of macrolide antibiotics in Barcelona urban aquifers (Spain). These substances were detected in aquifers located below urbanized areas at average concentrations of 2.6 ng/L for clarithromycin, 190.2 ng/L for azithromycin and below the limit of detection for erythromycin. In contrast, azithromycin, clarithromycin and erythromycin were ubiquitous in the observation points of the shallow aquifer of the Besòs River Delta, with average concentration of 15.7, 10.7 and 28.7 ng/L, respectively. These concentrations were lower than those detected in the main recharge source of this aquifer, which is the polluted Besòs River (Jurado et al. 2017). The concentrations measured in this river were 133 ng/L for azithromycin, 101 ng/L for clarithromycin and 54.9 ng/L for erythromycin. This finding suggests that these substances might be naturally attenuated when river water infiltrates through the soil to the aquifer, but the actual remediation process has not been investigated to date. The occurrence of azithromycin and clarithromycin was also reported in the Baix Fluvià alluvial aquifer that is located in an agricultural area and thus, the main source of these substances is the application to crops of organic fertilizers, such as manure and slurry (Boy-Roura et al. 2018). The average concentrations for azithromycin and clarithromycin were low, being 0.93 and 0.34 ng/L, respectively.

Concerning other European groundwater bodies, clarithromycin and erythromycin were detected in French groundwater at maximum concentrations of 42 ng/L and 1872 ng/L, respectively (Lopez et al. 2015). Szekeres et al. (2018) also reported high concentrations of erythromycin in the city of Cluj-Napoca and the surrounding areas of the northwest of Romania. This substance was reported at concentrations ranging from 384 to 436.7 ng/L in urban areas, but the maximum concentration of 568.4 ng/L was found in an area located at 40 km from the urban centre. The potential pollution sources identified were faecal contamination from wildlife or domestic animals. High concentrations of azithromycin, up to 140 ng/L and of erythromycin, 392.4 ng/L, were also reported in Belgrade's groundwater (Serbia) and in a karst aquifer system in Germany (Grujić et al. 2009; Reh et al. 2013), respectively. In many other groundwater bodies of Europe such as those of Germany, Sweden, the Netherlands, Portugal and Serbia, some of these substances were not detected or were detected at concentrations below 50 ng/L (Table 2.4).

Generally, the frequency of detection of macrolide antibiotics was below the 20% with some exceptions such as azithromycin (97%) and clarithromycin (100%) in the urban groundwater of Barcelona (Spain) (López-Serna et al. 2013) and erythromycin in the groundwater bodies of Sweden (100%) and Romania (83%) (Rostvall 2017; Szekeres et al. 2018). Medium detection frequencies of these substances, between 42% and 67%, were reported in groundwater bodies of Spain and Sweden (Table 2.4).

The fluoroquinolone antibiotic ciprofloxacin was one of the most reported substances in European groundwater (Table 2.4). Several studies focused on the groundwater bodies of Spain where this substance was detected at concentrations above 100 ng/L (Cabeza et al. 2012; López-Serna et al. 2013; Candela et al. 2016; Boy-Roura et al. 2018). Its maximum concentration was 443 ng/L in the urban aquifers of Barcelona where the principal pollution source appeared to be leakage from the sewage network system. Similar concentrations were measured in the lower confined aquifer of the Llobregat River Delta, reaching levels up to 406 ng/L (Cabeza et al. 2012; Candela et al. 2016). Ciprofloxacin was also detected at somewhat lower but still high concentrations in the aquifers of an agricultural area of the Baix Fluvià, reaching values up to 298.3 ng/L (Boy-Roura et al. 2018). Similar concentrations were found in the shallow aquifer of the Besòs River Delta, ranging from 5.4 to 130 ng/L (López-Serna et al. 2013). The frequency of detection of ciprofloxacin in the urban aquifers of Barcelona was quite wide, from not detected to 94% (Table 2.4).

A study conducted in Romania evaluated the distribution of antibiotic residues and the microbial contamination in groundwater environments at various distances from urban areas (Szekeres et al. 2018). Ciprofloxacin was found at concentrations above 100 ng/L in two sites. The first one was located near a swine confinement facility where no special waste system was applied and manure was periodically spread onto crop fields. The second one was located in an urban area impacted by the wastewaters of two hospitals. This substance was also detected at low concentrations (maximum of 43 ng/L) or not detected in French groundwater bodies (Guillon et al. 2015; Lopez et al. 2015) and was not observed in groundwater from a riverbank filtration system in Berlin, Germany (Heberer et al. 2008) and in a groundwater body of Portugal used as water source for human consumption (de Jesus Gaffney et al. 2015).

Finally, the β-lactam antibiotic amoxicillin was frequently studied in groundwater, however it was not detected in Germany, Spain, Sweden and Serbia despite most of the limits of detection of the applied methods were below 15 ng/L (Table 2.4).

2.3 Fate of the Organic Micropollutants in the European Aquifers

2.3.1 Mobility of the Organic Micropollutants

The physico-chemical properties partly govern the mobility of pollutants in groundwater. Among these properties, water solubility (S_w) and the octanol-water partition coefficient (K_{ow}), typically expressed as log K_{ow}, are frequently used (Table 2.5). This partition coefficient provides information about the hydrophobic or hydrophilic character of the micropollutants. Overall, substances with log K_{ow} (log K_{ow} lower than 1, hydrophilic) are likely to be mobile, whereas substances with log K_{ow} values that range from 1 to 4 present medium mobility (Table 2.5). Conversely, the higher is the log K_{ow}, the more hydrophobic is the organic micropollutant (log K_{ow} higher than 4) and thus, they are classified as low mobility substances. Nevertheless, organic micropollutants might have ionized forms due to the presence of amino and carboxyl groups in their structures. Consequently, they can be present as anionic, cationic and/or zwitterion in different proportions. A zwitterion is a molecule with equivalence between the positively-charged functional group(s) and negatively-charged functional group(s).

The percentage of the ionized fraction depends on the pK_a of the organic function and the pH of the groundwater. Generally, the lower the pH value, the greater the concentration of protons in solution, prevailing the ionic forms of the organic micropollutants. The relationship between pH and pK_a is as follows:

(i) when pK_a and pH values are equal, there is 50% of both fractions in solution, ionized and non-ionized;
(ii) when the difference between pK_a and pH is equal to 0.5, the proportion of both fractions is 75% of ionized and 25% of non-ionized or 25% of ionized and 75% of non-ionized;
(iii) when the difference between pK_a and pH is higher than 1 the fraction of ionized or non-ionized is between 90% and 100%;
(iv) when the difference between pK_a and pH is higher than 2 the fraction of ionized or non-ionized is 99%.

Moreover, it is important to consider if the substance is an acid or a base. For acids: (i) high pK_a means that the species is predominantly non-ionized, is a poor proton donor, and a weak acid and (ii) low pK_a indicates that the species is predominantly ionized, is a good proton donor, and a strong acid. For bases: (i) high pK_a determines that the species is predominantly ionized, is a good proton acceptor, and a strong base and (ii) low pK_a means that the species is predominantly non-ionized, is a poor proton acceptor, and a weak base.

The parameter that embodies simultaneously the concepts of hydrophobicity and ionogenicity is the log D_{ow}, which is the pH-dependent n-octanol–water distribution. Thus, the log D_{ow} is more convenient than the log K_{ow} to foresee the mobility of the ionized forms of the organic micropollutants in the subsurface (Wells 2006).

Table 2.5 Physico-chemical properties of the steroid hormones, pesticides and pharmaceuticals included in the Watch List

Substance	Estrone	17-beta-estradiol	17-alpha-ethinylestradiol	Methiocarb	Metaflumizone
Molecular formula	$C_{18}H_{22}O_2$	$C_{18}H_{24}O_2$	$C_{20}H_{24}O_2$	$C_{11}H_{15}NO_2S$	$C_{24}H_{16}F_6N_4O_2$
Chemical structure	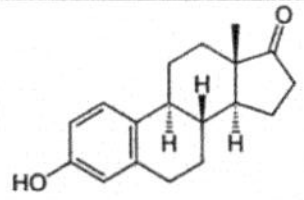	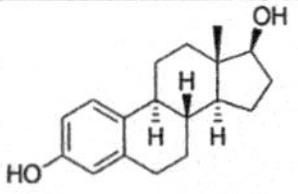	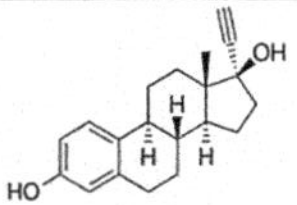	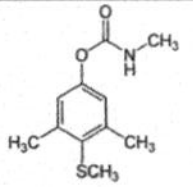	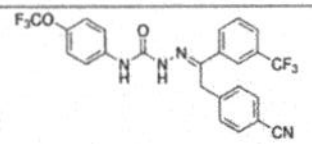
CAS number	53-16-7	50-28-2	57-63-3	2032-65-7	139968-49-3
Sw^a (mg/L)	30	3.9	11.3	27	0.0018
Log K_{ow} [b]	4.31	3.75	3.9	3.13	6.86
pK_a[b] (a,b)	10.33 (a)	10.33 (a)	10.33 (a)	14.77 (a)	11.17 (a)
Log D_{ow} [b]	4.31	3.74	3.9	3.13	6.86
Substance	Acetamiprid	Clothianidin	Imidacloprid	Thiacloprid	Thiamethoxam
Molecular formula	$C_{10}H_{11}ClN_4$	$C_6H_8ClN_5O_2S$	$C_9H_{10}ClN_5O_2$	$C_{10}H_9ClN_4S$	$C_8H_{10}ClN_5O_3S$
Chemical structure	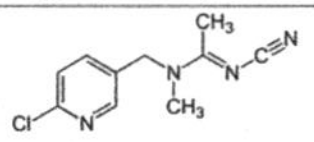	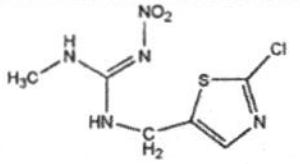	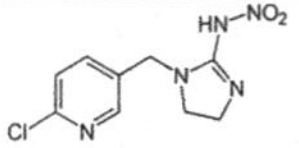	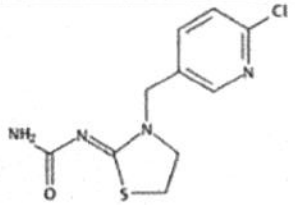	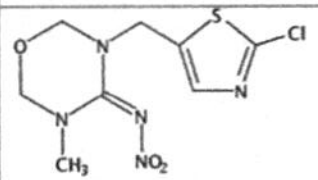
CAS number	135410-20-7	210880-92-5	105827-78-9, 138261-41-3	111988-49-9	153719-23-4
Sw^a (mg/L)	2950	340	610	184	4100
Log K_{ow} [b]	1.11	0.54	0.87	2.06	1.29
pK_a[b] (a,b)	4.16 (b)	15.59 (a), 0.44 (b)	10.6 (a), 6.75 (b)	1.62 (b)	3.01 (b)
Log D_{ow} [b]	1.11	0.54	0.78	2.06	1.29
Substance	Azithromycin	Clarithromycin	Erythromycin	Amoxicillin	Ciprofloxacin

Molecular formulat	$C_{38}H_{72}N_2O_{12}$	$C_{38}H_{69}NO_{13}$	$C_{37}H_{67}NO_{13}$	$C_{16}H_{19}N_3O_5S$	$C_{17}H_{18}FN_3O_3$
Chemical structure	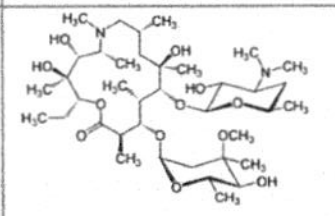	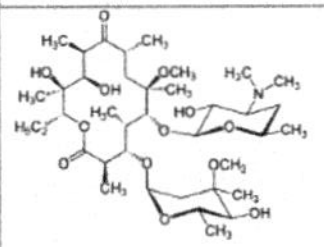	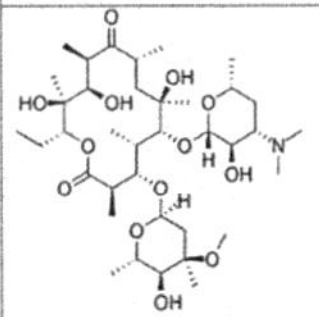	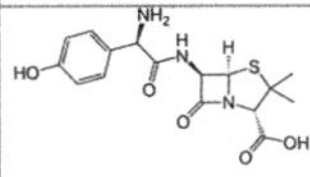	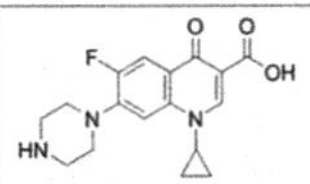
CAS number	83905-01-5	81103-11-9	114-07-8	26787-78-0	85721-33-1
Sw^a (mg/L)	2.4	1.7	4.2	3430	30,000
Log $K_{ow}{}^b$	2.44	3.24	2.6	−2.31	-0.85
$pK_a{}^b$ (a,b)	12.43 (a), 9.57 (b)	12.46 (a), 8.38 (b)	12.45 (a), 8.39 (b)	3.23 (a), 7.22 (b)	5.56 (a), 8.68 (b)
Log $D_{ow}{}^b$	−1.23	2.22	1.57	−2.67	-0.85

Note that log D_{ow} is evaluated at pH = 7.4

pK_a: acid dissociation constant, (a): strongest acidic pK_a, (b): strongest basic pK_a

[a]Data from Hazardous Substances Database (HSDB) and Pesticides Properties Data Base (PPDB)

[b]Calculated using Advanced Chemistry Development Inc. (© 1998–2020 ChemAxon Ltd.)

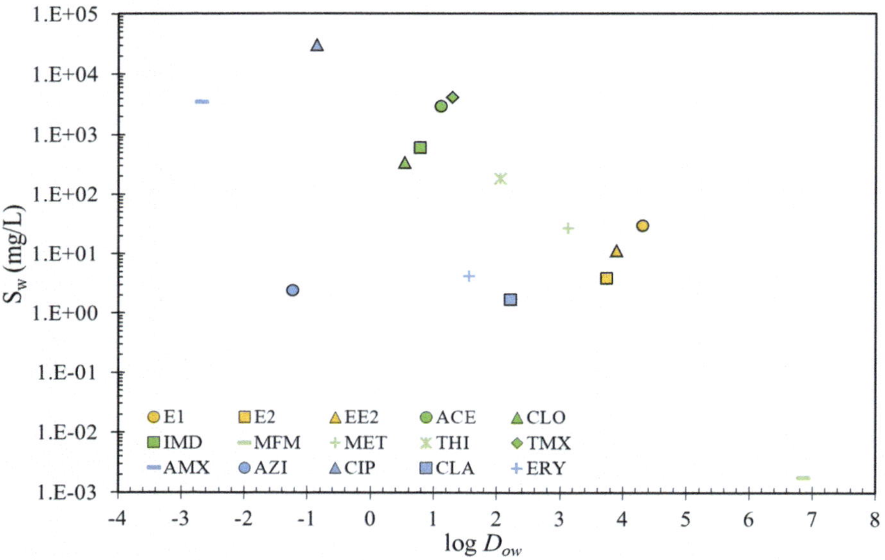

Fig. 2.4 Log D_{ow} values versus water solubility (S_w, mg/L) of the target substances. Note that steroid hormones are displayed in yellow, pesticides in green and pharmaceuticals in blue. (S_w data was extracted from Hazardous Substances Database (HSDB) and Pesticides Properties Data Base (PPDB) and log D_{ow} was calculated using Advanced Chemistry Development Inc. (© 1998–2020 ChemAxon Ltd.); abbreviations: *ACE* acetamiprid, *AMX* amoxicillin, *AZI* azithromycin, *CIP* ciprofloxacin, *CLA* clarithromycin, *CLO* clothianidin, *E1* estrone, *E2* 17-beta-estradiol, *EE2* 17-alpha-ethinylestradiol, *ERY* erythromycin, *IMD* imidacloprid, *MFM* metaflumizone, *MET* methiocarb, *THI* thiacloprid, *TMX* thiamethoxam)

Overall, the higher the water solubility, the lower the log D_{ow} value (Fig. 2.4). The most hydrophobic substance among the pesticides is metaflumizone, which log D_{ow} is 6.86 and water solubility is 0.0018 mg/L, whereas the pharmaceuticals ciprofloxacin, which log D_{ow} is −0.85 and water solubility is 30,000 mg/L and amoxicillin, which log D_{ow} is −2.67 and water solubility is 3430 mg/L, are the most hydrophilic substances included in the Watch List (EC 2018).

Moreover, a parameter frequently used to rank pesticides for their potential to move towards groundwater is the Groundwater Ubiquity Score (GUS) (Gustafson 1989). The GUS index is an experimentally computed value that links the pesticides' soil half-life (DT_{50}) and the partition coefficient between soil organic carbon and water (K_{oc}). This parameter is estimated as follows:

$$\mathrm{GUS} = \log\left(\mathrm{DT}_{50}\right) \times \left(4 - \log\left(K_{oc}\right)\right) \tag{2.1}$$

If the GUS index is higher than 2.8 indicates that a given pesticide is classed as "probable leacher" and GUS index lower than 1.8 shows that the potential for movement towards groundwater is low and thus, these pesticides are "non-leachers". Pesticides with values of GUS index between 1.8 and 2.8 are categorized as

Table 2.6 Groundwater Ubiquity Score (GUS) estimated for the target pesticides using the values of the partition coefficient between soil organic carbon and water (K_{oc}) and the soil half-life (DT_{50}) (Pesticides are classified as "non-leachers" when GUS index is lower than 1.8, "transition" when GUS index is between 1.8 and 2.8 and "probable leachers" if GUS is higher than 2.8; data from Hazardous Substances Database (HSDB) and Pesticides Properties Data Base (PPDB))

Group	Substance	log K_{oc}	DT_{50} typical/field study (days)	GUS index	Classification
Carbamates	Methiocarb	2.96	2.94/35	1.60	Non-leacher
Semicarbazone	Metaflumizone	4.49	141/13.8	−0.56	Non-leacher
Neonicotinoids	Acetamiprid	2.3	1.6/3	0.81	Non-leacher
	Clothianidin	2.09	545/121	3.98	Leacher
	Imidacloprid	2.19–2.90	191/174	4.05	Leacher
	Thiacloprid	3.04	0.88/8.1	0.87	non-leacher
	Thiamethoxam	1.74	50/39	3.60	Leacher

"moderate leachers" (Table 2.6). It is important to note that this parameter is solely an indicator of the potential movement of pesticides to groundwater; however, it does not take into account the local environmental conditions such as redox potential, soil organic matter content and other technical parameters such as the field application rate and the application period of a given pesticide. Generally, pesticides that have low water solubility (S_w lower than 50 mg/L), short half-life (DT_{50} lower than 30 days) and high sorption potential might have low tendency to move through soil. Conversely, pesticides with high water solubility (S_w higher than 500 mg/L), long half-life (DT_{50} higher than 100 days) and low sorption potential may have greater tendency to move through soil.

Most pesticides are neutral substances in the environment. As an example, Fig. 2.5 shows the species distribution of methiocarb at pH range from 5 to 8 and this substance is present as a neutral form, meaning that log K_{ow} and log D_{ow} values are the same. An exception is the neonicotinoid imidacloprid that has a strongest basic dissociation constant (pK_a) of 6.75, indicating that this substance is charged at groundwater pH ranges. Based on the GUS index (Eq. (2.1)), clothianidin, imidacloprid and thiamethoxam are likely to be potential leachers to groundwater because of their GUS indexes are above 3.6 (Table 2.6). Additionally, their water solubility is moderate for clothianidin and high for imidacloprid and thiamethoxam and their log K_{ow} values indicate that they hold a hydrophilic character (Table 2.5). Conversely, acetamiprid, thiacloprid, methiocarb and metaflumizone are non-leachers, especially the latter with a GUS index of −0.56, which never was detected in European and non-European surface water and groundwater bodies (Pietrzak et al. 2019). The GUS indexes for thiacloprid, methiocarb and metaflumizone are in good agreement with their low-moderate solubility and medium-low mobility but acetamiprid, albeit classified as non-leacher, shows high solubility (S_w equal to 2950 mg/L) and medium mobility (log K_{ow} equal to 1.11), pointing out its moderate hydrophilic character.

Despite that some of these pesticides are fairly hydrophilic; they were barely detected in European groundwaters (Table 2.3). For instance, the neonicotinoids

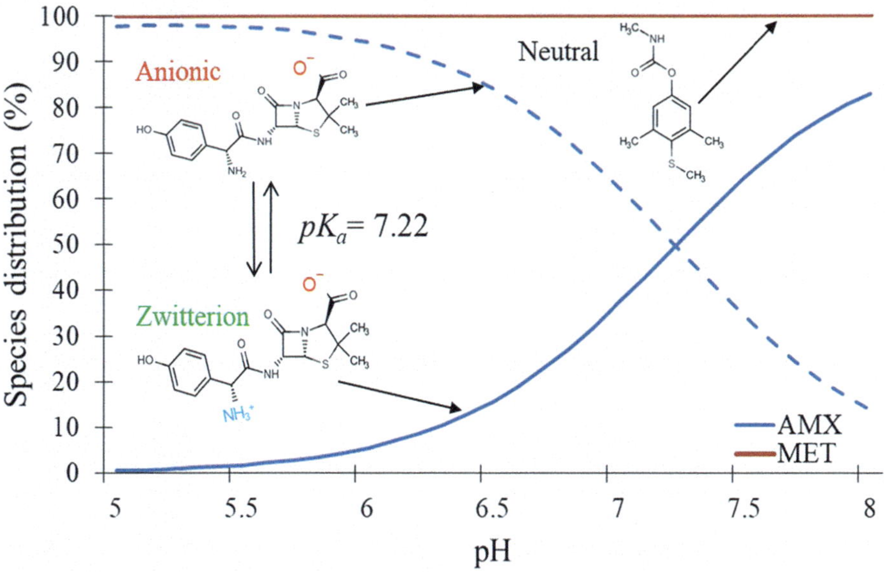

Fig. 2.5 Amoxicillin (AMX) and methiocarb (MET) species distribution (%) at pH ranging from 5 to 8 (methiocarb is a neutral substance and amoxicillin is present in anionic form and as zwitterion; note that at pK_a equal to 7.22 occurs the dissociation of the amino group; data was extracted from Advanced Chemistry Development Inc. (© 1998–2020 ChemAxon Ltd.))

imidacloprid and thiamethoxam, which are two of the most investigated pesticides in European groundwater bodies, were eventually detected, with a frequency up to 35% in the groundwater from Spain and Portugal (Melo et al. 2012; Pitarch et al. 2016).

Concerning pharmaceuticals, macrolide antibiotics have low water solubility (lower than 5 mg/L) and their medium lipophilicity (log K_{ow} higher than 2.4) indicate that their mobility is moderate (Table 2.5). Conversely, the antibiotics amoxicillin and ciprofloxacin have high water solubility and log K_{ow} values lower than −0.85, indicating their hydrophilic character and thus they are likely to be mobile substances. It is important to note that amoxicillin, ciprofloxacin and the macrolide antibiotics have ionized groups at groundwater pH ranges and, consequently, different values of log K_{ow} are observed with respect to log D_{ow} (Table 2.5). As an example, Fig. 2.5 shows the species distribution of amoxicillin at pH from 5 to 8. Amoxicillin has two ionized groups: the carboxyl and the amino groups. The dissociation of the carboxyl group is given by the strongest acidic pK_a (pK_a equal to 3.23) whereas the strongest basic pK_a (pK_a equal to 7.22) leads to the dissociation of the amino group. Thus, amoxicillin is present in zwitterion and anionic forms at groundwater pH ranges (Fig. 2.5). Ciprofloxacin pK_a values are 5.56 for the carboxylic acid group and 8.68 for the nitrogen of the piperazinyl ring (Torniainen et al. 1996). Hence, at environmental pH, ciprofloxacin exists as zwitterion because carboxyl acid is deprotonated and the nitrogen is protonated. Macrolide antibiotics are weak bases with the strongest basic pK_a ranging from 8.38 to 9.57 (Table 2.5). They exist as cationic and neutral species in groundwater at pH 7.4. The ionized forms of these pharmaceuticals result in very different values of log K_{ow} and log D_{ow} for azithromycin (2.44 vs. −1.23), moderate different values for clarithromycin (3.24 vs. 2.22) and erythromycin (2.6 vs. 1.57) and slightly different value for amoxicillin (−2.31 vs. −2.67) (Table 2.5). Summarizing, the log D_{ow} appears to be more convenient than log K_{ow} to predict the mobility of these organic micropollutants in environmental conditions.

Finally, the steroid hormones estrone, 17-beta-estradiol and 17-alpha-ethinylestradiol are present in neutral forms in the environment. These substances have low solubility in water, which is lower or equal to 30 mg/L, and high log K_{ow} values and therefore, they can be considered hydrophobic substances (Table 2.5). In fact, these hormones were not detected or measured at very low concentrations, which were equal or lower than 9 ng/L in the groundwater bodies of Europe despite being frequently investigated (Fig. 2.3 and Table 2.2).

It is important to mention that the physico-chemical properties of the organic micropollutants do not always predict correctly their mobility in groundwater bodies. For example, some hydrophilic substances and "potential leachers", such as the neonicotinoids clothianidin and imidacloprid were eventually detected in European groundwater bodies (Tables 2.3, 2.5 and 2.6). On the other hand, the macrolide antibiotics erythromycin and clarithromycin had high detection frequencies, 97% and 100%, respectively, in Barcelona urban aquifers (Spain) despite being considered moderate hydrophobic substances based on their physico-chemical properties (Tables 2.4 and 2.5). Moreover, azithromycin was detected in these aquifers at average concentrations above 100 ng/L (López-Serna et al. 2013).

Similarly, many substances included in the Watch List often exceeded the concentration of 100 ng/L in the groundwater bodies of Europe for a wide range of log D_{ow} values (Fig. 2.6, highlighted in red). This is the case of the pesticides imidacloprid and methiocarb and the pharmaceuticals azithromycin, ciprofloxacin and erythromycin. Consequently, local environmental conditions such as the aquifer heterogeneity, hydraulic properties, groundwater residence time, microbiological conditions and concentration of the organic micropollutants in the aquifer pollution sources as well as their consumption have to be considered to gain reliable insight on the fate of these substances in groundwater.

2.3.2 Processes Controlling the Fate of the Organic Micropollutants in Groundwater

Some studies reported concentrations of a number organic micropollutants in surface waters and in groundwater (Rodríguez-Mozaz et al. 2004; Zuehlke et al. 2004; Heberer et al. 2008; López-Serna et al. 2013; Pignotti et al. 2017; Boy-Roura et al. 2018; ISPRA 2018) (Fig. 2.7). Overall, the average concentrations reported in

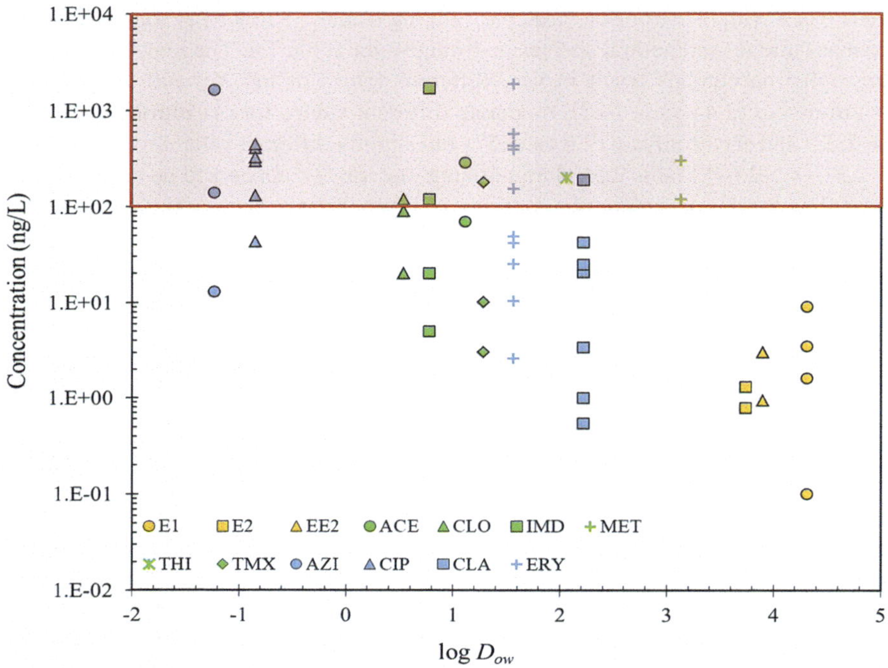

Fig. 2.6 Log D_{ow} vs. maximum concentrations (ng/L) of the substances listed in the Decision 2018/840/EU reported in European groundwater bodies (the concentrations exceeding the threshold of 100 ng/L are highlighted in red; steroid hormones are displayed in yellow, pesticides in green and pharmaceuticals in blue; Log D_{ow} was extracted from Advanced Chemistry Development Inc. (© 1998–2020 ChemAxon Ltd.); abbreviations: *ACE* acetamiprid, *AZI* azithromycin, *CIP* ciprofloxacin, *CLA* clarithromycin, *CLO* clothianidin, *E1* estrone, *E2* 17-beta-estradiol, *EE2* 17-alpha-ethinylestradiol, *ERY* erythromycin, *IMD* imidacloprid, *MET* methiocarb, *THI* thiacloprid, *TMX* thiamethoxam)

surface waters are higher than those in groundwater for steroid hormones and antibiotics (Fig. 2.7a). When surface water infiltrates through the soil, the aquifer sediments act as a natural filter and the concentration of some organic micropollutants is reduced likely due to physical and chemical processes such as sorption, mixing and biodegradation, which may provide substantial improvements in water quality.

Once the organic micropollutants reach groundwater environment, their fate depends on the transport and chemical processes that simultaneously occur in the subsurface. The main processes driving the organic micropollutants fate during subsurface migration are sorption to organic matter and clay minerals, ion exchange in the soil and aquifer, biological degradation and abiotic transformation (Lapworth et al. 2012). However, the assessment of the biogeochemical processes involving organic micropollutants at field conditions is not an easy task. For example, transformation processes at field scale will most probably be difficult to evaluate because of the low concentrations of these contaminants in groundwater (Table 2.2). The

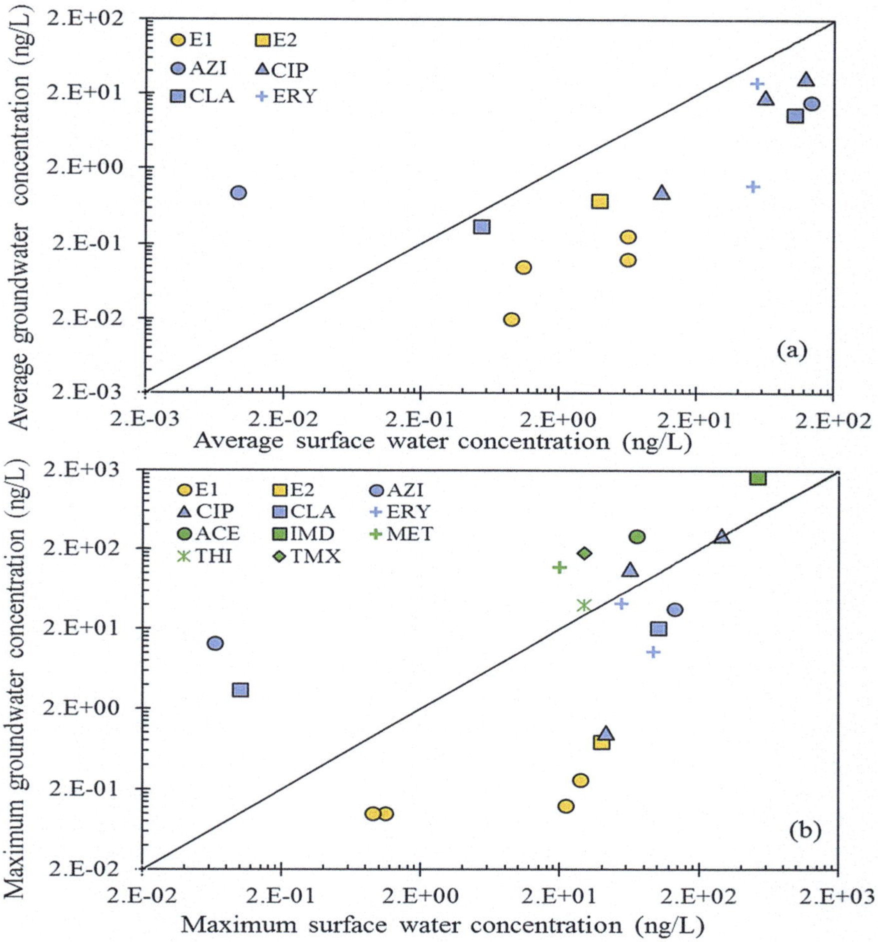

Fig. 2.7 (**a**) Average and (**b**) maximum measured concentrations (ng/L) of selected organic micropollutants in surface waters vs. concentrations reported in European groundwater bodies (note that steroid hormones are displayed in yellow, pesticides in green and pharmaceuticals in blue; abbreviations: *ACE* acetamiprid, *AZI* azithromycin, *CIP* ciprofloxacin, *CLA* clarithromycin, *E1* estrone, *E2* 17-beta-estradiol, *ERY* erythromycin, *IMD* imidacloprid, *MET* methiocarb, *THI* thiacloprid, *TMX* thiamethoxam; source: Rodríguez-Mozaz et al. 2004; Zuehlke et al. 2004; Heberer et al. 2008; López-Serna et al. 2013; Pignotti et al. 2017; Boy-Roura et al. 2018; ISPRA 2018)

attenuation of these pollutants occurs mainly through microbial degradation, because adsorption is generally reversible and, therefore, only retards the contaminants' transport (Greskowiak et al. 2017). Biodegradation of some organic micropollutants is described as a redox-dependent process (Massmann et al. 2008a; Burke et al. 2014), but the redox state of groundwater is not usually described in field studies. Even more, for an evaluation in field conditions, competing bacterial communities and their involvement on degradation processes are difficult to determine in comparison to laboratory controlled conditions (Imfeld et al. 2008). Sorption processes depend on the ionogenicity of the organic micropollutants and the nature of the surface. Negative substances have more affinity with positively charged surfaces such as iron oxides, positive charged substances tend to sorb onto negatively charged surfaces such as clay and neutral substances tend to sorb onto soil organic matter (Valhondo et al. 2015).

On the other hand, some substances presented higher maximum concentrations in groundwater than in surface water, such as some pesticides, i.e., acetamiprid, imidacloprid, methiocarb and thiamethoxam, and the antibiotics azithromycin and clarithromycin (Fig. 2.7b). This observation might be due to the desorption of a given pollutant from low permeable aquifer sediments and/or their storage in stagnant water zones, e.g. unsaturated zone, from where they can be slowly released with time and come into solution again. Therefore, a process-related characterization of subsurface heterogeneity, i.e., relevant zonation of porous media types with respect to hydraulic and geochemical properties, is required to better understand the fate of the organic micropollutants in groundwater.

A wide body of literature focused on the ability of managed aquifer recharge to degrade waterborne contaminants such as steroid estrogens and a vast array of pharmaceuticals over the last two decades (Heberer et al. 2001; Maeng et al. 2011; Regnery et al. 2017; Valhondo et al. 2020). For example, a field study conducted in a managed aquifer recharge site in Berlin (Germany) showed that 17-beta-estradiol and 17-alpha-ethinylestradiol were not detected in the surface water and estrone was removed more than 80% at a monitoring well located close to the lake shore (Zuehlke et al. 2004). In a similar study, 17-beta-estradiol was not detected in the surface water but it was detected in one groundwater sample of the riverbank in the Grand Duchy of Luxembourg (Banzhaf et al. 2013). The authors postulated that this rare occurrence was probably linked to sorption process but its quantification was difficult to evaluate at field scale. Thus, the assessment of the processes driving the fate of steroid hormones during soil and/or aquifer passage have been done at laboratory scale (Mansell et al. 2004; Carr et al. 2011; Patterson et al. 2010, 2011; Maeng et al. 2013), where experimental conditions are controlled.

Mansell et al. (2004) conducted laboratory studies to investigate the removal mechanisms behind the behaviour of 17-beta-estradiol during soil aquifer passage using reclaimed water. The findings revealed that 17-beta-estradiol adsorption to porous media was the main removal mechanism, but additional attenuation occurred due to biodegradation. Biodegradation process took place regardless of the redox conditions, i.e., aerobic vs. anoxic, and type of organic carbon of the matrix (Mansell et al. 2004). Similarly, Maeng et al. (2013) investigated the fate of 17-beta-estradiol and 17-alpha-ethinylestradiol in batch and column studies simulating managed aquifer recharge. Batch experiments showed that 17-beta-estradiol and 17-alpha-ethinylestradiol were mainly attenuated via adsorption onto soil particles. These authors further studied the removal of 17-alpha-ethinylestradiol in soil column experiments under abiotic and biotic conditions, being 67% and 87%, respectively. These results suggested that adsorption was the major removal mechanism of 17-alpha-ethinylestradiol; however, biodegradation also contributed.

The evaluation of biodegradation under oxic conditions of steroid estrogens was carried out by Carr et al. (2011) using batch experiments. 17-beta-estradiol degraded with the shortest half-life under both saturated and unsaturated conditions. Half-lives for 17-beta-estradiol ranged from 1.5 to 4 days; 66–97% was removed from the soils. Estrone and 17-alpha-ethinylestradiol exhibited slower rates of biological transformation under saturated and draining conditions, with half-life values from 27.5 to 56.8 days and from 22.6 to 207 days, respectively. Similarly, Patterson et al. (2010, 2011) assessed the potential natural attenuation of 17-beta-estradiol and 17-alpha-ethinylestradiol during aquifer passage associated with managed aquifer recharge through columns that were packed with aquifer sediment under natural aerobic and anaerobic geochemical conditions. The degradation of 17-beta-estradiol was observed under aerobic and anaerobic conditions with half-lives below 1 and 6 days, respectively (Patterson et al. 2010, 2011). In contrast, 17-alpha-ethinylestradiol degradation was observed neither in natural aerobic conditions nor in anaerobic with long half-lives of more than 50 days (Patterson et al. 2011). These results suggest that the biodegradation of these substances is driven by the redox conditions in the soil-aquifer system.

Research on managed aquifer recharge schemes shows that pharmaceuticals attenuation is primarily consequence of the microbial activity at field scale and the removal efficiency depends on the redox state (Schmidt et al. 2007; Wiese et al. 2011). According to Heberer et al. (2008), river bank filtration is a very effective process for the mitigation of antibiotics such as clarithromycin and anhydroerythromycin, which is a degradation product of erythromycin, as well as ciprofloxacin. These substances were removed during soil passage after 2–4 months of travel time. Macrolide antibiotics were removable by bank filtration under oxic and anoxic conditions, but clarithromycin showed slightly higher degradation rate under oxic conditions and anhydroerythromycin under anoxic ones (Heberer et al. 2008).

Similarly, Schmidt et al. (2007) investigated the efficiency of riverbank filtration in the removal of clarithromycin and anhydroerythromycin at four bank filtration sites located along the Rhine River, the Elbe River and the Ruhr River (Germany). These substances were removed to an extent above 70% under aerobic, suboxic and anaerobic conditions. Wiese et al. (2011) investigated the removal of these substances in managed aquifer recharge sites in Germany, reporting removal frequencies above 96% for clarithromycin and 97% for anhydroerythromycin under different redox conditions, ranging from aerobic to iron-reducing conditions. The removal of erythromycin and clarithromycin were successfully achieved after soil aquifer treatment in agricultural fields irrigated with secondary effluents in Braunschweig (Germany, Ternes et al. 2007). This study suggested that these substances were removed due to degradation rather than sorption through soil passage based on their affinity to sorb to organic carbon, e.g., log K_d equal to −1.12 for clarithromycin. This assumption agrees with results obtained in laboratory experiments for the degradation of erythromycin in soils, reporting a half-life of 20 days (Schlüsener and Bester 2006; Salvia et al. 2014).

Based on the aforementioned studies, it can be concluded that aquifers can effectively remove the target organic micropollutants. Generally, the removal of these substances gradually increases as residence time and travel distance increase, however, the redox state of an aquifer has to be regarded as the key-factor driving the removal of the organic micropollutants because redox conditions have a more pronounced effect than residence time (Schmidt et al. 2007). Further, accurate knowledge about the groundwater flow pattern in an aquifer is of utmost importance for reliable calculation of contaminants' removal (Massmann et al. 2008b). So far, most research on the biodegradation rate constants, adsorption coefficients and half-lives of the organic micropollutants has been determined at laboratory scale (Ghattas et al. 2017; Greskowiak et al. 2017), which misses potential synergetic effects associated with the heterogeneous and complex hydrochemical conditions that are inherent in aquifers. Moreover, scale-up laboratory experiments to field conditions may be unsuccessful due to scale dependencies as highlighted by Regnery et al. (2017). For example, laboratory scale experiments often employ organic micropollutants' concentrations higher than naturally occurring ones, which may result in degradation rate constants or removal rates not representative of field conditions.

Numerical models are used to characterize the flow and the conservative or reactive transport parameters. They are a useful tool to assess the residence time distributions and the mixing rates to obtain degradation and sorption coefficients (Massmann et al. 2008b). However, Boy-Roura et al. (2018) claimed that antibiotic transport modelling at field scale is likely unsuccessful due to difficulties in predicting the coefficients in the sorption equations, i.e., K_{oc} of zwitterion substances, and the variation of retardation factors in orders of magnitude. One reason might be the already mentioned challenge to parameterize a model in the required detail concerning the spatial distribution of hydraulic and geochemical zones that essentially determine the fate of the organic micropollutants in groundwater.

These uncertainties might be the reason why so far only few studies have successfully evaluated the hydrogeochemical processes controlling the fate of the organic micropollutants in groundwater at field scale by means of numerical modelling. Further research should be devoted to this relevant topic as highlighted by the recurrent occurrence of some pesticides and antibiotics in groundwater at high concentrations in European groundwater bodies (Tables 2.3 and 2.4, Fig. 2.6).

2.4 Risk Assessment of the Organic Micropollutants

2.4.1 Human Health Risk Assessment

Although organic micropollutants' concentrations reported in European groundwater bodies are in the ng/L to low µg/L range, an increasing concern exists about the potential human health impact of these substances. This issue is of outmost importance in the future water scarcity scenario boosted by climate change. To increase the freshwater availability, the replenishment of aquifers through managed aquifer recharge is an excellent strategy, but depending on the contamination extent of the source water used, increased contamination can be reached into the aquifer. Considering that groundwater is the main contributor for drinking water production, humans are ultimately exposed by ingestion and diet.

To address this issue, human health risk characterization is performed and expressed as risk quotients (RQs). This assessment is estimated for different life stages to increase the accuracy (de Jesus Gaffney et al. 2015). The life stages considered are 0–6 months, 6–12 months, 1–2 years, 2–3 years, 4–8 years, 9–13 years, 14–18 years and adults (Table 2.7).

The risk quotients were assessed for pesticides, pharmaceuticals and 17-alpha-ethinylestradiol by calculating the ratio between the maximum concentrations measured in groundwater (C_{max}) (Fig. 2.3) and the Drinking Water Equivalent Level (DWEL) of each substance through Eq. (2.2):

$$RQ = {C_{max}}/{\text{DWEL}} \tag{2.2}$$

DWEL values (ng/L) are estimated according to Eq. (2.3) (US EPA 2006) where ADI is the acceptable daily intake (μg/kg day), BW represents the 50th percentile values of body weight for the different life stages (kg), 1000 corresponds to the unit conversion from μg/L to ng/L and DWI is the daily ingestion rate of water (L/day). The age-specific values considered for DWI are those set by the European Food Safety Authority (EFSA 2010) and are listed in Table 2.7. ADI values represent the concentration of a given pollutant in water that can be ingested on a daily basis and should not cause any adverse effect in a potentially exposed population (Schwab et al. 2005). These values range from 0.017 μg/kg day for 17-alpha-ethinylestradiol to 97 μg/kg day for clothianidin (Schwab et al. 2005; Kumar et al. 2010; Leung et al. 2013; Lewis et al. 2016).

$$DWEL = {\text{ADI} \times \text{BW} \times 1000}/{\text{DWI}} \tag{2.3}$$

Maximum measured concentrations are used to foresee the worst-case scenario approach. A commonly used risk ranking criteria is applied: substances with a risk quotient lower than 0.1 mean minimal risk, substances with a risk quotient between

Table 2.7 Selected age groups and respective 50th percentile body weights (BW, kg) and drinking water intakes (DWI, L/day) (Data on body weight was extracted from the WHO website and the DWI doses were those published by European Food Safety Authority (EFSA 2010))

Age groups	BW (kg)	DWI (L/day)
0–6 months	7.6	0.68
6–12 months	8.6	1
1–2 years	10.6	1.2
2–3 years	13	1.3
4–8 years	20.4	1.6
9–13 years	35.4	2
14–18 years	58	2.25
Adults	62	2.25

0.1 and 1 mean median risk and substances with a risk quotient value higher than 1 could potentially affect human health (Martin et al. 2011).

The risk quotients estimated for the 11 target organic micropollutants across the eight age intervals considered range from 0.000045 to 0.11 (Fig. 2.8) which are 1–2 orders of magnitude lower than their respective DWELs (Table 2.8 and Fig. 2.8). Most of the substances have risk quotients far below 0.1 for all life stages considered and, therefore, posing minimal risk for human health. The pesticides clothianidin, tamoxifen and acetamiprid are the substances with the lowest risk quotients

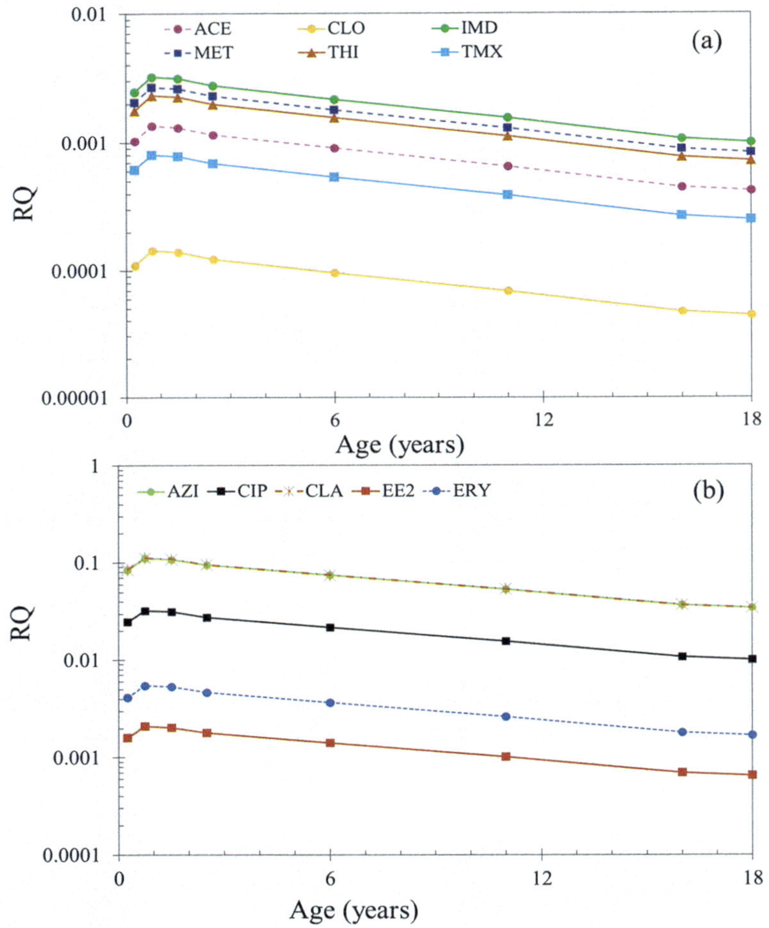

Fig. 2.8 Human health life-stage risk quotient (RQ) profile for (**a**) pesticides and (**b**) pharmaceuticals and 17-alpha-ethinylestradiol in European groundwater bodies (adults' age adopted is 18 years old; abbreviations: *ACE* acetamiprid, *AZI* azithromycin, *CIP* ciprofloxacin, *CLA* clarithromycin, *CLO* clothianidin, *EE2* 17-alpha-ethinylestradiol, *ERY* erythromycin, *IMD* imidacloprid, *MET* methiocarb, *THI* thiacloprid, *TMX* thiamethoxam)

Table 2.8 Estimated drinking water equivalent levels (DWELs in ng/L) for the organic micropollutants per each age group

DWEL (ng/L)								
	Age groups							
Substance	0–6 months	6–12 months	1–2 years	2–3 years	4–8 years	9–13 years	14–18 years	Adults
Acetamiprid	2.8E + 05	2.1E + 05	2.2E + 05	2.5E + 05	3.2E + 05	4.4E + 05	6.4E + 05	6.9E + 05
Azithromycin	1.9E + 04	1.5E + 04	1.5E + 04	1.7E + 04	2.2E + 04	3.0E + 04	4.4E + 04	4.7E + 04
Ciprofloxacin	1.8E + 04	1.4E + 04	1.4E + 04	1.6E + 04	2.0E + 04	2.8E + 04	4.1E + 04	4.4E + 04
Clarithromycin	2.2E + 03	1.7E + 03	1.8E + 03	2.0E + 03	2.6E + 03	3.5E + 03	5.2E + 03	5.5E + 03
Clothianidin	1.1E + 06	8.3E + 05	8.5E + 05	9.7E + 05	1.2E + 06	1.7E + 06	2.5E + 06	2.7E + 06
17-alpha-ethinylestradiol	1.9E + 03	1.4E + 03	1.5E + 03	1.7E + 03	2.1E + 03	3.0E + 03	4.3E + 03	4.6E + 03
Erythromycin	4.5E + 05	3.4E + 05	3.5E + 05	4.0E + 05	5.1E + 05	7.1E + 05	1.0E + 06	1.1E + 06
Methiocarb	1.5E + 05	1.1E + 05	1.1E + 05	1.3E + 05	1.7E + 05	2.3E + 05	3.3E + 05	3.6E + 05
Imidacloprid	6.7E + 05	5.1E + 05	5.3E + 05	6.0E + 05	7.7E + 05	1.1E + 06	1.5E + 06	1.7E + 06
Thiacloprid	1.1E + 05	8.6E + 04	8.8E + 04	1.0E + 05	1.3E + 05	1.8E + 05	2.6E + 05	2.8E + 05
Thiamethoxam	2.9E + 05	2.2E + 05	2.3E + 05	2.6E + 05	3.3E + 05	4.6E + 05	6.7E + 05	7.2E + 05

(Fig. 2.8a). Only the pharmaceuticals azithromycin and clarithromycin have risk quotients higher than 0.1 at early life stages (Fig. 2.8b). For all the target organic micropollutants, the life stage at higher risk is the infants group from 6 to 12 months, followed by the children from 1 to 3 years old (Fig. 2.8). Greater water intake with respect to body weight in early life stages results in higher exposure levels and higher risks than in adults. Thus, the risk quotients estimated for the entire target substances in infants from 6 to 12 months are threefolds higher than the risk quotients calculated for the adolescence (14–18 years old) and adulthood stages.

Summarizing, according to the data available, the human health risk associated to the ingestion of drinking water from groundwater source is low. Nevertheless, as pharmaceuticals consumption is just expected to increase (i.e., population increase and ageing), further research should be devoted to that issue. To this end, toxicity assessment of metabolites and transformation products of the target organic micropollutants, as well as of their mixtures, should be performed (Houtman et al. 2014; Qin et al. 2020) to achieve a reliable and up to date human health risk characterization. Further, chronic effects on health from the continuous intake of the organic micropollutants even at low dosage are unknown, as well as such effects that emerge from an exposure to a cocktail of organic micropollutants.

2.4.2 Environmental Risk Assessment

The assessment of the ecotoxicological risk posed by the target organic micropollutants on aquatic ecosystems is performed following the European Agency for the Evaluation of Medicinal Products guidelines (EMEA/CHMP 2006) through the estimation of the Hazard Quotients (HQs) defined by Eq. (2.4):

$$HQ = {}^{\mathrm{MEC}}\!/_{\mathrm{PNEC}} \tag{2.4}$$

where MEC is the measured environmental concentration of the substance and PNEC corresponds to the predicted no-effect concentration against selected organisms in surface waters. When these toxicity data are not available, estimated PNECs are derived as follows (Eq. (2.5)):

$$PNEC = {}^{\mathrm{EC_{50}\ or\ LC_{50}}}\!/_{1000} \tag{2.5}$$

where EC_{50} is the half maximal effective concentration and LC_{50} is the lethal concentration lethal concentration that kills 50% of the tested organisms during an observation period. EC_{50} or LC_{50} values are divided by an uncertainty factor of 1000 (Sanderson et al. 2004). EC_{50}, LC_{50} and PNEC values for different organisms such as the fish *Oncorhynchus mykiss, Danio rerio* and *Pimephales promelas*, the aquatic invertebrates *Daphnia magna, Ceriodaphnia dubia, Americamysis bahia, Mysidopsis bahia, Thamnocephalus platyurus* and *Brachionus calyciflorus*, the

Table 2.9 Predicted no-effect concentrations (PNECs in mg/L), half maximal effective concentrations (EC_{50} in mg/L) and lethal concentration for half of the population (LC_{50} in mg/L) for pesticides considering different taxonomic groups (PNEC is evaluated as EC_{50} and LC_{50} with an uncertainty factor of 1000)

Taxonomic group	Species	Exposure time	Substance	EC_{50} or LC_{50} (mg/L)	PNEC (mg/L)
Fish	Oncorhynchus mykiss[a]	96 h	Acetamiprid	> 100	1.0E – 01
			Clothianidin	> 104.2	1.0E – 01
			Imidacloprid	> 83	8.3E – 02
			Methiocarb	0.65	6.5E – 04
			Thiacloprid	> 90.1	9.0E – 02
			Thiamethoxam	> 125	1.3E – 01
Aquatic invertebrates	*Americamysis bahia*[a]	96 h	Imidacloprid	0.034	3.4E – 05
	Daphnia magna[b]	48 h	Acetamiprid	49.8	5.0E – 02
			Clothianidin	> 40	4.0E – 02
			Imidacloprid	85	8.5E – 02
			Methiocarb	0.008	8.0E – 06
			Thiacloprid	> 85.1	8.5E – 02
			Thiamethoxam	> 100	1.0E – 01
	Mysidopsis bahia[a]	96 h	Thiacloprid	0.0006	6.0E – 07
Aquatic plants	*Lemna gibba*[b]	7 days	Acetamiprid	1	1.0E – 03
			Clothianidin	> 121	1.2E – 01
			Thiacloprid	95.4	9.5E – 02
			Thiamethoxam	> 90	9.0E – 02
Algae	*Pseudokirchneriella subcapitata*[b]	72 h	Clothianidin	55	5.5E – 02
			Thiamethoxam	> 100	1.0E – 01
	Scenedemus subspicatus[b]	72 h	Acetamiprid	> 98.3	9.8E – 02
			Imidacloprid	> 10	1.0E – 02
			Methiocarb	2.2	2.2E – 03
	Selenastrum capricornutum[b]	72 h	Thiacloprid	60.6	6.1E – 02

Data of EC_{50} and LC_{50} is extracted from Lewis et al. (2016)
[a]PNEC is assessed using LC_{50}
[b]PNEC is estimated using EC_{50}

aquatic plants *Lemna gibba* and *Lemna minor* and the algae *Scenedemus subspicatus, Pseudokirchneriella subcapitata, Selenastrum capricornutum* and *Microcystis aeruginosa* are shown in Table 2.9 for pesticides and in Table 2.10 for pharmaceuticals and the steroid 17-alpha-ethinylestradiol. EC_{50} and LC_{50} values were extracted from the literature (Halling-Sørensen et al. 1998; Sanderson et al. 2003; Isidori et al. 2005; Robinson et al. 2005; Ando et al. 2007; Lewis et al. 2016).

In this assessment, the maximum detected concentrations of the organic micropollutants in European groundwater bodies were used to consider the worst-case scenario and when more than one EC_{50} value was reported for a given pharmaceutical and species, the lowest value is selected.

Table 2.10 Predicted no-effect concentrations (PNECs in mg/L) and half maximal effective concentrations (EC_{50} in mg/L) for pharmaceuticals and 17-alpha-ethinylestradiol considering different taxonomic groups (PNEC is evaluated as EC_{50} divided by an uncertainty factor of 1000; data of EC_{50} was extracted from Halling-Sørensen et al. (1998), Sanderson et al. (2003), Isidori et al. (2005), Robinson et al. (2005) and Ando et al. (2007), abbreviations: *CLA* clarithromycin, *ERY* erythromycin; – not specified)

Taxonomic group	Species	Exposure time	Substance	EC_{50} (mg/L)	PNEC (mg/L)
Fish	*Danio rerio*	96 h	CLA/ERY	< 1000	1.0E + 00
	Pimephales promelas	–	Ciprofloxacin	246,000	2.5E + 02
		–	17-alpha-ethinylestradiol	2.1	2.1E – 03
		–	Erythromycin	61	6.1E – 02
Aquatic invertebrates	*Brachionus calyciflorus*	24 h	Clarithromycin	35.46	3.5E – 02
		24 h	Erythromycin	27.53	2.8E – 02
	Ceriodaphnia dubia	48 h	Clarithromycin	18.66	1.9E – 02
		48 h	Erythromycin	10.23	1.0E – 02
	Daphnia magna	24 h	Clarithromycin	25.72	2.6E – 02
		–	Ciprofloxacin	991	9.9E – 01
		–	17-alpha-ethinylestradiol	5.7	5.7E – 03
		–	17-alpha-ethinylestradiol	2.1	2.1E – 03
		24 h	Erythromycin	22.45	2.2E – 02
		–	Erythromycin	7.8	7.8E – 03
	Thamnocephalus platyurus	24 h	Clarithromycin	33.64	3.4E – 02
		24 h	Erythromycin	17.68	1.8E – 02
Aquatic plants	*Lemna minor*	24 h	Ciprofloxacin	0.203	2.0E – 04
Algae	–	–	17-alpha-ethinylestradiol	0.84	8.4E – 04
	Cyanobacteria	144 h	Erythromycin	0.023–0.43	2.3E – 05
	Microcystis aeruginosa	24 h	Ciprofloxacin	0.017	1.7E – 05
	Pseudokirchneriella subcapitata	24 h	Ciprofloxacin	18.7	1.9E – 02
	Selenastrum capricornutum	–	Ciprofloxacin	938	9.4E – 01
		–	17-alpha-ethinylestradiol	2	2.0E – 03
		–	Erythromycin	4.3	4.3E – 03

Similarly to the risk quotients, when hazard quotients are higher than 1 the substances pose a high ecotoxicological risk; if hazard quotients are 0.1 and 1 the organic substance indicate a medium risk and hazard quotients below 0.1 indicate minimal associated risk (Hernando et al. 2006).

It is worthy to highlight that the hazard quotient was specifically developed to assess the risk in surface water bodies because the PNEC is estimated solely for surface water organisms. However, the hazard quotient evaluation is particularly relevant when groundwater interact with surface water, specifically when groundwater is the main recharge source of surface waters, for instance in long droughts episodes. Hence, hazard quotients' evaluation presented in the following might be considered as an indicator because it does not reflect the real potential effect of the organic micropollutants on subsoil biodiversity and microbiodiversity.

Figure 2.9 shows the hazard quotients estimated for pesticides, pharmaceuticals and 17-alpha-ethinylestradiol against several aquatic species considering the lowest EC_{50} or LC_{50} values reported in the literature (Tables 2.9 and 2.10). Most of the organic micropollutants at the concentrations found in groundwater are not likely to pose a risk for the aquatic target species, showing hazard quotients far below 1. Specifically, none of the pesticides represent a risk for fish, plants and algae (Fig. 2.9a). However, some pesticides pose a high risk for invertebrates, showing very high hazard quotients; methiocarb (38), imidacloprid (49) and thiacloprid (333). Concerning pharmaceuticals, they do not pose a high risk for invertebrates and fish. However, high hazard quotient for ciprofloxacin (2.2), ciprofloxacin (26) and erythromycin (81) were estimated for plants and algae (Fig. 2.9b), indicating fairly high ecological risks to these taxonomic groups. The hazard quotients calculated for 17-alpha-ethinylestradiol were always far below 0.1, suggesting that this hormone is not expected to pose any risk for the selected species at the concentrations observed (Fig. 2.9b).

Summarizing, most of the organic micropollutants do not constitute a risk for the taxonomic groups considered at the concentrations measured in European groundwater bodies. Nevertheless, 13% of the calculated hazard quotients are higher than 1 for the pesticides methiocarb, imidacloprid and thiamethoxam and the antibiotics ciprofloxacin and erythromycin. Moreover, 17% of the estimated hazard quotients are between 0.1 and 1 for the aforementioned substances plus acetamiprid, indicating medium ecological risk. To the light of these values, increased efforts should be devoted to evaluate the potential risk that the organic micropollutants included in the Watch List (EC 2018) can pose to the living organisms in aquatic ecosystems as a result of the widespread consumption of human and veterinary antibiotics and pesticides usage. Furthermore, these substances are likely to reach groundwater as demonstrated by their reported occurrence level and frequency of detection in the groundwater bodies across Europe (Tables 2.2, 2.3, and 2.4).

2.5 Conclusions and Prospects

This chapter summarizes important issues related to the organic micropollutants listed in the surface water Watch List for European Union monitoring defined in the Decision 2018/840/EU (EC 2018) concerning groundwater bodies of all over Europe, together with their geochemical behaviour in subsurface water bodies and potential harm for humans and the environment. The target pollutants are known to reach the aquifers because of the close interaction of groundwater with surface waters, mainly rivers, impacted by discharges of effluents from wastewater treatment plants. Current wastewater treatments are not intended to remove the organic

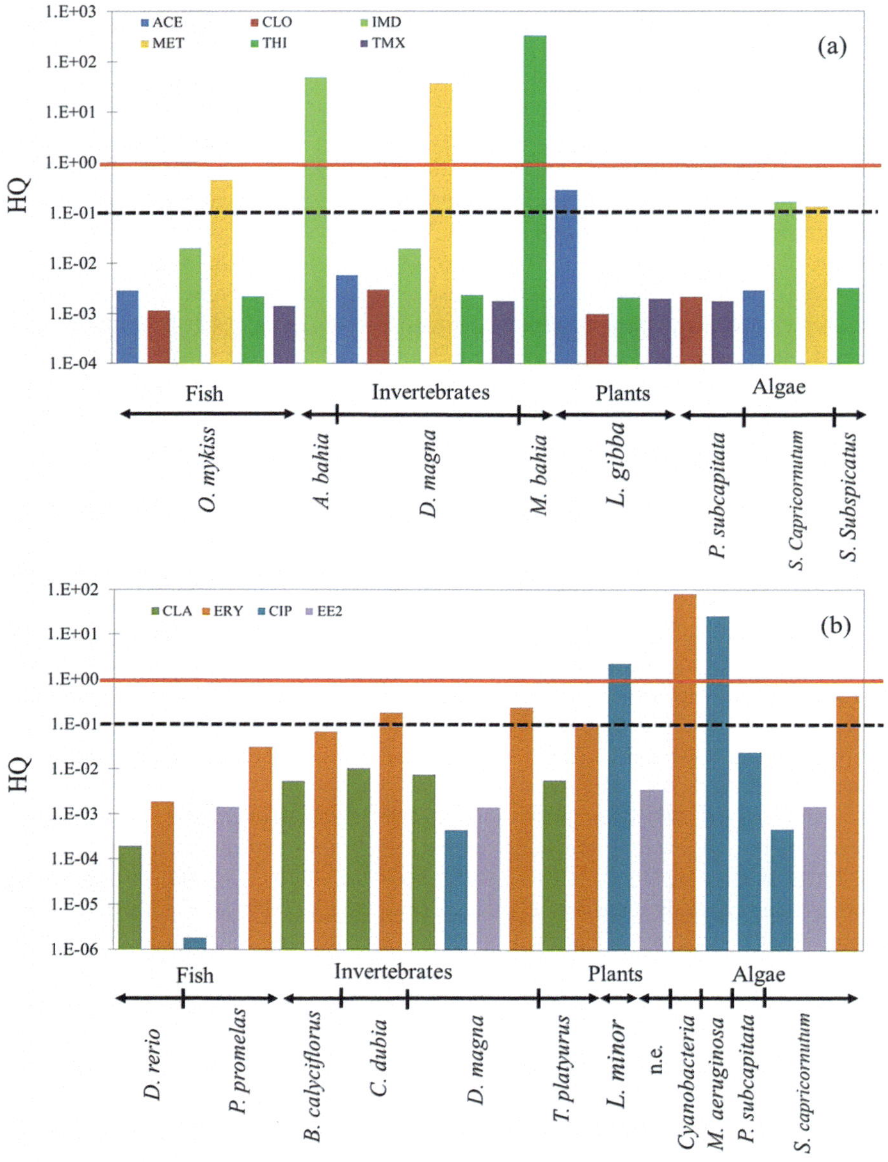

Fig. 2.9 Hazard quotients (HQ) for (**a**) pesticides and (**b**) pharmaceuticals and 17-alpha-ethinylestradiol for different species of fish, invertebrates, plants, algae and crustaceans (abbreviations: *ACE* acetamiprid, *CIP* ciprofloxacin, *CLA* clarithromycin, *CLO* clothianidin, *EE2* 17-alpha-ethinylestradiol, *ERY* erythromycin, *IMD* imidacloprid, *MET* methiocarb, *THI* thiacloprid, *TMX* thiamethoxam; HQ equal to 1 are indicated with a red line and HQ equal to 0.1 are indicated with a black dashed line; n.e.: not specified)

micropollutants and therefore, these facilities are considered the main source of rivers' contamination loads.

Among the target contaminants, the most investigated and frequently detected substances are the antibiotics erythromycin and clarithromycin. Some pesticides, such as imidacloprid, have been extensively studied but most of them were barely detected in European groundwater bodies. For instance, metaflumizone has never been detected in either groundwater or surface water (Pietrzak et al. 2019). Concerning the steroid hormones estrone, 17-beta-estradiol and 17-alpha-ethinylestradiol, they have been widely studied but rarely reported in European groundwater (Fig. 2.3). Nonetheless, some of these substances have been observed at concentrations higher than 100 ng/L in several European groundwater bodies. This is the case of the pesticides imidacloprid (1670 ng/L) and methiocarb (300 ng/L) and the pharmaceuticals azithromycin (1620 ng/L), ciprofloxacin (443 ng/L) and erythromycin (1872 ng/L). Hence, it is of paramount importance to define groundwater quality standards for the non-regulated substances such as antibiotics. Notice that values in brackets are the maximum concentrations reported in the groundwater bodies of Europe but, generally, these organic micropollutants have been usually reported at concentrations below 50 ng/L in European groundwater, causing a potential risk for chronic health concerns as a result of constant exposure.

The behaviour of the selected organic micropollutants in aquifers is highly uncertain. Once these pollutants reach groundwater, their fate is driven by simultaneous transport and chemical processes, i.e., sorption, transformation and biodegradation, taking place in the subsurface as well as on their physico-chemical properties. On the one hand, the physico-chemical properties command their mobility since many organic micropollutants may be as ionized forms as a consequence of the protonation or deprotonation of the functional groups in their molecule. Therefore, the use of log D_{ow} may provide a more reliable approach to assess the potential adsorption onto soil surface considering that negative substances tend to interact with positively charged surfaces, such as metal oxides, positively charged substances are prone to sorb onto negatively charged surfaces such as clays and neutral substances sorb onto the solid organic matter. On the other hand, the attenuation of these pollutants in groundwater occurs mainly through microbial degradation; because of adsorption is usually a reversible process, which only retards the transport of the organic micropollutants in the subsurface. Biodegradation of some organic micropollutants is described as redox-dependent process but the redox state of groundwater is not addressed in many field studies. Moreover, although the long residence time of water in aquifers might result in strong attenuation of some organic micropollutants, others are persistent in groundwater due to absent beneficial geochemical conditions for degradation. These processes require an appropriate understanding of all these factors controlling the fate of these pollutants at field scale, however, so far, most research has been performed at laboratory scale, missing knowledge on the relevance of complexity in field-scale models. Transfer laboratory experiments to field conditions can be unsuccessful due to scale issues as highlighted in many studies. Moreover, laboratory-scale experiments often employ organic micropollutants' concentrations higher than naturally occurring, which can result in degradation rate constants or removal rates not representative for field conditions.

A sound evaluation of health and environmental risks is difficult due to missing information on field concentrations, the aforementioned lack of understanding of the behaviour of the organic micropollutants in the field and the absence of studies that evaluate long-term exposure on human or environments. Yet, to acquire a first-order approximation, the maximum concentrations in European groundwater bodies were used to evaluate the human health and the environmental risks posed by the target organic micropollutants through the calculation of risk and hazard quotients, respectively. Concerning human health risk assessment, the estimated risk quotients of pesticides, pharmaceuticals and 17-alpha-ethinylestradiol showed that risk to human health due to the ingestion of groundwater as source of water for human consumption is low. For all the substances, the life stage presenting the highest risk is the group of infants from 6 to 12 months, since their risk quotients are up to three-folds higher than those corresponding to the adolescence and the adulthood. Therefore, greater drinking water ingestion with respect to the body weight in early life stages result in higher exposure levels and higher risks than in adults. According to the estimated hazard quotients estimated for pesticides, pharmaceuticals and 17-alpha-ethinylestradiol, most of these substances do not pose a threat for the aquatic environment. However, a few organic micropollutants, i.e., imidacloprid, thiacloprid and methiocarb, constitute a high risk against the invertebrates *A. bahia, M. bahia* and *D. magna.* Similarly, high hazard quotients values for ciprofloxacin against the aquatic plants *L. minor* and ciprofloxacin and erythromycin for the alga *M. aerugionosa* were calculated. These results underline that few aquatic ecosystems might be threatened by the organic micropollutants at trace levels of ng/L and it should furthermore be noted that the approximations are done conservatively and might in fact be higher at specific occasions or for chronic exposure.

Despite the fact there is an increasing number of studies targeting the organic micropollutants in groundwater bodies of Europe, there are still many knowledge gaps that should further studied. Based on the data presented in this chapter, following issues deserve much attention and should be addressed in future investigations:

1. Field studies should focus on the occurrence and fate of a wider range of organic micropollutants in groundwater or aquifer ecosystems. For instance, approximately 1500 active pharmaceutical ingredients are currently in use, however, the occurrence in aquifers and toxicity data only exist for a few of them (Oldenkamp 2019);
2. Global contamination of European groundwater bodies remains largely unknown because most of these substances are not routinely monitored since they were not included in the Water Framework Directive. Water Catchment and/or Federal Agencies should recommend regular monitoring groundwater sampling programs for organic micropollutants to propose a Groundwater Watch List of priority pollutants, similarly to surface waters, that will help to set up groundwater quality standards of these yet unregulated substances;

3. Field studies that investigate in further detail the factors controlling the extent of natural removal of the organic micropollutants, including sorption, biodegradation and transformation processes. This knowledge will allow the development of modelling approaches for predicting the behaviour of the organic micropollutants in groundwater and thus, will facilitate the management of groundwater resources;
4. More research should be devoted to characterize the hazard and to estimate the risk posed by certain organic micropollutants listed in the second Watch List. Nevertheless, for a proper evaluation of the risk, metabolites and other transformation products of the selected organic micropollutants should be integrated in next surveys. Besides, as inevitably several substances co-exist in groundwater, the ecotoxicity assessment of mixtures of organic micropollutants and their derivatives should also be object of study because additivity, synergy and antagonism have been described in the ecotoxic behaviour of mixtures of pollutants. Finally, it might be interesting to evaluate the PNEC values for subsurface biodiversity in order to better assess the hazard quotients in groundwater because so far these values have only been evaluated for surface water organisms.

Acknowledgements Anna Jurado gratefully acknowledges the financial support from the Excellence Initiative of the German Federal and State Governments trough the "Open Topic Postdoc Position" Fellow at the Technische Universität Dresden and the support from the Secretary for Universities and Research of the Ministry of Economy and Knowledge of the Government of Catalonia and the Marie Sklodowska-Curie COFUND of the programme H2020 (BP3, contract number 801370). M. Silvia Díaz-Cruz acknowledges the Generalitat de Catalunya (Water and Soil Quality Unit 2017-SGR-1404) and the Spanish Ministry of Science and Innovation (Project CEX2018-000794-S). IDAEA-CSIC is a Centre of Excellence Severo Ochoa (Spanish Ministry of Science and Innovation, Project CEX2018-000794-S).

References

Ando T, Nagase H, Eguchi K, Hirooka T, Nakamura T, Miyamoto K, Hirata K (2007) A novel method using cyanobacteria for ecotoxicity test of veterinary antimicrobial agents. Environ Toxicol Chem 26:601–606. https://doi.org/10.1016/S0378-4274(03)00257-1

Arnold KE, Brown AR, Ankley GT, Sumpter JP (2014) Medicating the environment: assessing risks of pharmaceuticals to wildlife and ecosystems. Philos Trans R Soc B 369:20130569. https://doi.org/10.1098/rstb.2013.0569

Banzhaf S, Krein A, Scheytt T (2013) Using selected pharmaceutical compounds as indicators for surface water and groundwater interaction in the hyporheic zone of a low permeability riverbank. Hydrol Process 27:2892–2902. https://doi.org/10.1002/hyp.9435

Barbosa MO, Moreira NFF, Ribeiro AR, Pereira MR, Silva AMT (2016) Occurrence and removal of organic micropollutants: an overview of the watch list of EU Decision 2015/495. Water Res 94:257–279. https://doi.org/10.1016/j.watres.2016.02.047

Barceló D, Chiron S, Fernandez-Alba A, Valverde A, Alpendurada MF (1996) Monitoring pesticides and metabolites in surface waters and groundwaters in Spain. In: Meyer MT (ed) Herbicides-metabolites in surface and groundwater. Am Chem Soc, Washington, pp 237–253. https://doi.org/10.1021/bk-1996-0630.ch018

Boy-Roura M, Mas-Pla J, Petrovic M, Gros M, Soler D, Brusi D, Menció A (2018) Towards the understanding of antibiotic occurrence and transport in groundwater: findings from the Baix Fluvià alluvial aquifer (NE Catalonia, Spain). Sci Total Environ 612:1387–1406. https://doi.org/10.1016/j.scitotenv.2017.09.012

Brüsch W, Rosenbom AE, Badawi N, Olsen P (2016) Monitoring of pesticide leaching from cultivated fields in Denmark. Geol Surv Denmark Greenland Bull 35:17–22

Burke V, Richter D, Hass U, Duennbier U, Greskowiak J, Massmann G (2014) Redox-dependent removal of 27 organic trace pollutants: compilation of results from tank aeration experiments. Environ Earth Sci 71:3685–3695. https://doi.org/10.1007/s12665-013-2762-8

Cabeza Y, Candela L, Ronen D, Teijon G (2012) Monitoring the occurrence of emerging contaminants in treated wastewater and groundwater between 2008 and 2010. The Baix Llobregat (Barcelona, Spain). J Hazard Mater 239:32–39. https://doi.org/10.1016/j.jhazmat.2012.07.032

Camacho-Muñoz D, Martin J, Santos JL, Alonso E, Aparicio I, De la Torre T, Rodriguez C, Malfeito JJ (2012) Effectiveness of three configurations of membrane bioreactors on the removal of priority and emergent organic compounds from wastewater: comparison with conventional wastewater treatments. J Environ Monit 14:1428–1436. https://doi.org/10.1039/c2em00007e

Campo J, Masiá A, Blasco C, Picó Y (2013) Occurrence and removal efficiency of pesticides in sewage treatment plants of four Mediterranean River Basins. J Hazard Mater 263:146–157. https://doi.org/10.1016/j.jhazmat.2013.09.061

Candela L, Tamoh K, Vadillo I, Valdes-Abellan J (2016) Monitoring of selected pharmaceuticals over 3 years in a detrital aquifer during artificial groundwater recharge. Environ Earth Sci 75:244. https://doi.org/10.1007/s12665-015-4956-8

Carr DL, Morse AN, Zak JC, Anderson TA (2011) Biological degradation of common pharmaceuticals and personal care products in soils with high water content. Water Air Soil Pollut 217:127–134. https://doi.org/10.1007/s11270-010-0573-z

Carvalho AM, Cardoso VV, Rodrigues A, Ferreira E, Benoliel MJ, Duarte EA (2015) Occurrence and analysis of endocrine-disrupting compounds in a water supply system. Environ Monit Assess 187:139. https://doi.org/10.1007/s10661-015-4374-0

Collado N, Rodriguez-Mozaz S, Gros M, Rubirola A, Barceló D, Comas J, Rodriguez-Roda I, Buttiglieri G (2014) Pharmaceuticals occurrence in a WWTP with significant industrial contribution and its input into the river system. Environ Pollut 185:202–212. https://doi.org/10.1016/j.envpol.2013.10.040

Corada-Fernández C, Candela L, Torres-Fuentes N, Pintado-Herrera MG, Paniw M, González-Mazo E (2017) Effects of extreme rainfall events on the distribution of selected emerging contaminants in surface and groundwater: the Guadalete River basin (SW, Spain). Sci Total Environ 605:770–783. https://doi.org/10.1016/j.scitotenv.2017.06.049

Corso M, Galey C, Seux R, Beaudeau P (2018) An assessment of current and past concentrations of trihalomethanes in drinking water throughout France. Int J Environ Res Public Health 15:1669. https://doi.org/10.3390/ijerph15081669

Damstra T, Barlow S, Bergman A, Kavlock R, Van Der Kraak G (2002) Global assessment of the sate-of-the-science of endocrine disruptors. WHO publication no. WHO/PCS/EDC/02.2. World Health Organisation, Geneva

de Jesus Gaffney V, Almeida CMM, Rodrigues A, Ferreira E, Benoliel MJ, Cardoso VV (2015) Occurrence of pharmaceuticals in a water supply system and related human health risk assessment. Water Res 72:199–208. https://doi.org/10.1016/j.watres.2014.10.027

De Liguoro M, Dalla Bona M, Gallina G, Capolongo F, Gallocchio F, Binato G, Di Leva V (2014) A monitoring of chemical contaminants in waters used for field irrigation and livestock watering in the Veneto region (Italy), using bioassays as a screening tool. Environ Sci Pollut Res Int 21:3546–3557. https://doi.org/10.1007/s11356-013-2357-7

De Stefano L, Fornés JM, López-Geta JA, Villarroya F (2015) Groundwater use in Spain: an overview in light of the EU Water Framework Directive. Int J Water Resour Dev 31:640-656. doi:https://doi.org/10.1080/07900627.2014.938260

Diamanti-Kandarakis E, Bourguignon JP, Giudice LC, Hauser R, Prins GS, Soto AM, Zoeller RT, Gore AC (2009) Endocrine-disrupting chemicals: an endocrine society scientific statement. Endocr Rev 2(30):293–342. https://doi.org/10.1210/er.2009-0002

Dolar D, Gros M, Rodriguez-Mozaz S, Moreno J, Comas J, Rodriguez-Roda I, Barcelo D (2012) Removal of emerging contaminants from municipal wastewater with an integrated membrane system, MBR-RO. J Hazard Mater 239-240:64–69. https://doi.org/10.1016/j.jhazmat.2012.03.029

Dujaković N, Grujić S, Radišić M, Vasiljević T, Laušević M (2010) Determination of pesticides in surface and ground waters by liquid chromatography-electrospray-tandem mass spectrometry. Anal Chim Acta 678:63–72. https://doi.org/10.1016/j.aca.2010.08.016

Eamus D, Fu BH, Springer AE, Stevens LE (2016) Groundwater dependent ecosystems: classification, identification techniques and threats. In: Jakeman AJ, Barreteau O, Hunt RJ, Rinaudo JD, Ross A (eds) Integrated groundwater management. Springer, Berlin, pp 313–346. https://doi.org/10.1007/978-3-319-23576-9_13

EFSA Panel on Dietetic Products; Nutrition and Allergies (2010) Scientific opinion on dietary reference values for water. EFSA J 8:1459. https://doi.org/10.2903/j.efsa.2010.1459

EMEA/CHMP-European Medicines Evaluation Agency/Committee for Medicinal Products for Human Use (2006) Guideline on the environmental risk assessment of medicinal products for human use. Doc Ref EMEA/CHMP/SWP/4447/00, 12 p

Ernst M, Hein A, Asmin J, Krauss M, Fink G, Hollender J, Ternes T, Jørgensen C, Jekel M, McArdell C (2012) Chapter 12: water quality analysis: detection, fate and behaviour of selected trace organic pollutants at managed aquifer recharge sites. In: Kazner C, Wintgens T, Dillon P (eds) Advances in water reclamation technologies for safe managed aquifer recharge. IWA Publishing, London, pp 197–225

European Commission (2006) Directive 2006/118/EC of the European Parliament and the Council of 12 December 2006 on the protection of ground water against pollution and deterioration. Off J Eur Union L 372/19

European Commission (2008) Directive 2008/105/EC of the European Parliament and of the Council of 16 December 2008 on environmental quality standards in the field of water policy, amending and subsequently repealing Council Directives 82/176/EEC, 83/513/EEC, 84/156/EEC, 84/491/EEC, 86/280/EEC and amending Directive 2000/60/EC of the European Parliament and of the Council. Off J Eur Union L 348/84

European Commission (2013) Directive 2013/39/EU of the European Parliament and of the Council of 12 August 2013 amending Directives 2000/60/EC and 2008/105/EC as regards priority substances in the field of water policy. Off J Eur Union L226/1

European Commission (2014) Directive 2014/80/EU of 20 June 2014 amending Annex II to Directive 2006/118/EC of the European Parliament and of the Council on the Protection of Groundwater Against Pollution and Deterioration. Off J Eur Union L182/52

European Commission (2015) European Commission Implementation Decision 2015/495/EC of 20 March 2015establishing a watch list of substances for Union-wide monitoring in the field of water policy pursuant to Directive 2008/105/EC of the European Parliament and of the Council. Off J Eur Union L78/40

European Commission (2018) European Commission Implementation Decision 2018/840 of 5 June 2018 establishing a watch list of substances for Union-wide monitoring in the field of water policy pursuant to Directive 2008/105/EC of the European Parliament and of the Council and repealing Commission Implementing Decision (EU) 2015/495. Off J Eur Union L141/9

Fahrenfeld N, Ma Y, O'Brien M, Pruden A (2013) Reclaimed water as a reservoir of antibiotic resistance genes: distribution system and irrigation implications. Front Microbiol 4–130:1–10. https://doi.org/10.3389/fmicb.2013.00130

Garcia-Becerra FY, Ortiz I (2018) Biodegradation of emerging organic micropollutants in nonconventional biological wastewater treatment: a critical review. Environ Eng Sci 35:1012–1036. https://doi.org/10.1089/ees.2017.0287

Ghattas A, Fischer F, Wick A, Ternes TA (2017) Anaerobic biodegradation of (emerging) organic contaminants in the aquatic environment. Water Res 116:268–295. https://doi.org/10.1016/j.watres.2017.02.001

Glineur A, Barbera B, Nott K, Carbonnelle P, Ronkart S, Lognay G, Tyteca E (2018) Trace analysis of estrogenic compounds in surface and groundwater by ultra high performance liquid chromatography-tandem mass spectrometry as pyridine-3-sulfonyl derivatives. J Chromatogr A 1534:43–54. https://doi.org/10.1016/j.chroma.2017.12.042

Greskowiak J, Hamann E, Burke V, Massmann G (2017) The uncertainty of biodegradation rate constants of emerging organic compounds in soil and groundwater – a compilation of literature values for 82 substances. Water Res 126:122-133. https://doi.org/10.1016/j.watres.2017.09.017

Grujić S, Vasiljević T, Laušević M (2009) Determination of multiple pharmaceutical classes in surface and ground waters by liquid chromatography-ion trap-tandem mass spectrometry. J Chromatogr A 1216:4989-5000. https://doi.org/10.1016/j.chroma.2009.04.059

Guillon A, Noyon N, Gogot C, Robert S, Bruchet A, Esperanza M (2015) Study on veterinary and human antibiotics in raw and treated water from a French basin. Water Sci Technol Water Supply 15:1275–1284. https://doi.org/10.2166/ws.2015.094

Gustafson DI (1989) Ground water ubiquity score: a simple method for assessing pesticide leachability. Environ Toxicol Chem 8:339–357. https://doi.org/10.1002/etc.5620080411

Halling-Sørensen B, Nor Nielsen S, Lanzky PF, Ingerslev F, Holten-Lützhøft HC, Jøbrgensen SE (1998) Occurrence, fate and effects of pharmaceutical substances in the environment. A review. Chemosphere 36:357–393. https://doi.org/10.1016/S0045-6535(97)00354-8

Hayes TB, Collins A, Lee M, Mendoza M, Noriega N, Stuart AA, Vonk A (2002) Hermaphroditic demasculinized frogs after exposure to the herbicide atrazine at low ecologically relevant doses. Proc Natl Acad Sci 99:5476–5480. https://doi.org/10.1073/pnas.082121499

Heberer T, Verstraeten IM, Meyer MT, Mechlinski A, Reddersen K (2001) Occurrence and fate of pharmaceuticals during bank filtration-preliminary results from investigations in Germany and the United States. J Contemp Water Res Educ 120:4–17

Heberer T, Massmann G, Fanck B, Taute T, Dünnbier U (2008) Behaviour and redox sensitivity of antimicrobial residues during bank filtration. Chemosphere 73:451–460. https://doi.org/10.1016/j.chemosphere.2008.06.056

Hernando MD, Mezcua M, Fernández-Alba AR, Barceló D (2006) Environmental risk assessment of pharmaceutical residues in wastewater effluents, surface waters and sediment. Talanta 69:334–342. https://doi.org/10.1016/j.talanta.2005.09.037

Hohenblum P, Gans O, Moche W, Scharf S, Lorbeer G (2004) Monitoring of selected estrogenic hormones and industrial chemicals in groundwaters and surface waters in Austria. Sci Total Environ 333:185–193. https://doi.org/10.1016/j.scitotenv.2004.05.009

Houtman CJ, Kroesbergen J, Lekkerkerker-Teunissen K, van der Hoek JP (2014) Human health risk assessment of the mixture of pharmaceuticals in Dutch drinking water and its sources based on frequent monitoring data. Sci Total Environ 496:54–62. https://doi.org/10.1016/j.scitotenv.2014.07.022

Iancu VI, Radu GL (2018) Occurrence of neonicotinoids in waste water from the Bucharest treatment plant. Anal Methods 10:2691–2700. https://doi.org/10.1039/C8AY00510A

Imfeld G, Vuilleumier S (2012) Measuring the effects of pesticides on bacterial communities in soil: a critical review. Eur J Soil Biol 49:22–30. https://doi.org/10.1016/j.ejsobi.2011.11.010

Imfeld G, Nijenhuis I, Nikolausz M, Zeiger S, Paschke H, Drangmeister J, Grossmann J, Richnow HH, Weber S (2008) Assessment of in situ degradation of chlorinated ethenes and bacterial community structure in a complex contaminated groundwater system. Water Res 42:871–882. https://doi.org/10.1016/j.watres.2007.08.035

Isidori M, Lavorgna M, Nardelli A, Pascarella L, Parrella A (2005) Toxic and genotoxic evaluation of six antibiotics on non-target organisms. Sci Total Environ 345:87–89. https://doi.org/10.1016/j.scitotenv.2004.11.017

ISPRA—Italian National Institute for Protection and Environmental Research (2018) National report on pesticides in water Italian monitoring 2016. Report 289/2018, 25 p

Jurado A, Vázquez-Suñé E, Carrera J, López de Alda M, Pujades E, Barceló D (2012) Emerging organic contaminants in groundwater in Spain: a review of sources, recent occurrence and fate in a European context. Sci Total Environ 440:82–94. https://doi.org/10.1016/j.scitotenv.2012.08.029

Jurado A, Vázquez-Suñé E, Pujades E (2017) Potential uses of pumped urban groundwater: a case study in Sant Adrià del Besòs (Spain). Hydrogeol J 25:1745–1758. https://doi.org/10.1007/s10040-017-1575-3

Jurado A, Walther M, Díaz-Cruz MS (2019) Occurrence, fate and environmental risk assessment of the organic micropollutants included watch list set by EU Decision 2015/495 in the groundwater of Spain. Sci Total Environ 663:285–296. https://doi.org/10.1016/j.scitotenv.2019.01.270

Kharbouche L, García MG, Lozano A, Hamaizi H, Galera MM (2019) Solid phase extraction of pesticides from environmental waters using an MSU-1 mesoporous material and determination by UPLC-MS/MS. Talanta 199:612–619. https://doi.org/10.1016/j.talanta.2019.02.092

Kløve B, Ala-aho P, Bertrand G, Boukalova Z, Ertürk A, Goldscheider N, Ilmonen J, Karakaya N, Kupfersberger H, Kværner J, Lundberg A, Mileusnić M, Moszczynska A, Muotka T, Preda E, Rossi P, Siergieiev D, Šimek J, Wachniew P, Widerlund A (2011) Groundwater dependent ecosystems: part I-hydroecological status and trends. Environ Sci Pol 14:770–781. https://doi.org/10.1016/j.envsci.2011.04.002

Kovačević S, Radišić M, Laušević M, Dimkić M (2017) Occurrence and behavior of selected pharmaceuticals during riverbank filtration in the Republic of Serbia. Environ Sci Pollut Res 24:2075–2088. https://doi.org/10.1007/s11356-016-7959-4

Kumar A, Chang B, Xagoraraki I (2010) Human health risk assessment of pharmaceuticals in water: issues and challenges ahead. Int J Environ Res Public Health 7:3929–3953. https://doi.org/10.3390/ijerph7113929

Kuster M, Díaz-Cruz S, Rosell M, López de Alda MJ, Barceló D (2010) Fate of selected pesticides, estrogens, progestogens and volatile organic compounds during artificial aquifer recharge using surface waters. Chemosphere 79:880–886. https://doi.org/10.1016/j.chemosphere.2010.02.026

Lapworth DJ, Baran N, Stuart ME, Ward RS (2012) Emerging contaminants: a review of occurrence, sources and fate in groundwater. Environ Pollut 163:287–303. https://doi.org/10.1016/j.envpol.2011.12.034

Lapworth DJ, Lopez B, Laabs V, Kozel R, Wolter R, Ward R, Vargas Amelin E, Besein T, Claessens J, Delloye F, Ferretti E (2018a) Developing a groundwater watch list for substances of emerging concern: a European perspective. Environ Res Lett 14:035004. https://doi.org/10.1088/1748-9326/aaf4d7

Lapworth DJ, Crane EJ, Stuart ME, Talbot JC, Besien T, Civil W (2018b) Micro-organic contaminants in groundwater in England: summary results from the Environment Agency LC-MS and GC-MS screening data. British Geological Survey Open Report, OR/18/052, 54p

Leung HW, Jin L, Wei S, Tsui MMP, Zhou B, Jiao L, Cheung PC, Chun YK, Murphy MB, Lam PKS (2013) Pharmaceuticals in tap water: human health risk assessment and proposed monitoring framework in China. Environ Health Perspect 121:839–846. https://doi.org/10.1289/ehp.1206244

Lewis KA, Green A, Tzilivakis J, Warner D (2016) An international database for pesticide risk assessments and management. Hum Ecol Risk Assess 22:1050–1064. https://doi.org/10.1080/10807039.2015.1133242

Lopez B, Ollivier P, Togola A, Baran N, Ghestem JP (2015) Screening of French groundwater for regulated and emerging contaminants. Sci Total Environ 518:562–573. https://doi.org/10.1016/j.scitotenv.2015.01.110

López-Serna R, Jurado A, Vázquez-Suñé E, Carrera J, Petrović M, Barceló D (2013) Occurrence of 95 pharmaceuticals and transformation products in urban groundwaters underlying the metropolis of Barcelona, Spain. Environ Pollut 174:305–315. https://doi.org/10.1016/j.envpol.2012.11.022

Luo Y, Guo W, Ngo HH, Nghiem LD, Hai FI, Zhang J, Liang S, Wang XC (2014) A review on the occurrence of micropollutants in the aquatic environment and their fate and removal

during wastewater treatment. Sci Total Environ 473:619–641. https://doi.org/10.1016/j.scitotenv.2013.12.065
Maeng SK, Sharma SK, Lekkerkerker-Teunissen K, Amy GL (2011) Occurrence and fate of bulk organic matter and pharmaceutically active compounds in managed aquifer recharge: a review. Water Res 45:3015–3033. https://doi.org/10.1016/j.watres.2011.02.017
Maeng SK, Sharma SK, Lee JW, Amy GL (2013) Fate of 17β-estradiol and 17α-ethinylestradiol in batch and column studies simulating managed aquifer recharge. J Water Supply Res Technol 62:409–416. https://doi.org/10.2166/aqua.2013.202
Mansell J, Drewes JE, Rauch T (2004) Removal mechanisms of endocrine disrupting compounds (steroids) during soil aquifer treatment. Water Sci Technol 50:229–237. https://doi.org/10.2166/wst.2004.0132
Martín J, Camacho-Muñóz D, Santos JL, Aparicio I, Alonso E (2011) Monitoring of pharmaceutically active compounds on the Guadalquivir river basin (Spain): occurrence and risk assessment. J Environ Monit 13:2042–2049. https://doi.org/10.1039/c1em10185d
Martínez-Galera M, Garrido-Frenich A, Martinez-Vidal JL, Parrilla Vasquez PJ (1998) Resolution of imidacloprid pesticide and its metabolite 6-chloronicotinic acid using cross-sections of spectrochromatograms obtained by high-performance liquid chromatography with diode-array detection. J Chromatogr A 799:149–154. https://doi.org/10.1016/S0021-9673(97)01086-8
Massmann G, Duennbier U, Heberer T, Taute T (2008a) Behaviour and redox sensitivity of pharmaceutical residues during bank filtration – investigation of residues of phenazone-type analgesics. Chemosphere 71:1476–1485. https://doi.org/10.1016/j.chemosphere.2007
Massmann G, Sültenfuß J, Dünnbier U, Knappe A, Taute T, Pekdeger A (2008b) Investigation of groundwater residence times during bank filtration in Berlin: a multi-tracer approach. Hydrol Process 22:788–801. https://doi.org/10.1002/hyp.6649
Meffe R, de Bustamante I (2014) Emerging organic contaminants in surface water and groundwater: a first overview of the situation in Italy. Sci Total Environ 481:280–295. https://doi.org/10.1016/j.scitotenv.2014.02.053
Melo A, Pinto E, Aguiar A, Mansilha C, Pinho O, Ferreira IM (2012) Impact of intensive horticulture practices on groundwater content of nitrates, sodium, potassium and pesticides. Environ Monit Assess 184:4539–4551. https://doi.org/10.1007/s10661-011-2283-4
Moreira NF, Orge CA, Ribeiro AR, Faria JL, Nunes OC, Pereira MF, Silva MTA (2015) Fast mineralization and detoxification of amoxicillin and diclofenac by photocatalytic ozonation and application to an urban wastewater. Water Res 87:87–96. https://doi.org/10.1016/j.watres.2015.08.059
Mostafalou S, Abdollahi M (2017) Pesticides: an update of human exposure and toxicity. Arch Toxicol 91:549–599. https://doi.org/10.1007/s00204-016-1849-x
Oldenkamp R (2019) The boomerang effect–environmental exposure to pharmaceuticals. Sustain Chem Pharm 12:100128. https://doi.org/10.1016/j.scp.2019.01.005
Palace WP, Evans RE, Wautier K, Baron C, Vandenbylllardt L, Vandersteen W, Kidd K (2002) Induction of vitellogenin and histological effects in wild fathead minnows from a lake experimentally treated with the synthetic estrogen, ethynylestradiol. Water Qual Res J Can 37:637–650. https://doi.org/10.2166/wqrj.2002.042
Patterson BM, Shackleton M, Furness AJ, Pearce J, Descourvieres C, Linge KL, Busetti F, Spadek T (2010) Fate of nine recycled water trace organic contaminants and metal(loid)s during managed aquifer recharge into a anaerobic aquifer: column studies. Water Res 44:1471–1481. https://doi.org/10.1016/j.watres.2009.10.044
Patterson BM, Shackleton M, Furness AJ, Bekele E, Pearce J, Linge KL, Busetti F, Spadek T, Toze S (2011) Behaviour and fate of nine recycled water trace organics during managed aquifer recharge in an aerobic aquifer. J Contam Hydrol 122:53–62. https://doi.org/10.1016/j.jconhyd.2010.11.003
Pereira AMPT, Silva LJG, Meisel LM, Lino CM, Pena A (2015) Environmental impact of pharmaceuticals from Portuguese wastewaters: geographical and seasonal occurrence, removal and risk assessment. Environ Res 136:108–119. https://doi.org/10.1016/j.envres.2014.09.041

Petrie B, Barden R, Kasprzyk-Hordern B (2015) A review on emerging contaminants in wastewaters and the environment: current knowledge, understudied areas and recommendations for future monitoring. Water Res 72:3–27. https://doi.org/10.1016/j.watres.2014.08.053

Pietrzak D, Kania J, Malina G, Kmiecik E, Wątor K (2019) Pesticides from the EU first and second watch lists in the water environment. Clean Soil Air Water 47:1800376. https://doi.org/10.1002/clen.201800376

Pignotti E, Farré M, Barceló D, Dinelli E (2017) Occurrence and distribution of six selected endocrine disrupting compounds in surface-and groundwaters of the Romagna area (North Italy). Environ Sci Pollut Res 24:21153–21167. https://doi.org/10.1007/s11356-017-9756-0

Pitarch E, Cervera MI, Portolés T, Ibáñez M, Barreda M, Renau-Pruñonosa A, Morell I, López F, Albarrán F, Hernández F (2016) Comprehensive monitoring of organic micro-pollutants in surface and groundwater in the surrounding of a solid-waste treatment plant of Castellón, Spain. Sci Total Environ 548:211–220. https://doi.org/10.1016/j.scitotenv.2015.12.166

Postigo C, Barceló D (2015) Synthetic organic compounds and their transformation products in groundwater: occurrence, fate and mitigation. Sci Total Environ 503:32–47. https://doi.org/10.1016/j.scitotenv.2014.06.019

Public Service of Wallonia (SPW), Belgium (2020) State of the aquifer of Wallonia (État des nappes d'eau souterraine de Wallonie), 64 p

Qin LT, Pang XR, Zeng HH, Liang YP, Mo LY, Wang DQ, Dai JF (2020) Ecological and human health risk of sulfonamides in surface water and groundwater of Huixian karst wetland in Guilin, China. Sci Total Environ 708:134552. https://doi.org/10.1016/j.scitotenv.2019.134552

Radovic T, Grujic S, Petkovic A, Dimkic M, Lausevic M (2015) Determination of pharmaceuticals and pesticides in river sediments and corresponding surface and ground water in the Danube River and tributaries in Serbia. Environ Monit Assess 187:4092. https://doi.org/10.1007/s10661-014-4092-zT

Ratola N, Cincinelli A, Alves A, Athanasios K (2012) Occurrence of organic micro contaminants in the wastewater treatment process. A mini review. J Hazard Mater 239–240:1–18. https://doi.org/10.1016/j.jhazmat.2012.05.040

Regnery J, Gerba CP, Dickenson ERV, Drewes JE (2017) The importance of key attenuation factors for microbial and chemical contaminants during managed aquifer recharge: a review. Crit Rev Environ Sci Technol 47:1409–1452. https://doi.org/10.1080/10643389.2017.1369234

Reh R, Licha T, Geyer T, Nödler K, Sauter M (2013) Occurrence and spatial distribution of organic micro-pollutants in a complex hydrogeological karst system during low flow and high flow periods, results of a two-year study. Sci Total Environ 443:438–445. https://doi.org/10.1016/j.scitotenv.2012.11.005

Robinson AA, Belden JB, Lydy MJ (2005) Toxicity of fluoroquinolone antibiotics to aquatic organisms. Environ Toxicol Chem 24:423–430. https://doi.org/10.1897/04-210R.1

Rodriguez-Mozaz S, López de Alda MJ, Barceló D (2004) Monitoring of estrogens, pesticides and bisphenol A in natural waters and drinking water treatment plants by solid-phase extraction-liquid chromatography–mass spectrometry. J Chromatogr A 1045:85–92. https://doi.org/10.1016/j.chroma.2004.06.040

Roose-Amsaleg C, Laverman AM (2016) Do antibiotics have environmental side-effects? Impact of synthetic antibiotics on biogeochemical processes. Environ Sci Pollut Res 23:4000–4012. https://doi.org/10.1007/s11356-015-4943-3

Ross A (2016) Groundwater governance in Australia, the European Union and the Western USA. In: Jakeman AJ, Barreteau O, Hunt RJ, Rinaudo JD, Ross A (eds) Integrated groundwater management: concepts, approaches and challenges. Springer, Cham, pp 145–171. https://doi.org/10.1007/978-3-319-23576-9_6

Rostvall A (2017) Evaluation of sorption materials for the removal of organic micropollutants in domestic wastewater and their potential infiltration in groundwater. MSc thesis, Swedish University of Agricultural Sciences, 119 p

Sacher F, Lange FT, Brauch HJ, Blankenhorn I (2001) Pharmaceuticals in groundwaters: analytical methods and results of a monitoring program in Baden-Württemberg, Germany. J Chromatogr A 938:199–210. https://doi.org/10.1016/S0021-9673(01)01266-3

Salvia MV, Experton J, Geandel C, Cren-Olivé C, Vuilliet E (2014) Fate of pharmaceutical compounds and steroid hormones in soil: study of transfer and degradation in soil columns. Environ Sci Pollut Res 21:10525–10535. https://doi.org/10.1007/s11356-014-3038-x

Sanderson H, Johnson DJ, Wilson CJ, Brain RA, Solomon KR (2003) Probabilistic hazard assessment of environmentally occurring pharmaceuticals toxicity to fish, daphnids and algae by ECOSAR screening. Toxicol Lett 144:383–395. https://doi.org/10.1016/S0378-4274(03)00257-1

Sanderson H, Johnson DJ, Reitsma T, Brain RA, Wilson CJ, Solomon KR (2004) Ranking and prioritization of environmental risks of pharmaceuticals in surface waters. Regul Toxicol Pharmacol 39:158–183. https://doi.org/10.1016/j.yrtph.2003.12.006

Santos LH, Araújo AN, Fachini A, Pena A, Delerue-Matos C, Montenegro MCBSM (2010) Ecotoxicological aspects related to the presence of pharmaceuticals in the aquatic environment. J Hazard Mater 175:45–95. https://doi.org/10.1016/j.jhazmat.2009.10.100

Schlüsener MP, Bester K (2006) Persistence of antibiotics such as macrolides, tiamulin and salinomycin in soil. Environ Pollut 143:565–571. https://doi.org/10.1016/j.envpol.2005.10.049

Schmidt CK, Lange FT, Brauch HJ (2007) Characteristics and evaluation of natural attenuation processes for organic micropollutant removal during riverbank filtration. Water Sci Technol Water Supply 7:1–7. https://doi.org/10.2166/ws.2007.060

Schwab BW, Hayes EP, Fiori JM, Mastrocco FJ, Roden NM, Cragin D, Meyerhoff RD, D'Aco VJ, Anderson PD (2005) Human pharmaceuticals in US surface waters: a human health risk assessment. Regul Toxicol Pharmacol 42:296–312. https://doi.org/10.1016/j.yrtph.2005.05.005

Senta I, Kostanjevecki P, Krizman-Matasic I, Terzic S, Ahel M (2019) Occurrence and behavior of macrolide antibiotics in municipal wastewater treatment: possible importance of metabolites, synthesis by products and transformation products. Environ Sci Technol 53:7463–7472. https://doi.org/10.1021/acs.est.9b01420

Sjerps R, Kooij PJ, van Loon A, Van Wezel AP (2019) Occurrence of pesticides in Dutch drinking water sources. Chemosphere 235:510–518. https://doi.org/10.1016/j.chemosphere.2019.06.207

Stolker AAM, Niesing W, Hogendoorn EA, Versteegh JFM, Fuchs R, Brinkman UAT (2004) Liquid chromatography with triple-quadrupole or quadrupole-time of flight mass spectrometry for screening and confirmation of residues of pharmaceuticals in water. Anal Bioanal Chem 378:955–963. https://doi.org/10.1007/s00216-003-2253-y

Stuart M, Lapworth D, Crane E, Hart A (2012) Review of risk from potential emerging contaminants in UK groundwater. Sci Total Environ 416:1–21. https://doi.org/10.1016/j.scitotenv.2011.11.072

Sui Q, Cao X, Lu S, Zhao W, Qiu Z, Yu G (2015) Occurrence, sources and fate of pharmaceuticals and personal care products in the groundwater: a review. Emerg Contam 1:14–24. https://doi.org/10.1016/j.emcon.2015.07.001

Szekeres E, Chiriac CM, Baricz A, Szőke-Nagy T, Lung I, Soran ML, Rudi K, Dragos N, Coman C (2018) Investigating antibiotics, antibiotic resistance genes and microbial contaminants in groundwater in relation to the proximity of urban areas. Environ Pollut 236:734–744. https://doi.org/10.1016/j.envpol.2018.01.107

Tanwar S, Di Carro M, Ianni C, Magi E (2014) Occurrence of PCPs in natural waters from Europe. In: Díaz-Cruz M, Barceló D (eds) Personal care products in the aquatic environment, The handbook of environmental chemistry, volume 36. Springer, Cham, pp 37–71. https://doi.org/10.1007/698_2014_276

Ternes TA, Bonerz M, Herrmann N, Teiser B, Andersen HR (2007) Irrigation of treated wastewater in Braunschweig, Germany: an option to remove pharmaceuticals and musk fragrances. Chemosphere 66:894–904. https://doi.org/10.1016/j.chemosphere.2006.06.035

Tong L, Wang YX, Hermo MP, Barron D, Barbosa J (2014) Simultaneous determination and toxicological assessment of penicillins in different water matrices. Ecotoxicology 23:2005–2013. https://doi.org/10.1007/s10646-014-1336-1

Torniainen K, Tammilehto S, Ulvi V (1996) The effect of pH, buffer type and drug concentration on the photodegradation of ciprofloxacin. Int J Pharm 132:53–61. https://doi.org/10.1016/0378-5173(95)04332-2

US EPA (2006) Setting standards for safe drinking water. Available from: http://www.epa.gov/ogwdw/standard/setting.html

Valhondo C, Carrera J, Ayora C, Tubau I, Martínez-Landa L, Nodler K, Licha T (2015) Characterizing redox conditions and monitoring attenuation of selected pharmaceuticals during artificial recharge through a reactive layer. Sci Total Environ 512:240–250. https://doi.org/10.1016/j.scitotenv.2015.01.030

Valhondo C, Martínez-Landa L, Carrera J, Díaz-Cruz MS, Amalfitano S, Levantesi C (2020) Six artificial recharge pilot replicates to gain insight into water quality enhancement processes. Chemosphere 240:124826. https://doi.org/10.1016/j.chemosphere.2019.124826

van den Brink C, Wuijts S (2016) Towards an effective protection of groundwater resources: putting policy into practice with the drinking water protection file. Water Policy 18:19. https://doi.org/10.2166/wp.2015.197

Vázquez–Suñé E, Carrera J, Tubau I, Sánchez–Vila X, Soler A (2010) An approach to identify urban groundwater recharge. Hydrol Earth Syst Sci 14:2085–2097. https://doi.org/10.5194/hess-14-2085-2010

Vulliet E, Wiest L, Baudot R, Grenier-Loustalot MF (2008) Multi-residue analysis of steroids at sub-ng/L levels in surface and ground-waters using liquid chromatography coupled to tandem mass spectrometry. J Chromatogr A 1210:84–91. https://doi.org/10.1016/j.chroma.2008.09.034

Wells MJM (2006) Log DOW: key to understanding and regulating wastewater derived contaminants. Environ Chem 3:439–449. https://doi.org/10.1071/EN06045

Wiese B, Massmann G, Jekel M, Heberer T, Dünnbier U, Orlikowski D, Grützmacher G (2011) Removal kinetics of organic compounds and sum parameters under field conditions for managed aquifer recharge. Water Res 45:4939–4950. https://doi.org/10.1016/j.watres.2011.06.040

Zhang K, Fent K (2018) Determination of two progestin metabolites (17α-hydroxypregnanolone and pregnanediol) and different classes of steroids (androgens, estrogens, corticosteroids, progestins) in rivers and wastewaters by high-performance liquid chromatography-tandem mass spectrometry (HPLC–MS/MS). Sci Total Environ 610–611:1164–1172. https://doi.org/10.1016/j.scitotenv.2017.08.114

Zhao YY, Pei YS (2012) Risk evaluation of groundwater pollution by pesticides in China: a short review. Procedia Environ Sci 13:1739–1747. https://doi.org/10.1016/j.proenv.2012.01.167

Zuehlke S, Duennbier U, Heberer T, Fritz B (2004) Analysis of endocrine disrupting steroids: investigation of their release into the environment and their behavior during bank filtration. Ground Water Monit Remidiat 24:78–85. https://doi.org/10.1111/j.1745-6592.2004.tb00715.x

Chapter 3
Emergence of a Pathogenic Fungus Resistant to Triazole Antifungal Drugs

Steffi Rocchi, Chloé Godeau, Grégorio Crini, and Eveline Snelders

Abstract Over the past 20 years, emerging antifungal resistance has been observed in the human pathogen, *Aspergillus fumigatus*, and is responsible for serious infections. Resistance in this pathogenic fungus is mainly due to the unintended impact of triazole fungicides widely used in the environment to protect crops and materials from destroying fungi. *A. fumigatus* is naturally present in the environment and will thus be widely exposed to fungicides, which, through selection pressure, will favour strains that have become resistant. These fungicidal compounds have a chemical structure similar to that of medical compounds, thus rendering them inactive for the treatment of an infection in humans with a resistant strain. *A. fumigatus* does not cause infections in crops yet fungicidal compounds with activity against this widely present fungus select for an unintended cross-resistance. Therefore, it is important to understand the phenomena leading to resistance in the environment.

In this chapter, we first reviewed resistance mechanisms that have been described, both occurring under medical treatments and those rather attributed to environmental fungicidal use. We have detailed its spread from its probable origin in Europe to its current global distribution. Then, a focus on environmental contributing factors leading to resistance was done and in particular the cross-disciplinary use of azoles. We defined what can be considered as a hotspot of resistance selection and discussed

S. Rocchi (✉)
Service de Parasitologie-Mycologie, CHRU Jean Minjoz, Besançon, France

Laboratoire Chrono-environnement, UMR 6249, UFR Santé, Université Bourgogne Franche-Comté, Besançon, France
e-mail: steffi.rocchi@univ-fcomte.fr

C. Godeau · G. Crini
Laboratoire Chrono-environnement, UMR 6249, UFR Sciences et Techniques, Université Bourgogne Franche-Comté, Besançon, France
e-mail: chloe.godeau@univ-fcomte.fr; gregorio.crini@univ-fcomte.fr

E. Snelders
Laboratory of Genetics, Wageningen University, Wageningen, The Netherlands
e-mail: eveline.snelders@wur.nl

N. Morin-Crini et al. (eds.), *Emerging Contaminants Vol. 1*, Environmental Chemistry for a Sustainable World 65,
https://doi.org/10.1007/978-3-030-69079-3_3

the possible abiotic and biotics factors in soils implicated in the resistance emergence. Finally, we have considered various possibilities to stop the increase of this phenomenon: limit the uncontrolled spread of fungicides in the environment on the one hand and to reduce the exposure of fungicides on the other hand.

Keywords *Aspergillus fumigatus* · Antifungal resistance · Azoles · Agricultural fungicide use · Fungal life-cycle · Evolution

3.1 Introduction

The problem of resistance to biocides, and more particularly to antimicrobials, concerns all infectious organisms: bacteria, viruses, fungi and parasites. Resistance can occur naturally, but abuse and/or misuse of antimicrobials leads to resistance. The most well-known case is bacterial resistance, where the misuse of antibiotics in animal and human health has led to multidrug-resistance. In contrast, multidrug-resistant pathogenic fungi are much less well-known, even though it represents a considerable threat to the control of plant, animal and human diseases. Over the past 20 years, emerging antifungal resistance in human fungal pathogens has been observed but remains neglected in the overall context of antimicrobial resistance and is not currently the subject of most official surveillance programmes or addressed in for example any of the World Health Organisation programs.

Antimicrobial resistance threatens the heart of modern medicine and the long-term sustainability of an effective global public health response to the constant threat of infectious diseases. The World Health Organization adopted a global plan of action against antimicrobial resistance to maintain the capacity to treat and prevent infectious diseases as long as possible, through effective drugs, responsibly used and accessible to all those who need them. The Global Antimicrobial Resistance Surveillance System (GLASS) was launched in 2015 by the World Health Organization to support a standardized approach to global antimicrobial resistance data collection, analysis and reporting to support decision-making, motivate local, national and regional actions, and provide the evidence base for action and advocacy (Tornimbene et al. 2018). Antibiotic resistance is now recognized as a public health issue with concrete implemented actions: surveillance/prevention programs and directives such as prohibiting the use of certain antibiotics as additives in feeding stuffs. On the contrary, for 20 years, an emerging antifungal resistance has been observed but remains neglected in the overall antimicrobial resistance context. Three fungi of concern were however mentioned in the report of the second Global Antimicrobial Resistance Surveillance System meeting (5–16 December 2016, Geneva, Switzerland), namely *Candida, Aspergillus,* and fungi causing mucormycosis (GLASS 2016).

One particular fungus of concern for human health is *Aspergillus fumigatus* (Fig. 3.1). This ubiquitous fungus is responsible for a wide spectrum of pulmonary pathologies, ranging from chronic or immuno-allergic forms to severe invasive

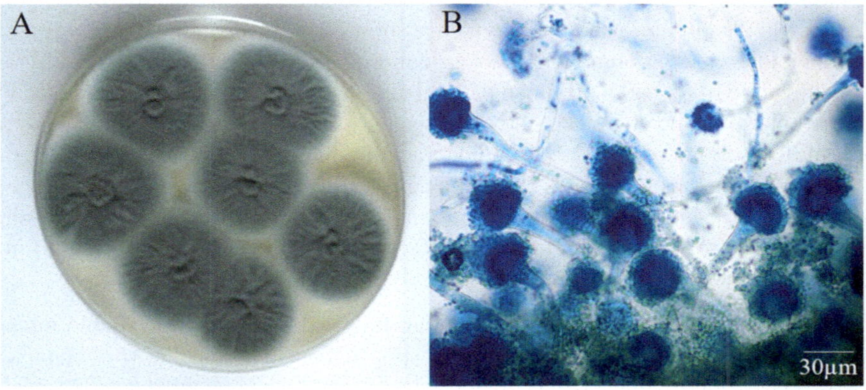

Fig. 3.1 Macroscopic (**a**) and microscopic (**b**) observations of *Aspergillus fumigatus*

infection with high mortality rates. In fact, *A. fumigatus* is the major etiological agent of opportunistic fungal infection in immunocompromised patients. Invasive aspergillosis, the most severe manifestation of *Aspergillus* disease, is difficult to diagnose and manage, has a mortality rate that can reach 65%, and is limited in treatment options (Verweij et al. 2009; Van der Linden et al. 2011).

Antifungal treatments for humans are limited to four classes of compounds: polyenes, echinocandins, pyrimidine analogue and triazoles. The development of triazole antifungal drugs, particularly voriconazole, in the early 2000s, has greatly improved invasive aspergillosis survival rate (Schwartz et al. 2005; Kourkoumpetis et al. 2012). Two other more recently available triazole molecules play an important role in the management of these diseases: isavuconazole, a first-line therapy for treating invasive aspergillosis, similar to voriconazole, and posaconazole, mostly used as a prophylactic treatment (Ullmann et al. 2018). However, the clinical advances made possible using medical triazoles could be threatened by the emergence of triazole-resistant *A. fumigatus* strains with a 20% lower survival rate in patients with culture-positive voriconazole-resistant invasive aspergillosis compared to azole-susceptible infections (Lestrade et al. 2019; Resendiz-Sharpe et al. 2019).

Antifungal resistance in *A. fumigatus* is however not widely considered to be an urgent public health problem for the following three reasons:

1. Fungal diseases are less known and there is a general lack of awareness despite their ubiquity and with global mortality rates exceeding those for malaria and breast cancer (Fisher et al. 2018);
2. Emerging antifungal resistance affects a number of patients that may seem limited, but there is a substantial underdiagnoses of fungal infections preventing accurate estimates. Also, during the congress of the European Society of Respiratory Diseases (September 17, 2018, Paris, France) it was recalled that aspergillosis is in fact a public health issue, with three million affected people worldwide;

3. An increasing part of the antifungal resistance is not due to misuse of antifungals in medicine but more and more evidence shows that it is due to the heavy use of fungicides in environment. Most agricultural research is focussed on phytopathogenic fungi and not human pathogenic fungi with a ubiquitous presence in our environment. Therefore, the antifungal cross-resistance development needs a multidisciplinary approach to maintain a durable use of triazoles in both agriculture as well as medicine. This is however easier said than done with the economic interests at stake and the necessity of intensive food production in our society. Therefore, substantial funding for interdisciplinary research on antifungal resistance is essential but unfortunately greatly under rewarded.

Triazoles molecules are used to treat human fungal diseases with available molecules that have activity against *Aspergillus* spp. such as itraconazole, voriconazole, posaconazole and isavuconazole. But triazoles are also used in animal health as well as fungicidal application in the environment for crop protection, with over 31 available molecules. In the past, it has been shown that using similar molecule structures for human health as well as animal health can select for multidrug cross-resistant bacteria. The use of large quantities of antibiotics in livestock for the purpose of growth enhancement is believed to contribute greatly to the emergence and spread of antibiotics resistant bacteria (Rohr et al. 2019). Similar to the massive use of antibiotics that has led to more and more frequent resistance in bacteria, the large use of fungicides has now led to antifungal resistance in fungi, for application of both crops and human health.

The time has therefore come to mobilize and better understand the sources of antifungal cross-resistance. At the European level, the European Centre for Disease Prevention and Control (ECDC) published a full report in 2013 that provides an overview of the threat situation (Kleinkauf et al. 2013). Researchers were encouraged to conduct studies by implementing large-scale surveillance of antifungal resistance to better identify populations and areas at risk and develop means of prevention. This need for monitoring has also been reported by international experts (International Society for Human & Animal Mycology, ISHAM, Resendiz-Sharpe et al. 2018) and by some agencies with, for example, a report on airborne antimicrobial resistance written by the UK Environment Agency and published in February 2020 (Jones 2020).

The emerging resistance problem in all its complexity, as well as an inventory of the phenomenon and the possible solutions are described in the following paragraphs.

3.2 Triazoles: Mode of Action and Mechanisms of Resistance

Azoles including imidazoles and triazoles are characterized by the presence of an azole nucleus, a pentavalent ring containing two (imidazoles) or three (triazoles) nitrogen atoms. In human and animal medicine historically, there are three generations of azoles: First generation azoles are imidazoles: clotrimazole, enilconazole, parconazole and miconazole. The first three molecules are available in veterinary medicine, clotrimazole and miconazole in cosmetics. The second generation is ketoconazole, marketed for humans and animals. Finally, the triazoles with itraconazole, fluconazole, voriconazole, posaconazole and isavuconazole are the third generation. Fungicide products are however not only used in medicine. Triazoles are largely used in agriculture to protect flower, cereal and wine crops from phytopathogenic molds such as *Botrytis* and *Fusarium* but are also employed in the processing of vegetable crops and to preserve a variety of materials, such as wood (Verweij et al. 2009; Gisi 2014; Berger et al. 2017).

Triazoles affect an important step of the ergosterol biosynthetic pathway, a predominant sterol and essential membrane component. They bind at 14-alpha-lanosterol demethylase encoded by the *cyp51A* gene and stop the conversion of lanosterol to ergosterol, resulting in the accumulation of 14-alpha methyl sterols in the fungal cell. Thus, triazoles disrupt cell transport and fungal membrane structure (Verweij et al. 2009; Berger et al. 2017). Resistance to triazoles can be intrinsic, as is the case with some cryptic species of *A. fumigatus* belonging to the *Fumigati* section or *A. fumigatus* complex (Lamoth 2016). Intrinsic resistance is demonstrated for example for *Aspergillus lentulus*, *Aspergillus thermomutatus*, *Aspergillus udagawae* and *Aspergillus oerlinghausensis*. However, there is also acquired resistance which is the result of adaptation and selection processes. In fact, as for most "icide" products synthesized to counter biological pathogens (bactericides, fungicides, parasiticides) resistance can be selected for over time.

The use of azoles in agriculture leads to resistance selection, in crop pathogens such as *Penicillium digitatum* destroying citrus or *Mycosphaerella graminicola* prevalent on wheat crops (Leroux and Walker 2011; de Ramón-Carbonell and Sánchez-Torres 2020), but also in non-target organisms such as *A. fumigatus* that co-exist in the same environments. This is probably even more favoured for azoles that are single site inhibitors, and above targeting pathogens that have very important spores production and short generation times. Acquired resistance is defined when a fungal strain belonging to a species naturally sensitive to the antifungal agent, is able to tolerate significantly higher concentrations of this antifungal molecule than those that inhibit the *in vitro* growth of the majority of so-called susceptible strains of the same species. From a microbiological point of view, the acquired resistance results in an increase in minimal inhibitory concentration. In the clinical situation, this resistance can lead to therapeutic failure during a treatment under standard conditions including dose, frequency and administration.

The first descriptions of rare cases of triazole resistance in *A. fumigatus* have been reported in the 90's, but since 2000 more and more publications became available on triazole resistance. Two searches for publications were done in the PubMed database (2020, April 7) to assess the number of peer-reviewed publications per year on the following queries: (1) "*Aspergillus*" and "resistance"; (2) "*Aspergillus*" and "resistance" and "(tri)azole". For the first search, 3171 publications were found

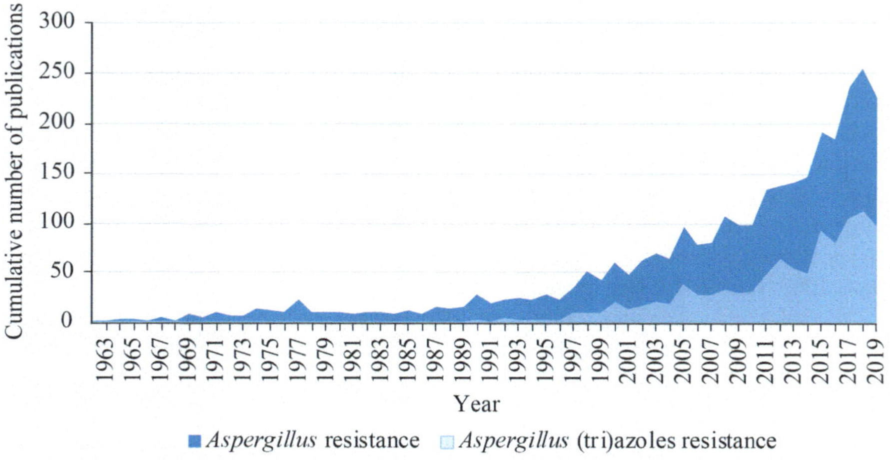

Fig. 3.2 Increasing number of publications on *Aspergillus* resistance and *Aspergillus* (tri)azole resistance

and 1084 for the second search. Results showing the increasing number of publications are presented in Fig. 3.2.

Mechanisms of triazole-resistance can be divided in two groups: *cyp*51A and non-*cyp*51A-mediated. In *A. fumigatus,* resistance mechanisms linked to mutations of the *cyp*51A gene (single nucleotide polymorphisms) and its promoter (tandem repeats) are the most commonly described. More than thirty single nucleotide polymorphisms have been identified in the *A. fumigatus cyp*51A gene, but whether they all have a significant impact on resistance has yet to be determined. In fact, single nucleotide polymorphisms could be classified in three categories (Fig. 3.3):

1. Those whose associated resistance could be explained by the analysis of the gene structure (modification of the active site for example) and by replacement of a wild type allele for the mutant allele and thereby conferring azole resistance;
2. Mutations that are associated with resistant phenotype but the mechanisms of which have not yet been studied;
3. Mutations that have only a small impact in phenotype (small increase in minimal inhibitory concentration) or that have been found in both susceptible and resistant strains and can therefore not be considered to be associated with azole resistance.

There are also mutations that do not confer resistance which are not presented in Fig. 3.3.

The most common mutated loci, reported to be mutation hotspots in triazole-resistant phenotypes are G54, M220 and G448. They are frequently found in patients with chronic aspergillosis who have been treated for long periods with triazole antifungals (Wiederhold and Patterson 2015; Meis et al. 2016; Garcia-Rubio et al. 2017) but also in the environment for G54 (Sharma et al. 2015; Tangwattanachuleeporn et al. 2017) and M220 (Bader et al. 2015). These *cyp*51A gene single nucleotide

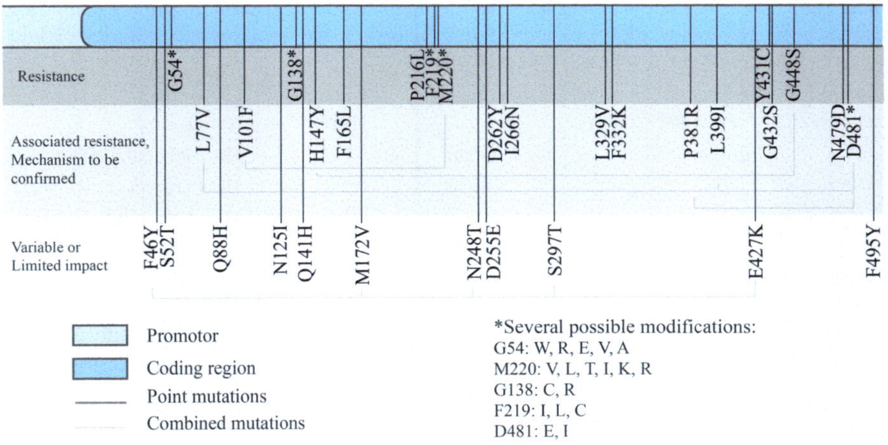

Fig. 3.3 Major single nucleotide polymorphisms found in *cyp51A* gene (position of mutations are represented by vertical lines in the coding region of *cyp51A* gene; the first letter is the amino acid on the wild gene, the number the position on the gene and the second letter the amino acid resulting from the mutated codon; the asterisk indicates that more than one amino acid modification is possible)

polymorphisms affect the protein's affinity for the azoles by modifying the binding site of azoles and restricting their entry. Table 3.1 summarizes the first descriptions of isolates with single nucleotide polymorphisms conferring azole resistance, by country and mutation.

Tandem repeats coupled or not with single nucleotide polymorphism(s), cause overexpression of the *cyp*51A gene and confer multi-triazole-resistant phenotypes. So far, six tandem repeat mutations have been described: TR_{34}, TR_{46} (with the commonly reported mutations TR_{34}/L98H and TR_{46}/Y121F/T289A), TR_{46}^{3}, TR_{46}^{4} (Zhang et al. 2017), TR_{53} (Hodiamont et al. 2009; van der Linden et al. 2013; Alvarez-Moreno et al. 2017) and TR_{120} (Hare et al. 2019). Table 3.2 summarizes the first descriptions of multi-triazoles-resistant isolates with mutations coupling both single nucleotide polymorphism(s) and tandem repeat, by country and mutation.

Other *cyp51A*-independent resistant mechanisms have been described and are also thought to be responsible for therapeutic failure (Wiederhold and Patterson 2015; Garcia-Rubio et al. 2017). *A. fumigatus* genome encodes for two paralogs genes of the P450 cytochrome, *cyp*51A and *cyp*51B (Mellado et al. 2001). Even if azole resistance in this fungus seems more depending on *cyp*51A mutations, *cyp*51B has already been described as resistance mediator in other species. The direct role of *cyp*51B ortholog on *A. fumigatus* azole resistance is not yet totally understand. No mutation on this gene been described, but a significant overexpression of it has been reported in some resistant strains without *cyp*51A mutation. In this case, *cyp*51B seems to act a replacement of *cyp*51A functions to ensure the membrane integrity (Kano et al. 2015).

Efflux pumps are a big mediator of resistance in many microbiological species. Efflux pumps are an energy-dependant resistance mechanism that consist in moving

Table 3.1 First descriptions of isolates with single nucleotide polymorphism(s) conferring resistance ordered by mutation and country (publications listed are those reporting the earliest isolate in time (year), not necessarily first published)

cyp51A mutations[a]	Continents	Country	Source of strain	Isolation year	References
G54 W/R/E/ V/A	Europe	UK	Clinical	1992	Howard et al. (2009)
		Spain	Clinical	2002	Diaz-Guerra et al. (2003)
		The Netherlands	Clinical	2007–2009	Van der Linden et al. (2011)
		Germany	Clinical	2011–2012	Bader et al. (2013)
			Environmental	2012	Bader et al. (2015)
		France	Clinical	2010	Burgel et al. (2012)
		Romania	Environmental	2013–2014	Sharma et al. (2015)
		Switzerland	Environmental	2015	Riat et al. (2018)
	America	USA	Clinical	2003	Mann et al. (2003)
		Argentina	Clinical	2013	Leonardelli et al. (2017)
		Peru	Clinical	2016–2017	Bustamante et al. (2020)
	Asia	Japan	Clinical	1998–2001	Tashiro et al. (2012)
		China	Clinical	2002/2003	Chen et al. (2005)
		India	Clinical	2011–2014	Chowdhary et al. (2015)
			Environmental	2013–2014	Sharma et al. (2015)
		Thailand	Environmental	2014–2015	Tangwattanachuleeporn et al. (2017)
	Oceania	Australia	Clinical	2011	Kidd et al. (2015)
	Africa	Tanzania	Environmental	2013–2014	Sharma et al. (2015)
M220 V/L/T/I/K	Europe	UK	Clinical	1992	Howard et al. (2009)
		France	Clinical	1995	Snelders et al. (2008)
		Sweden	Clinical	1998	
		The Netherlands	Clinical	2004	
		Spain	Clinical	2004	Mellado et al. (2004)
		Denmark	Clinical	2007	Mortensen et al. (2011)
		Germany	Clinical	2010	Fischer et al. (2014)
			Environmental	2012	Bader et al. (2015)
	Asia	China	Clinical	2002	Chen et al. (2005)
	America	USA	Clinical	2001	Wiederhold et al. (2015)
		Peru	Clinical	2016–2017	Bustamante et al. (2020)
G448S	Europe	UK	Clinical	1992–2007	Howard et al. (2009)
		France	Clinical	2004	Bellete et al. (2010)
		Spain	Clinical	2011	Pelaez et al. (2012)
	Oceania	Australia	Clinical	2013	Kidd et al. (2015)
	Asia	Japan	Clinical	2013	Toyotome et al. (2016)
G138C	Europe	UK	Clinical	2004	Howard et al. (2006)
		The Netherlands	Clinical	1994–2008	Snelders et al. (2010)
	America	USA	Clinical	2012	Wiederhold et al. (2015)

(continued)

Table 3.1 (continued)

cyp51A mutations[a]	Continents	Country	Source of strain	Isolation year	References
F219 L/I/C	Europe	The Netherlands	Clinical	2008	Camps et al. (2012)
		Germany	Clinical	Unspecified	Bader et al. (2013)
	America	USA (Maryland)	Clinical	2014	Wiederhold et al. (2015)
P216L	Europe	The Netherlands	Clinical	2008	Camps et al. (2012)
		UK	Clinical	2005	Howard et al. (2009)
		France	Environmental	2016	Jeanvoine et al. (2017)
Y431C	Europe	UK	Clinical	2004	Albarrag et al. (2011)

[a]For mutations, the first letter is the amino acid on the wild gene, the number the position on the gene and the second letter the amino acid resulting from the mutated codon. When several letters are listed, more than one amino acid is possible

a toxic substance (here the antifungal treatment) out of the cell, preventing the antifungal from playing its role. In molds, two main types of efflux pumps are described: ATP binding cassette transporters and major facilitator superfamily transporters. ATP binding cassette transporters permit a unidirectional transport of molecules with hydrolyzing ATP as energy source whereas major facilitator superfamily transporters use proton exchange to translocate substrates. In *A. fumigatus*, 45 ATP binding cassette transporters and 275 major facilitator superfamily transporters are described and overexpression of genes (*AtrI*; *AtrF*; *Cdr1b*) encoding efflux pumps have already been observed as in environmental than in clinical strains (Hagiwara et al. 2016; Meneau et al. 2016; Berger et al. 2017). In other studies, the deletion of certain genes (*mdrA, Cdr1b*) has led to an increase triazole susceptibility (Meneau et al. 2016; Hagiwara et al. 2016).

HapE is a transcriptional factor, it's a part of a three subunits complex named CCAT binding complex. This complex is involved in the *cyp*51A gene expression as a repressor. A point mutation identified as P88L in *HapE* gene seems to affect the repressor effect of the complex. Once mutated, *HapE* induced a decreased affinity of the CCAT binding complex for its target, inducing a decreased of repressor effect and so an overexpression of *cyp*51A gene. This phenomenon is a theoretical implication in azole resistance since it has not been described in environmental or clinical strains but only in laboratory conditions (Camps et al. 2012; Gsaller et al. 2016).

3.3 Origin and Evolution of Environmental Azole Resistance

Since 1998, multi-triazole-resistance in *A. fumigatus* has rapidly spread in the Netherlands (Snelders et al. 2008) as well as globally (Table 3.2). Initially the rapidly spreading azole resistance phenotype was predominantly associated with two co-occurring mutations in the target gene of the azoles, *cyp*51A: (1) a tandem repeat

Table 3.2 First descriptions of isolates with tandem repeat (TR) and single nucleotide polymorphism(s) conferring multi-triazole resistance ordered by mutation and country (publications listed are those reporting the earliest isolate in time (year), not necessarily first published)

cyp51A mutations[a]	Continents	Country	Source of strain	Isolation year	References
TR_{34}/L98H also with S297T/F495I	Europe	The Netherlands	Clinical	1998	Snelders et al. (2008)
			Environmental	2002	Snelders et al. (2009)
		Italy	Clinical	1998	Lazzarini et al. (2016)
			Environmental	2011–2012	Prigitano et al. (2014)
		Turkey	Clinical	2000	Ozmerdiven et al. (2015)
		Norway	Clinical	2001	Snelders et al. (2008)
		Spain	Clinical	2003	Chowdhary et al. (2013)
		UK	Clinical	2007	Howard et al. (2010)
			Environmental	2009–2011	Bromley et al. (2014)
		Denmark	Clinical	2007	Howard et al. (2009)
			Environmental	2010	Mortensen et al. (2010)
		Portugal	Clinical	2010–2016	Pinto et al. (2018)
		France	Clinical	2010–2011	Burgel et al. (2012)
			Environmental	2012	Rocchi et al. (2014a)
		Germany	Clinical	2011	Fischer et al. (2014)
			Environmental	2012–2013	Bader et al. (2015)
		Belgium	Clinical	2011	Vermeulen et al. (2015)
		Poland	Clinical	2011	Nawrot et al. (2018)
		Romania	Environmental	2013–2014	Sharma et al. (2015)
		Switzerland	Clinical	2015	Riat et al. (2018)
			Environmental		

Asia	China	Clinical	2008	Lockhart et al. (2011)
		Environmental	2010–2015	Chen et al. (2016)
	India	Clinical	2008	Chowdhary et al. (2012a)
		Environmental	2012–2013	Chowdhary et al. (2014b)
	Iran	Clinical	2005	Seyedmousavi et al. (2013)
		Environmental	2013	Badali et al. (2013)
	Taiwan	Clinical	2011	Wu et al. (2015)
		Environmental	2014–2016	Wang et al. (2018)
	Kuwait	Environmental	Unspecified	Ahmad et al. (2014)
		Clinical	Unspecified	Ahmad et al. (2015)
	Thailand	Environmental	2014–2015	Tangwattanachuleeporn et al. (2017)
	Japan	Clinical	2016	Toyotome et al. (2017)
		Environmental	2016	Onishi et al. (2017)
	Korea	Clinical	Unspecified	Lee et al. (2018)
America	USA	Clinical	2008	Wiederhold et al. (2015)
		Environmental	2015	Hurst et al. (2017)
	Mexico	Clinical	2014–2017	Gonzalez-Lara et al. (2019)
	Colombia	Environmental	2015	Le Pape et al. (2016)
	Brazil	Clinical[a]	2015	Pontes et al. (2020)
	Peru	Clinical	2016–2017	Bustamante et al. (2020)
Oceania	Australia	Clinical	2004	Kidd et al. (2015)
	New Zealand	Environmental	Unspecified	Korfanty et al. (2019)
Africa	Tanzania	Environmental	2013–2014	Chowdhary et al. (2014b)
	Kenya	Environmental	Unspecified	Kemoi et al. (2018)

(continued)

Table 3.2 (continued)

cyp51A mutations[a]	Continents	Country	Source of strain	Isolation year	References
TR_{46}/Y121F/T289A	Europe	The Netherlands	Clinical	2009	van der Linden et al. (2013)
			Environmental	2013	
		Portugal	Clinical	2010–2016	Pinto et al. (2018)
		Belgium	Clinical	2011	Vermeulen et al. (2015)
		Germany	Clinical	2012	Fischer et al. (2014)
			Environmental	2013	Bader et al. (2015)
		Spain	Clinical	2014	Pelaez et al. (2015)
		Denmark	Clinical	2014	Astvad et al. (2014)
		UK	Clinical	2016	Moore et al. (2017)
			Environmental	2014–2016	Sewell et al. (2019a)
		France	Clinical	2014	Lavergne et al. (2015)
			Environmental	2014	Lavergne et al. (2017)
	Asia	India	Environmental	2012–2013	Chowdhary et al. (2013)
		Japan	Clinical	2013	Hagiwara et al. (2016)
		China	Clinical	2013	Chen et al. (2016)
			Environmental	2016	Ren et al. (2017)
		Iran	Environmental	2017–2018	Ahangarkani et al. (2020)
	America	USA	Clinical	2008	Wiederhold et al. (2015)
		Argentina	Clinical	2009	Isla et al. (2018)
		Colombia	Environmental	2015	Le Pape et al. (2016)
	Africa	Tanzania	Environmental	2014	Chowdhary et al. (2014b)
TR_{53}	Europe	Netherlands	Clinical	2008	Hodiamont et al. (2009)
	America	Colombia	Environmental	2015	Le Pape et al. (2016)
TR_{120}	Europe	Denmark	Clinical	2013	Hare et al. (2019)
TR_{46}^{3} or TR_{46}^{4} Y121F/M172I/T289A/G448S	Europe	The Netherlands	Environmental	2012	Zhang et al. (2017)
			Clinical	2012–2015	

[a]The index number corresponds to the number of base pairs (= motif) that are repeated in the promotor; the superscript number is the number of times the motif is repeated (nothing is indicated when it is only one more pattern repetition); the first letter is the amino acid on the wild gene, the number the position on the gene and the second letter the amino acid resulting from the mutated codon; when several letters are listed, more than one amino acid is possible

(TR) of 34 base pairs in the promoter region and (2) a leucine to histidine substitution in codon 98 (L98H) (TR_{34}/L98H). *A. fumigatus* harbouring these mutations significantly complicates patient management and recent studies show a 20% excess mortality in patients with azole-resistant invasive aspergillosis compared with azole-susceptible infection (van Paassen et al. 2016). Interestingly and importantly, in 2010 a new but similar mutation was described; TR_{46}/Y121F/T289A, that was recovered from patients and environmental samples like the TR_{34}/L98H (van der Linden et al. 2013) and has since then also been spreading on a global scale in both clinical as well as environmental *A. fumigatus* isolates.

Resistance of clinical *A. fumigatus* isolates has been reported at a low frequency in the past and is generally considered to be a rare phenomenon. Acquired resistance was only seen in *A. fumigatus* isolates cultured from rare cases of patients with aspergilloma during long-term treatment with azoles (Chen et al. 2005; Howard et al. 2006, 2009; Snelders et al. 2008). Evidence has however shown that the TR_{34}/L98H and TR_{46}/Y121F/T298A in *A. fumigatus* detected in humans are likely to be acquired from the environment rather than from selection within patients during azole therapy for four reasons:

1. Multi-azole-resistant isolates were found in azole-naïve patients with aspergillus disease (Verweij et al. 2007; Snelders et al. 2008) ruling out an azole resistance selection within the patient;
2. The tandem repeat azole resistance mechanism was previously described in azole-resistant phytopathogenic molds but not in clinical molds (Snelders et al. 2009);
3. Isolates with the specific TR_{34} and TR_{46} mutations are recovered from clinical samples as well as environmental samples (Snelders et al. 2009);
4. Medical-azole-resistant TR_{34} and TR_{46} isolates are cross-resistant to five agricultural azoles (Snelders et al. 2012). Environmental azoles are widely used as fungicides (including fruit preservation during storage, protecting wood, flower bulb dipping, and many more (Verweij 2017). However, they are not specifically directed against *A. fumigatus* but rather to protect crops from fungal pathogens such as *Botrytis* and *Fusarium*. Taken together, cross-resistance to medical azoles is likely to be an unintended side effect of the evolved resistance to environmental azole fungicides. Now, two possible routes for resistance development can be considered: through patient exposure; the "patient-route" and through exposure to azole compounds in the environment; the "environmental-route" (Fig. 3.4). On the left side in this figure, the patient-route of azole resistance development is shown. Patients at risk can be infected by an azole-susceptible isolate which only develops resistance after colonization and long-term azole treatment of the patient. On the right, the environmental-route of azole-resistance development is shown. *A. fumigatus* is airborne and present ubiquitously, however in the environment (Zhang et al. 2017) where *A. fumigatus* can grow and azole residues are present there can be selection and spread of resistant genotypes. Finally, patients can be infected by azole-resistant genotypes originating from the environment.

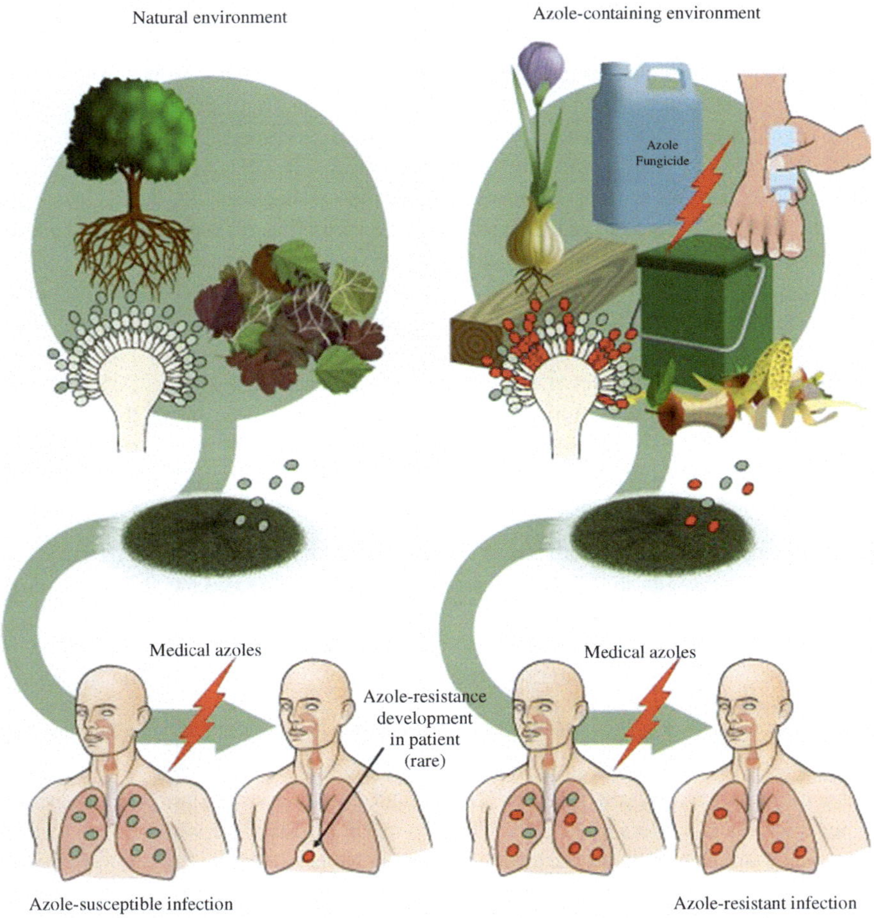

Fig. 3.4 Two routes of azole resistance development. (Source: Wageningen University & Research, The Netherlands)

The tandem repeat-genotypes of *A. fumigatus* have either spread or independently emerged on a global scale, however, newly emerged resistance mutations are generally thought to be associated with reduced growth rates in the absence of antifungals compared to wild type isolates. In order to persist in the absence of azoles and to not be outcompeted by sensitive isolates, resistant isolates need to mitigate this cost of resistance, and are thought to do so by acquiring second-site compensatory mutations. A fitness cost has however never been detected in clinical or environmental isolates. This can be explained because the compensation of fitness already took place or because there is no cost of fitness under standard laboratory conditions.

3.4 Global Spread of Azole Resistance

Today, the most affected country by tandem repeat azole resistance mechanism are the Netherlands, with a prevalence of more than 10% of resistance in clinical strains. Clinical and environmental cases of triazole resistance in *A. fumigatus* show inequalities in global distribution, with Europe at the centre of attention. High prevalence of triazole resistant *A. fumigatus* is observed in clinical practice and in the environment in the Netherlands but also in UK, Denmark, Germany and Eastern France. This prevalence varies between geographic regions and between hospitals, probably due to the differences in methods used to detect resistance and follow patients. However, there has recently been an emergence and extension of the phenomenon (also due to increased testing for antifungal resistance) to all countries in Europe, and to all continents, meaning that the phenomenon has reached worldwide prevalence.

A map showing the countries reporting resistance with TR_{34} and TR_{46} mutations, as it has been published before, did not seem informative to us since studies are not comparable in terms of sampling or techniques and often because the number of total strains isolated is not known. Thus, to generally illustrate global spread of tandem repeat azole resistance in *A. fumigatus*, we reported the number of papers published by year and by country in Fig. 3.5. We performed the following search on PubMed database: "TR34" or "TR46" and "*Aspergillus fumigatus*". Results were exported in .xml format and data analysed with RStudio with "maps" and "XML"

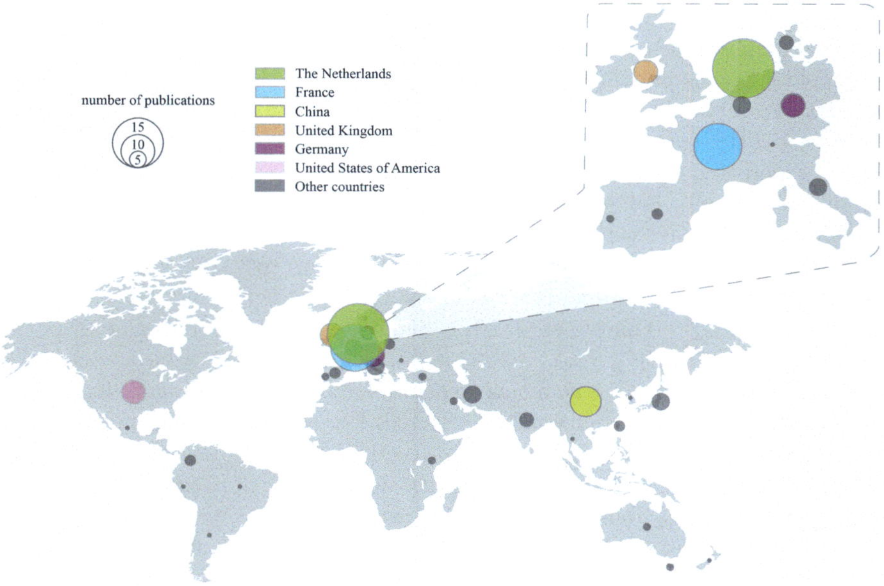

Fig. 3.5 Distribution by country of publications (source: PubMed database) reporting cases of resistant strains carrying tandem repeat mutations detected in clinical or environmental samples

packages. Reviews, original articles that develop diagnostic techniques on previously identified strains, as well as articles comparing several strains worldwide were excluded from the list. This resulting world map is a good illustration of disparities in resistance, but should be interpreted with caution, because papers does not reflect exact prevalence numbers and just because there's no publication does not mean the resistance is not there. One example is USA: in one of their publications on passive surveillance in 2015–2017, they reported 0.3% of resistant clinical isolates with TR_{34} mutation (Berkow et al. 2018) and only one study with TR_{34} mutation detected in the environment (Hurst et al. 2017). In the USA there are too few studies performed to have a representative picture of the situation in this vast country, especially since with the increasing use of azole fungicides, a potential increase in resistance can be expected to be observed similar to Europe (Lockhart et al. 2020).

In any case, tandem repeat azole resistance is increasingly present on all continents and it is expected to continue to spread. Previous studies in the Netherlands and UK have screened for environmental resistance of *A. fumigatus* isolates in plant waste material by using a single azole at a high concentration and thereby selectively enriching for highly resistant genotypes (Verweij 2017; Schoustra et al. 2019a). Rather easily, it is possible to detect environmental azole resistance, it is however unclear whether these tandem repeat type resistant isolates are genetically related or not. Are they arising frequently *de novo* in different environmental niches or have they only arisen a small number of times and subsequently spread to other geographic locations in a very efficient manner? It is also unclear, due to a lack of systematic surveillance programs, how many other tandem repeat resistance mechanisms potentially exist.

In parallel, with advances in modern medicine and cancer therapies, the number of people at risk is growing. In addition, the number of cases may affect a much broader category of patients than we think and not only immunocompromised patients. In 2018, the number of cases of invasive aspergillosis in patients with influenza has increased significantly. Triazole resistance is also an emerging concern for treating chronically colonized patients such as cystic fibrosis patients. Several European studies have been conducted and reported different prevalence of resistance depending on the country and the region, from 0 to 16% (Amorim et al. 2010; Mortensen et al. 2011; Burgel et al. 2012; Morio et al. 2012; Fischer et al. 2014; Prigitano et al. 2017; Guegan et al. 2018; Seufert et al. 2018; Godeau et al. 2019; Lavergne et al. 2019) (Fig. 3.6).

Initially, there seemed to be a clear separation between mutations called "clinically" selected and "environmentally" selected. The resistance mechanisms found in isolates that are thought to have become resistant through medical azole treatment consist of point mutations only, whereas the presence of a tandem repeat is an important mechanism found in plant pathogenic molds that developed resistance through exposure to azole fungicides. Therefore, the resistance mechanism seems to be indicative of the route of resistance development and therefore it implies that the TR_{34}/L98H and TR_{46}/Y121F/T298A have been selected in the environment rather than within patients. The tandem repeat resistant *A. fumigatus* isolates are also

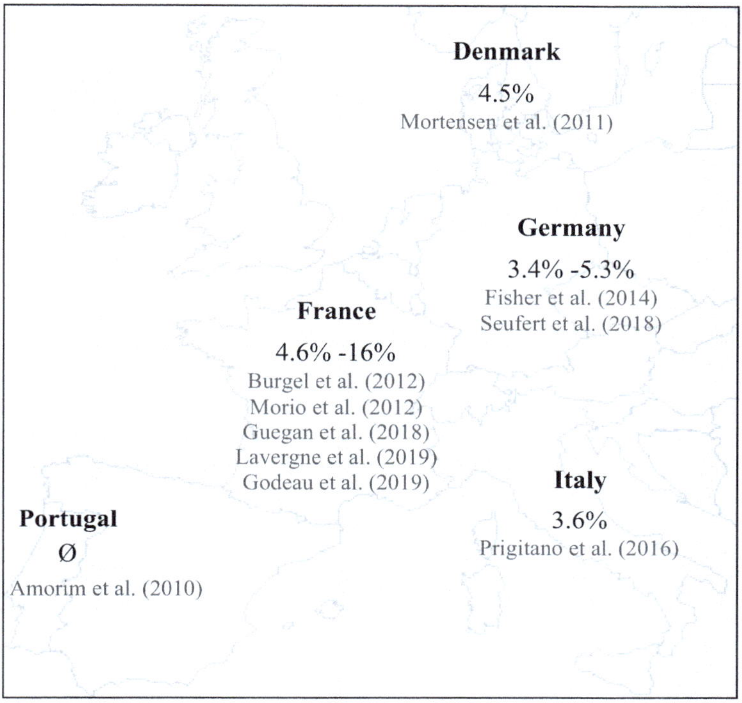

Fig. 3.6 European studies and prevalence of azole resistant *A. fumigatus* burden in cystic fibrosis patients (prevalence is given with the number of *A. fumigatus* positive culture)

widely distributed in the environment supporting this environmental-route of resistance development.

But, we soon also realized that the compartments may not be so distinct. For example, the G54 mutation initially described as "clinical" was also found in the environment in India, Tanzania, Romania, Germany (Bader et al. 2015; Sharma et al. 2015) (Table 3.1). A recent paper dealing with the TR_{120} mutation in a fatal case of aspergillosis is also interesting: A wild type and a TR_{120} isolate with the same genome (whole genome sequencing analysis) were detected in the same patient with azole long-term therapy (Hare et al. 2019). Thus, it questions the fact that only environmental exposure to fungicides can induce adaptations with tandem repeat mutations. The limits of the resistance acquisition pathways (patients *vs* environment) are not always distinct (Buil et al. 2019). It is therefore perhaps no longer necessary to speak of a compartment of emerging resistance but focus on environmental "conditions", whether it be the pulmonary environment or the agricultural outdoor environment and the capacity of the fungus to thrive and adapt.

Clearly, multi-azole-resistant *A. fumigatus* isolates are spreading rapidly across the globe, but there is an alarming paucity of fundamental research into this emerging problem. Therefore, a fundamental understanding of the causal factors of azole resistance selection is needed.

3.5 Contributing Factors in the Emergence of Azole Resistance

3.5.1 Cross-Disciplinary Use of Azoles

Impressive efforts have generally greatly increased living standards and conditions, which crucially includes intensified agriculture. However, with the intensification of agriculture, changes in land use that affects the ecological environment and the increase in human population density and concentration in urban areas, disease risk and epidemiology for plants, animals, and humans has similarly changed significantly. Fungal pathogens are a growing threat to human and animal health, ecosystems, food safety and the global economy, hence the importance of effectively controlling them. Triazole are used on a global scale for agriculture and horticulture, accounting for a quarter of all fungicides (Hollomon 2017), with 31 available molecules (Fisher et al. 2018). They are particularly used in Europe and Asia, where they represent one of the most commonly used pesticide classes. Forty percent of pesticides used in Europe are fungicides and more than a third of them are azole fungicides. This percentage could reach up to 23% in India and China (Chowdhary et al. 2013, 2015; Chen et al. 2016). Treatment with azoles fungicides would seem to be less important on the American continent where less than in 5% of the total cultivated crops azole fungicides are used (Verweij et al. 2009). However, azole fungicides are among the most predominant pesticides used in Colombia (Alvarez-Moreno et al. 2017) concluding that geographical differences of azole use are large.

A. fumigatus is a common saprobic fungus whose main habitat is decaying plant material. It also is the most prevalent airborne fungal pathogen causing fatal invasive infections in immunocompromised patients (Latgé 1999). Although *A. fumigatus* is not a phytopathogen, many environmental azoles were found to exhibit activity against *A. fumigatus* (Snelders et al. 2012). The fungicides are applied repeatedly over a long period of time and could thereby create a persistent pressure of azole compounds on saprophytic fungi. Several studies reported the presence of resistant strains (TR_{34}/L98H and TR_{46}/Y121F/T289A mutations) in industry, professional fields or other activities in link with triazole molecules. Due to the persistence of these molecules in the environment, resistant strains are not necessarily found in recently treated substrates. It takes at least 2 years after use for triazoles to be no longer detected in soils (Hollomon 2017). For example, in a study conducted in market gardens, a link has been identified between the presence azoles resistant *A. fumigatus* and the presence of difenoconazole in the soils of different farms, even 2 years after treatment of azoles (Rocchi et al. 2018). In a study in French sawmills (Jeanvoine et al. 2017), azole-resistant strains were isolated in sawmills in contact with azole fungicide either by dipping in a tank, by occasional fungicide spraying, by importing foreign wood treated with fungicides or by storing treated wood coming from a sawmill using azoles. Thus, even if there is no fungicide treatment on site, the storage of treated wood (and probably its leaching by rain) could be sufficient to provide an environmental azole pressure and find resistant strains in soils.

Some of the studies depicted in Table 3.3 combine the research of resistant strains with a dosage and application of triazole molecules. In each of the professional sites surveyed, resistant strains were found in soils with low to moderate triazole concentrations (on the scale of azole concentrations that can be found in agricultural fields). The highest concentrations probably result in complete elimination of all viable fungal forms while the lowest concentrations exert a selection pressure high enough for fungi to grow and adapt; to select for antifungal resistance. In sawmills study, propiconazole concentration in substrates were up to 50.000µg/g of substrate and resistance was found for samples with 0.3–60µg of propiconazole/g of substrate (Jeanvoine et al. 2017) but also less often in samples with greater quantity (one strain in soil with propiconazole concentration of 3.000µg/g). In market gardens study (Rocchi et al. 2018), concentrations ranged from 4.6 to 414.5 ng of difenoconazole/g of soil and resistance were found until 90 ng of difenoconazole /g of soil. Studies show that fungicide concentrations can vary greatly at one site and application of fungicides cannot always be done in a homogenous manner; therefore, some pockets of very high concentrations will exist similarly for pockets with intermediate, low, or no fungicidal pressure all at one site.

3.5.2 Hotspots of Azole Resistance

Triazole fungicides have now been largely blamed for the emergence of resistance, but in some cases, even if triazoles are used for crop protection, only susceptible strains are isolated (Rocchi et al. 2018). This indicates that not all (agricultural) environments have a high pressure of azole resistance and constitute as a hotspot.

Research has shown that azole-resistant *A. fumigatus* can accumulate and thrive in plant-waste containing environmental azoles (Zhang et al. 2017; Schoustra et al. 2019a). These locations can be defined as a hotspot that is characterized by an environment that supports the growth and reproduction of *A. fumigatus* and where azole fungicides with anti-*Aspergillus* activity are present. During adaptive evolution in such a hotspot, mutants arise spontaneously, and through a process of natural selection, the adaptive mutant expands in the population until it either gets fixed or replaced by an even fitter variant that arises. Natural populations of *A. fumigatus* are likely to have a complex environmental structure containing multiple resistant genotypes. The exact genetic composition will depend on the selection pressures, the frequency of mutations, and the population size in the hotspot.

Table 3.3 Examples of studies with detection of tandem repeat mutations in environmental samples and if described, the link with fungicides use

		Types of samples with resistant isolates (other samples without resistance are not mentioned)						
							Details of samples	
Country	Mutations	Bulb, seeds, flower beds/ fields	Compost	Culture fields	Green and wood chippings	Other	Fungicides used or residuals quantification	References
The Netherlands	TR_{34}	×	×				Fungicides used for plant production. Possible residues present in organic matter compost	Snelders et al. (2009)
	TR_{34} TR_{46}					×	Air in hospital and dwellings	van der Linden et al. (2013)
	TR_{34} TR_{46}		×				Quantifications: Prothioconazole-desthio, tebuconazole and thiabendazole. Concentrations ranging from 0.01 to 3.7 mg/kg	Personal communication
	TR_{34} TR_{46}	×			×		Quantifications: Prothioconazole, prothiconazoledestio, prochloraz, prochloraz-desimidazole-amino, difenoconazole, epoxiconazole, penconazole, propriconazole, azaconazole, tebuconazole. Concentrations ranged from 0.001 mg/kg to 6.4 mg/kg	Schoustra et al. (2019a)
UK	TR_{34}					×	Soil and air surrounding hospital	Dunne et al. (2017)
	TR_{34} TR_{46}	×				×	Flower beds outside hospitals, urban parks and remote forest	Sewell et al. (2019a)

France	TR_{34}			×			Cereals crops. Use of prothioconazole and epoxiconazole, amounts non available	Rocchi et al. (2014a)
	TR_{34}				×		Sawmills. Quantifications: Propiconazole, tebuconazole, difenoconazole, epoxiconazole, bromuconazole. Resistance until 3000µg/g of propiconazole	Jeanvoine et al. (2017)
	TR_{34}			×		×	Garden, indoor air from bedroom and bathroom	Lavergne et al. (2017)
	TR_{34}[a] TR_{46}			×			Carrots, celery and cabbage crops. Quantifications: Propiconazole, tebuconazole, difenoconazole, epoxiconazole, bromuconazole. Resistance until 90 ng/g of difenoconazole	Rocchi et al. (2018)
	TR_{34}					×	Electrostatic dust collector in dwellings	Dauchy et al. (2018)
	TR_{34}	×				×	Tulip bulbs coming from the Netherlands, air and dust in hospital	Godeau et al. (2019)
Norway	TR_{34}					×	Water	Snelders et al. (2008)
Denmark	TR_{34}	×					Surroundings hospital, flower beds in park	Mortensen et al. (2010)
Italy	TR_{34}	×		x			Cucurbit fields, apple orchards, rose pot composts. Use of azole fungicides in culture.	Prigitano et al. (2014)
	TR_{34}			×			Orchards, market gardens, vineyards	Prigitano et al. (2019)
Romania	TR_{34}			×			Not specified	Sharma et al. (2015)

(continued)

Table 3.3 (continued)

		Types of samples with resistant isolates (other samples without resistance are not mentioned)						
		Bulb, seeds, flower beds/			Green and		Details of samples	
Country	Mutations	fields	Compost	Culture fields	wood chippings	Other	Fungicides used or residuals quantification	References
Colombia	TR_{34} TR_{46} TR_{53}	×					Flower fields, tebuconazole and difenoconazole largely used	Le Pape et al. (2016)
	TR_{34} TR_{46} TR_{53}	×					Concentrations of penconazole (0.15 mg/kg), difenoconazole (0.41 mg/kg), tetraconazole (0.49 mg/kg) and tebuconazole (0.69 mg/kg)	Alvarez-Moreno et al. (2017)
	TR_{34} TR_{46} TR_{53}			×			Maize, potatoes, carrots, peas fields. Quantifications: Azaconazole, cyproconazole, difenoconazole, epoxiconazole, fluquinconazole, flusilazole, flutriafol, hexaconazole, penconazole, propiconazole, tebuconazole, tetraconazole, and triadimenol. No residuals.	Alvarez-Moreno et al. (2019)
India	TR_{34}	×				×	Air from hospital wards, flower beds of hospital garden, soils with bird droppings Paddy/rice/red chilly fields, tea garden, beneath cotton trees	Chowdhary et al. (2012b)
	TR_{34} TR_{46}	×		×			Potato, fenugreek, red chili, aubergine, mustard crops and rose bed soils	Chowdhary et al. (2014a)

China	TR_{34}[a] TR_{46}			x			Soils from market gardens, in greenhouses with *Luffa* and strawberry plants	Ren et al. (2017)
	TR_{34}[a] TR_{46}			x			Strawberry, vegetables and rice plants. Quantifications: Difenoconazole, tebuconazole, epoxiconazole, prochloraz, imazalil, tricyclazole.Soils with resistance: Prochloraz until 0.05 mg/kg, difénoconazole until 0.0385 mg/kg, tebuconazole until 0.0805 mg/kg.	Chen et al. (2020)
Kuwait	TR_{34}	×		×		×	Outdoor air and soil, seedlings or vegetables/ ornamental flowers	Ahmad et al. (2014)
	TR_{34}					×	Outdoor air	Ahmad et al. (2015)
Iran	TR_{34}	×					Surrounding the hospitals	Badali et al. (2013)
	TR_{46}		×				Not specified	Ahangarkani et al. (2020)
Tanzania	TR_{34} TR_{46}				×	×	Woody debris of tree trunk hollows, soil from hospital parking areas	Chowdhary et al. (2014b)
USA	TR_{34}		×				Compost pile containing peanut plant harvest debris from fields treated with azoles	Hurst et al. (2017)
Kenya	TR_{34}	×					Flower farming with extensive use of fungicides	Kemoi et al. (2018)

[a]TR34/L98H/S297T/F495I

When taking in mind the frequency of mutation, it is necessary to take into account the lifecycle of *A. fumigatus*. The exact mechanisms and circumstances by which tandem repeat genotypes have evolved in *A. fumigatus* are unknown. Whole-genome sequencing of *A. fumigatus* isolates sampled worldwide has so far not been able to determine whether the most common resistance genotype emerged once with subsequent global spread or emerged multiple times (Abdolrasouli et al. 2015). This is largely due to a lack of understanding of the population structure in *A. fumigatus,* the extent and the nature of standing genetic variation, and its origin by mutation and recombination. Importantly, to date, there have been no systematic studies to determine mutation rates. Similarly, recombination in *A. fumigatus* is not well understood and there is no recombination map available.

Selection and recombination of mutations can take place during asexual, sexual and/or parasexual reproduction, both of which have been documented for *A. fumigatus*. The sexual cycle has been demonstrated in a laboratory set-up in 2009 but its natural niche has not been described yet, therefore it is unknown to what extent the sexual cycle, and accompanying recombination, contributes to the population structure (O'Gorman et al. 2009). Sexual spores or ascospores are able to survive a 70 °C heat shock of 1 h in contrast to asexual spores that do not survive. A recent report by Schoustra et al. (2019b) showed that plant waste material *A. fumigatus* growth after heat shock indicating the presence of sexual spores. It has also been shown that completion of the sexual part of the life-cycle of *A. fumigatus* could facilitate the emergence of the tandem repeat in the promotor region (Zhang et al. 2017). Given this recent evidence, it seems important that *A. fumigatus* can complete its full life cycle in the selection of the tandem repeat genotypes.

Finally looking at the population size, it has become clear that certain environments contain a much higher load of *A. fumigatus* spores than others. A recent study showed that in particular flower bulb waste, green waste material, and wood chippings had high concentrations of *A. fumigatus* in contrast to animal manure, grain, maize silage, fruit storage or exotic fruit waste (Verweij 2017; Schoustra et al. 2019a). Therefore, sites that facilitate abundant and prolonged growth can be considered as a hotspot for azole resistance selection.

Taking in mind the selection pressure by azoles, the frequency of mutations, and the population size of *A. fumigatus* certain sites where conditions are favourable can be considered as hotspots for azole resistance selection (Verweij 2017). These hotspots need to facilitate abundant and prolonged growth of *A. fumigatus*, so that the fungus can fulfil its full life cycle to exploit its full capacity for genetic variation. Conditions that support the growth of *A. fumigatus* are plant biomass temperatures between 12 °C and 65 °C, high humidity (85%–100%) and an acidity between pH 3.7 and 7.6. For selection of azole resistant genotypes, azoles need to be present that are active against *A. fumigatus* and present at sufficient levels of concentration (Verweij 2017). Three sites were selected as hotspots by a study from Verweij et al. (2017): compost from flower bulb waste, green compost and wood chippings but more sites can be considered under the above-mentioned conditions of a hotspot. Summarizing a hotspot for azole resistance selection can be defined as a site where *A. fumigatus* can grow, complete its lifecycle and where azoles are present.

3.5.3 Abiotic and Biotic Factors Contributing to Azole Concentrations and Fungal Population

The advice given to reduce resistance in plant pathogenic molds is to rotate treatments and apply a high-dose treatment rather than several low-dose treatments (Azevedo et al. 2015). Nevertheless, for the problem with *A. fumigatus*, the amounts in soils or others hotspots such as composts will eventually be in the right concentration range to select for resistance. In fact, fungicides exert selection pressure at concentrations lower than those needed to kill fungi. Thus, fungicide residues in soils, wastewater, slurry or tulip bulbs could be the drivers of antifungal resistance, regardless of the quantity sprayed at the start. The pathways in which these more or less persistent molecules are degraded or eliminated are therefore very important in this matter. The most important thing is probably the persistence of azole molecules in soils, with half-life estimated between 20 and 200 days (Hollomon 2017). The physicochemical and biological properties of the soil are known to regulate the processes involved in pesticide dissipation in general (adsorption, transfer and biodegradation), thus conditioning the exposure of living organisms in the soil and the ecotoxicological consequences resulting from this exposure. In soils, the selection of resistance probably results from interactions between growing practices (crop rotations, ploughing, use of fertilizers, types and quantities of fungicides) and soil characteristics.

Azole fungicides (hydrophobic pesticides) are mainly accumulated in the top soil layer (0–20 cm depending on the molecule) but it has already been reported that their behaviour depends on the quantity and quality of organic matter. For example, tebuconazole is considered slightly mobile and stay in first 10 cm of soil, with higher retention recorded in soils amended with organic materials (Čadková et al. 2013). The addition of organic matter or surfactants to soils could also reduce the migration of penconazole (Komarek et al. 2010). Percolations experiments under controlled conditions have shown that difenoconazole adsorption also depended on the organic content of the soil, with a retention percentage that increases with the organic matter content (Léchenault-Bergerot et al. 2019). At soil characteristics levels, mineral elements are also important factors. For example, sandy soil (or soil with low organic matter content) retains less water and therefore less diluted triazoles due to their hydrophobicity (Komarek et al. 2010). Thus, triazoles may not be retained in this type of soil long enough to promote the selection of resistant strains (less contact time). This assumption of interaction was raised during a study of market gardeners where even if difenoconazole was used on crops, no fungicide was detected in the soil and only susceptible strains are isolated (Rocchi et al. 2018): The soil was so sandy that everything was probably drained very quickly.

It is also assumed that a transformation of pesticides in soils can occur without knowledge of the resulting compounds effects. Thus, "aging effect" can be another area of investigation. It is defined as the redistribution of an element into less available chemical fractions, which reduces its mobility and bioavailability. Aging of fungicides in soils can create stronger bonds between soil particles and fungicides with increased contact time that reduces the availability of pesticides for degradation and increases their half-life. The half-life of tetraconazole depends for example on the concentration at the time of treatment and soil texture ranging from 67 to 69 days to more than 1688 days (Sułowicz et al. 2016). But, global environmental factors such as climate would also be involved. Tetraconazole for example would tend to accumulate with successive annual precipitations (Sułowicz et al. 2016). Other fungicides like tebuconazole have half time between 40 and 170 days that would depend on the dose and average daily temperature (El Azhari et al. 2018). It is therefore likely that the nature of the soil plays a role in the dissipation of triazole residues and therefore indirectly in the development of resistance.

Agricultural work can also modify the soil structure (ploughing) or composition. For example, intensive agriculture can decrease the percentage of organic matter (Tilman et al. 2002) and the long-term use of fertilizers can also decrease pH (Bradley et al. 2006) with an impact of physicochemical and biological functioning of soils. It has been shown that vegetation cover (associated with high organic matter content) can reduce tebuconazole mobility (Dousset et al. 2010). The roots of plants will also modify the structural properties of the soil (compaction), draw water or secrete minerals, thus modifying very locally the bio-physical and chemical properties of the soil. There is a whole biological component (plant and microbial communities) to be considered.

The microbiological communities present will also have an impact on the soil, with perhaps a greater physical impact of fungal communities (filament networks) than bacteria. At the same time, the application of fungicides can affect the biomass of non-target organisms (others than *A. fumigatus*), change their biochemical, structural, functional activity, and genetic diversity. Depending on the treatment practices and molecules used, fungicides must have more or less marked toxicological effects, which will modify the biological communities present. Various controlled studies have shown, for example, that the impact of tebuconazole is even more negative when the soil has never been treated. For bacterial communities, a higher sensitivity of communities is found in soils with a treatment history, which can be explained by the lower diversity of communities (Sulowicz et al. 2016). It has also been shown that there is no effect of tebuconazole on the total abundance of bacterial communities, but rather on some taxa more than others. For example, beta and gamma proteobacteria increase in the most exposed soils. This can be explained by the fact that beta proteobacteria are capable of degrading organic compounds or that fungi exposed to fungicides reduce their ecological pressure on prokaryotic communities that may develop (El Azhari et al. 2018).

The fraction of bioavailable pesticides can also be visibly transformed by soil microorganisms. Environmental (bio)availability is therefore not homogeneous. Phenomena of solubility and mobility (availability) as well as absorption, metabolization, and bioaccumulation (bioavailability), related to the physico-chemical and biological structures of soils occur. Tebuconazole can be transformed by abiotic and biotic processes and give transformation products that can be recalcitrant and that are also involved in the transformation of lanosterol into ergosterol. Thirty-four transformation products for tebuconazole have been identified and little is known about their ecotoxicity on soil microbial communities (El Azhari et al. 2018).

In addition, there are phenomena of competition and/or interactions between microorganisms, which are currently only minimally studied in *A. fumigatus*. In soils, it has been estimated that *Aspergillus* compete with 4000 eukaryotic and prokaryotic species per gram of soil (Montesinos 2003). *A. fumigatus* is able to develop strategies to avoid predation by amoebae: intracellular germination in the predator, synthesis of inhibition factors that lead to host cell lysis (Hillmann et al. 2015). These strategies are the same as those used to counter the immune system during lung infections (van de Veerdonk et al. 2017). Cooperation phenomena have also been shown, where *A. fumigatus* spores can be transported by *Paenibacillus* vortex bacteria (also present in the Rhizosphere) to avoid areas that would be contaminated with triazoles (Ingham et al. 2011).

Fungicide levels that allow resistance selection in fungal populations are therefore not related to fungicide use alone. And to think that giving general recommendations for agricultural soil treatments would be sufficient is certainly utopian. It is the properties (physico-chemical and biological) of environments in modulating fungicide availability, which control the fungicide selection pressure on fungal populations, if they are able to prosper with these conditions.

3.6 Solutions to Limit the Risk of Resistance

The complexity of durable azole use lies in the fact that it is difficult to recommend that the use of all triazoles has to be discontinued in the environment and reserved for human therapeutics. The main challenge of agriculture is to feed the entire population, which is threatened by, among other things, crop infections, largely dominated by fungi. Therefore, often, the only solution is preventive action yet considering that organic farming is not sufficiently productive (−30%). Concerning treatment options, triazoles are preferred in agriculture because they specifically target fungi. An immediate ban on triazoles in the environment would lead to excessive economic losses.

Due to their use, which currently seems essential for the protection of agricultural or wood products, azole fungicides can be found in the various environmental compartments (water, soil, sediment, plants, air) and accumulate in organisms. The proven toxicity of pesticides in general (fungicides, insecticides, weed killers) is currently a growing public concern and a source of debate in the scientific community. It is necessary to limit their spread in ecosystems. The European Union has the strictest pesticide authorization and control system in the world with legislation and measures in place to ensure a reduced, more accurate and safe use of pesticides in agriculture. With regard to triazole resistance, it is essential to better characterize exactly which fungicides lead to emerging resistance and in what conditions. In any cases, industrial and agricultural sectors must stop releasing large quantities of triazoles if this is the case. Recent work conducted in sawmills has shown that even if the conditions for using treatment products (soaking time, drying) are respected, triazole residues are nevertheless released into the environment, around treatment tanks and near storage area (up to 45 mg of propiconazole per gram of substrate) (Jeanvoine et al. 2017).

3.6.1 Possible Environmental Remediation: A New Approach

A potential option to limit the emergence of resistance mechanisms in the environment would be to control the diffusion of triazole molecules in soils, before they accumulate in soils and end up in compost. Most of the time, those treatments are used to avoid the development of fungal infections on the aerial part of the plants. In those cases, there is no need for the fungicide to go through the soil, where they select resistant strains. Chemical components could be used to catch the fungicide before their penetration in the soil, and without affecting the treatment effectiveness on leaves.

Some conventional disposal methods can be used to remove this type of molecule (activated carbon or chemical oxidation, for example) from environmental compartments. However, despite their proven effectiveness, the latter have a number of disadvantages, including their high costs, and their non-ecological and energy-consuming aspects. Moreover, the significant production of waste is difficult to recycle, and the possibility of transferring technology from the laboratory scale to environmental sites is limited. It therefore appears necessary to propose new methods of eliminating these fungicidal molecules, which would be more chemically effective, greener and cost-effective.

Cyclodextrins are natural oligosaccharides obtained by enzymatic degradation of starch. These commercial molecules belong to the family of cage molecules due to their structure, which is composed of a hydrophobic cavity that can encapsulate other substances. These substances are also non-toxic, biodegradable, biocompatible, and produced at relatively low cost. Cyclodextrins find many applications from pharmaceuticals to cosmetics, food, textiles, agrochemicals, chromatography, biotechnology, catalysis, nanotechnology, and environment (Del Valle 2004; Crini et al. 2018). Chemical structures and dimensions of three native cyclodextrins, α-CD, β-CD and γ-CD, containing 6, 7 and 8 glucose units, respectively, are presented Fig. 3.7.

Their hydrophobic cavities make it possible to develop interactions as "host-guest" interaction type (Fig. 3.8). In medicine they are already commonly used as transporter or to increase solubility of triazole molecules (Peyton et al. 2015). In agriculture, some application has been described, especially in goal to gradually released a constant concentration of fungicide, which confirms their ability to encapsulate triazoles used in the environment (Balmas et al. 2006).

Applications in link with environment protection and soil and water decontamination are already described with organic and metallic molecules (Crini et al. 2018). Their potential application in this problem is by their ability to encapsulate azole

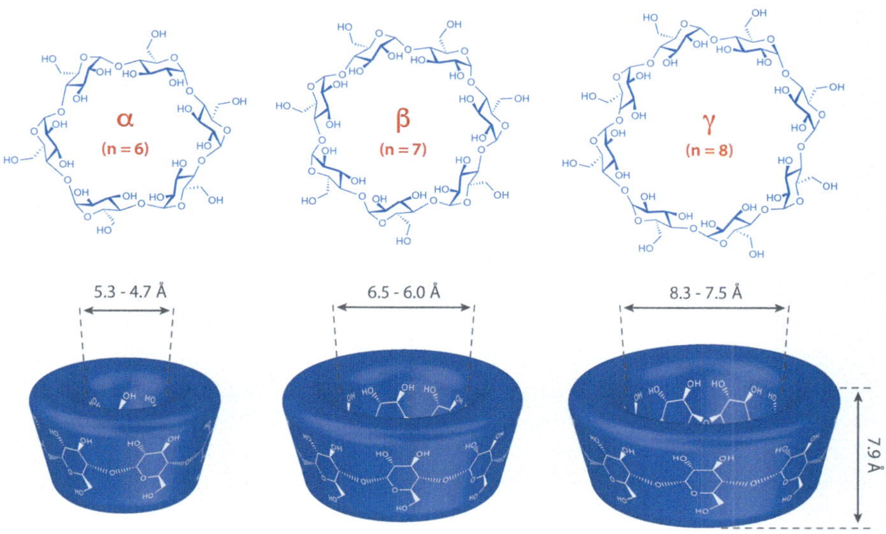

Fig. 3.7 Chemical structure and dimensions for the three native cyclodextrins: α-, β- and γ-cyclodextrin; (n = 6, 7 and 8 glucose units, respectively; source: with the kind courtesy of Marc Fourmentin, Dunkerque, France)

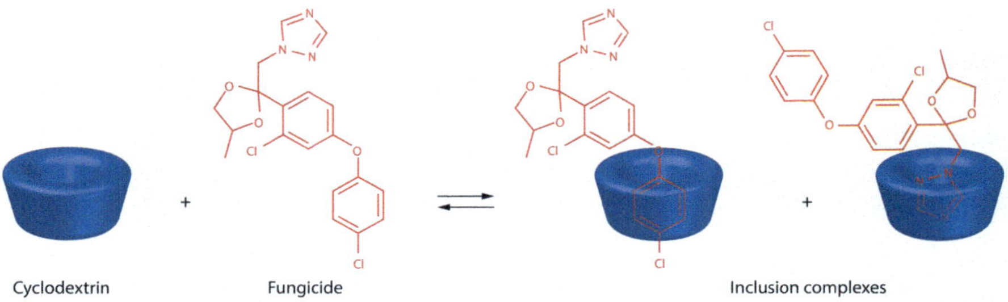

Fig. 3.8 Schematic representation of the inclusion phenomena between a cyclodextrin molecule (the host) and a solute (the guest) to form an inclusion complex (the example illustrates the two possibilities of interaction between difenoconazole fungicide and β-cyclodextrin; source: with the kind courtesy of Marc Fourmentin, Dunkerque, France)

molecules in the environment once they are, for example, grafted on a support which is currently under research. A moderate use of triazole molecules coupled with the development of that type of decontamination methods could permit to limit the dispersion of azole molecules and indirectly the phenomenon of resistance mechanisms selection. Recent *in vitro* studies, Rocchi et al. (2019) showed that in presence of cyclodextrins, triazole fungicides no longer had any action on *A. fumigatus* strains. Those results assumed that once trapped by cyclodextrins, triazole fungicides would probably have less impact leading to resistance. This underlines the interest of deepening the study on the interactions between cyclodextrins and triazole fungicides, which represent an interesting solution to limit the resistance emergence.

3.6.2 Changing Hotspots into Coldspots

Not all sites where azoles are present selection of azole resistance *A. fumigatus* takes place, nor sites that show abundant growth of *A. fumigatus* show selection of azole resistance. A hotspot of azole resistance selection is characterized with abundant and prolonged growth of *A. fumigatus*, the ability of the fungus to undergo a complete lifecycle and growth of the fungus in the presence of azoles. By identification and defining hotspots for azole resistance selection the question arises if it is possible to change these sites from hotspots into coldspots. Since prohibiting the use of azole in the environment will be difficult to achieve, preventing the prolonged growth of *A. fumigatus* in the presence of azole can be much more feasible with possibly highly satisfactory effect. Further research will need to show what the common characteristics are for both hot- and coldspots as well as the effect of possible interventions taken at former hotspots to observe if they can effectively be changed into coldspots.

Limiting the length of storage of waste stockpiles could potentially be an interesting intervention resulting in a lower environmental pressure of azole resistant *A. fumigatus* isolates. By simply not giving *A. fumigatus* the time to grow to substantial concentrations and not allowing the fungus to undergo its full lifecycle can potentially have great effect on azole resistance selection. Future research will first have to determine the exact parameters that can define an azole resistance selection hotspot and then in turn investigate what appropriate measures can be taken to turn these locations into hotspots.

3.6.3 Identify Places at Risk and Protect Sensitive Populations

The spread of resistant strains from their place of selection to at-risk populations is poorly understood, particularly as the strains are able to migrate across vast distances of land and sea due to environmental and anthropogenic factors (Ashu et al. 2017). Several common genotypes were found within the same region and country and could be carried by the wind, if topography and climate permit it. In other cases, there are up to 7500 km between two individuals with identical genotype, which could be explained by winds or humans and marketed products movements. Thus, there could be local and long-distance dispersals (Sewell et al. 2019b).

For human fungal pathologies, it is likely that a certain number of people can be colonized by resistant strains present in their workplace or at home, and that the transition to infection occurs in the hospital during immunosuppression. For others, nosocomial infection with resistant strains present in the hospital environment is also possible, although no publication has yet proved this. Various risk factors for invasive fungal infections during hospitalization of immunocompromised patients have already been mentioned for several years: weather conditions, construction work, plants, flowers, showers (Pagano et al. 2011). Given the increasing emergence of resistance, and because some resistant strains have been found at the entrance of intensive care haematology department (Godeau et al. 2019), there are probably simple steps that could be taken to further reduce the exposure of at-risk patients. For example, the bibliography points out that the presence of tulip bulbs is a source of soil contamination by resistant strains of *A. fumigatus* (Dunne et al. 2017; Godeau et al. 2019; Sewell et al. 2019a). These types of flowerbeds (but also composting soil) in area with intensive care units and departments with allogeneic hematopoietic stem cell transplantation should be forbidden, or alternative such as the planting of other types of flowers of controlled origin or of bulbs produced through organic farming methods must be proposed (Godeau et al. 2019). For example, the hospital in Besançon (Eastern France), which used to fill its flowerbeds with Dutch tulips, has changed its product reference and now orders tulips without azoles treatments from the same supplier.

In terms of patients' exposure outside hospitals, it is more complicated. It might seem that only patients in an environment using fungicides (work or housing surrounded by cultures) are at risk (Rocchi et al. 2014a; Jeanvoine et al. 2016), but exposure to resistant strains is much more complex. Indeed, resistant strains have also been isolated from urban housing, urban parks or forests far from potential hotspots (Dauchy et al. 2018; Sewell et al. 2019a). Therefore, since we do not know the possible extent of these strains, it could be important to follow at-risk patients' life courses and measure exposure at hospital, at home and at "at-risk" workplace. Some publications have moreover shown the presence of genetically related resistant isolates from gardens and dwellings of patients who have developed aspergillosis with resistant strains (van der Linden et al. 2013; Lavergne et al. 2017). Even without mentioning the problem of resistance, prevention in hospital and at home has been recognized as effective in apprehending the risk of invasive aspergillosis in immunocompromised patients (Reboux et al. 2014; Rocchi et al. 2014b). Following these works, a follow-up of immunocompromised patients with a home survey by an environmental counsellor to measure potential fungal sources is proposed to patients in the Burgundy-Franche-Comté region (Eastern France) undergoing induction chemotherapy or hospitalized for allogeneic hematopoietic stem cell transplantation. For sure, even if home fungal monitoring is implemented, it cannot always be possible to do something for invasive fungal infection prevention, above all if just some colonies of *A. fumigatus* are isolated. However, home visits represent a means of providing personalized prevention to patients and could elucidate sources mini hotspots. Incredible at-risk situations, such as immunocompromised patient living in a home decorated with plaited dried wheat in all rooms and associated with important *A. fumigatus* concentrations in air (Rocchi et al. 2014b), could not have been identified without the home visit, just with the general advice given to the patients by the doctor. At this time, resistance testing was not done. But, this kind of situation has to be mentioned to the clinician which can opt for a prophylactic treatment or give advices to patient to go in another dwelling temporarily, during at-risk periods.

3.7 Conclusion

To maintain the ability to control fungal infections with the effective use of antifungal compounds and taking into account human and ecosystem health, the fight against antifungal resistance must become a high priority. There are three priority actions with the following objectives: To better understand the conditions in which resistance develops, identify the characteristics of hotspots for azole resistance selection and link the spread of resistance from hotspots with patients. To improve techniques to better diagnose and cure patients from fungal infections; and finally to obtain a durable azole use for both environmental as well as medical purposes. A "One Health" approach (a concept combining human health, animal health and ecosystem health) is needed to understand the development and spread of these treatment-resistant infectious agents in order to ultimately protect the health of ecosystems (including wild and cultivated/cultivated/raised fauna and flora) and humans.

Acknowledgements The authors thank Dr. Marc Fourmentin (*Université du Littoral Côte-d'Opale*, Dunkerque, France) and Marc Maas (The Netherlands) for illustrations graphics. They are also grateful for Benoit Valot (*UniversitéBourgogne Franche-Comté*, France) for his PubMed data analysis for the world map.

References

Abdolrasouli A, Rhodes J, Beale MA, Hagen F, Rogers TR, Chowdhary A, Meis JF, Armstrong-James D, Fisher MC (2015) Genomic context of azole resistance mutations in *Aspergillus fumigatus* determined using whole-genome sequencing. MBio 6:e00536–e00536. https://doi.org/10.1128/mBio.00536-15

Ahangarkani F, Puts Y, Nabili M, Khodavaisy S, Moazeni M, Salehi Z, Laal Kargar M, Badali H, Meis JF (2020) First azole-resistant *Aspergillus fumigatus* isolates with the environmental TR46/Y121F/T289A mutation in Iran. Mycoses 63:430–436. https://doi.org/10.1111/myc.13064

Ahmad S, Khan Z, Hagen F, Meis JF (2014) Occurrence of triazole-resistant *Aspergillus fumigatus* with TR34/L98H mutations in outdoor and hospital environment in Kuwait. Environ Res 133:20–26. https://doi.org/10.1016/j.envres.2014.05.009

Ahmad S, Joseph L, Hagen F, Meis JF, Khan Z (2015) Concomitant occurrence of itraconazole-resistant and -susceptible strains of *Aspergillus fumigatus* in routine cultures. J Antimicrob Chemother 70:412–415. https://doi.org/10.1093/jac/dku410

Albarrag AM, Anderson MJ, Howard SJ, Robson GD, Warn PA, Sanglard D, Denning DW (2011) Interrogation of related clinical pan-azole-resistant *Aspergillus fumigatus* strains: G138C, Y431C, and G434C single nucleotide polymorphisms in cyp51A, upregulation of cyp51A, and integration and activation of transposon Atf1 in the cyp51A promoter. Antimicrob Agents Chemother 55:5113–5121. https://doi.org/10.1128/AAC.00517-11

Alvarez-Moreno C, Lavergne RA, Hagen F, Morio F, Meis JF, Le Pape P (2017) Azole-resistant *Aspergillus fumigatus* harboring TR34/L98H, TR46/Y121F/T289A and TR53 mutations related to flower fields in Colombia. Sci Rep 7:45631. https://doi.org/10.1038/srep45631

Alvarez-Moreno C, Lavergne RA, Hagen F, Morio F, Meis JF, Le Pape P (2019) Fungicide-driven alterations in azole-resistant *Aspergillus fumigatus* are related to vegetable crops in Colombia, South America. Mycologia 111:217–224. https://doi.org/10.1080/00275514.2018.1557796

Amorim A, Guedes-Vaz L, Araujo R (2010) Susceptibility to five antifungals of *Aspergillus fumigatus* strains isolated from chronically colonised cystic fibrosis patients receiving azole therapy. Int J Antimicrob Agents 35:396–399. https://doi.org/10.1016/j.ijantimicag.2009.12.007

Ashu EE, Hagen F, Chowdhary A, Meis JF, Xu J (2017) Global population genetic analysis of *Aspergillus fumigatus*. mSphere 2:e00019-17. https://doi.org/10.1128/mSphere.00019-17

Astvad KM, Jensen RH, Hassan TM, Mathiasen EG, Thomsen GM, Pedersen UG, Christensen M, Hilberg O, Arendrup MC (2014) First detection of TR46/Y121F/T289A and TR34/L98H alterations in *Aspergillus fumigatus* isolates from azole-naive patients in Denmark despite negative findings in the environment. Antimicrob Agents Chemother 58:5096–5101. https://doi.org/10.1128/AAC.02855-14

Azevedo MM, Faria-Ramos I, Cruz LC, Pina-Vaz C, Rodrigues AG (2015) Genesis of azole antifungal resistance from agriculture to clinical settings. J Agric Food Chem 63:7463–7468. https://doi.org/10.1021/acs.jafc.5b02728

Badali H, Vaezi A, Haghani I, Yazdanparast SA, Hedayati MT, Mousavi B, Ansari S, Hagen F, Meis JF, Chowdhary A (2013) Environmental study of azole-resistant *Aspergillus fumigatus* with TR34/L98H mutations in the *cyp51A* gene in Iran. Mycoses 56:659–663. https://doi.org/10.1111/myc.12089

Bader O, Weig M, Reichard U, Lugert R, Kuhns M, Christner M, Held J, Peter S, Schumacher U, Buchheidt D, Tintelnot K, Gross U (2013) cyp51A-based mechanisms of *Aspergillus fumigatus* azole drug resistance present in clinical samples from Germany. Antimicrob Agents Chemother 57:3513–3517. https://doi.org/10.1128/AAC.00167-13

Bader O, Tunnermann J, Dudakova A, Tangwattanachuleeporn M, Weig M, Gross U (2015) Environmental isolates of azole-resistant *Aspergillus fumigatus* in Germany. Antimicrob Agents Chemother 59:4356–4359. https://doi.org/10.1128/AAC.00100-15

Balmas V, Delogu G, Sposito S, Rau D, Migheli Q (2006) Use of a complexation of tebuconazole with beta-cyclodextrin for controlling foot and crown rot of durum wheat incited by Fusarium culmorum. J Agric Food Chem 54:480–484. https://doi.org/10.1021/jf0523014

Bellete B, Raberin H, Morel J, Flori P, Hafid J, Manhsung RT (2010) Acquired resistance to voriconazole and itraconazole in a patient with pulmonary aspergilloma. Med Mycol 48:197–200. https://doi.org/10.3109/13693780902717018

Berger S, El Chazli Y, Babu AF, Coste AT (2017) Azole resistance in *Aspergillus fumigatus*: a consequence of antifungal use in agriculture? Front Microbiol 8:1024. https://doi.org/10.3389/fmicb.2017.01024

Berkow EL, Nunnally NS, Bandea A, Kuykendall R, Beer K, Lockhart SR (2018) Detection of TR34/L98H CYP51A mutation through passive surveillance for azole-resistant *Aspergillus fumigatus* in the United States from 2015 to 2017. Antimicrob Agents Chemother 62:e02240-17. https://doi.org/10.1128/AAC.02240-17

Bradley K, Drijber RA, Knops J (2006) Increased N availability in grassland soils modifies their microbial communities and decreases the abundance of arbuscular mycorrhizal fungi. Soil Biol Biochem 38:1583–1595. https://doi.org/10.1016/j.soilbio.2005.11.011

Bromley MJ, van Muijlwijk G, Fraczek MG, Robson G, Verweij PE, Denning DW, Bowyer P (2014) Occurrence of azole-resistant species of *Aspergillus* in the UK environment. J Glob Antimicrob Res 2:276–279. https://doi.org/10.1016/j.jgar.2014.05.004

Buil JB, Hare RK, Zwaan BJ, Arendrup MC, Melchers WJG, Verweij PE (2019) The fading boundaries between patient and environmental routes of triazole resistance selection in *Aspergillus fumigatus*. PLoS Pathog 15:e1007858. https://doi.org/10.1371/journal.ppat.1007858

Burgel PR, Baixench MT, Amsellem M, Audureau E, Chapron J, Kanaan R, Honore I, Dupouy-Camet J, Dusser D, Klaassen CH, Meis JF, Hubert D, Paugam A (2012) High prevalence of azole-resistant *Aspergillus fumigatus* in adults with cystic fibrosis exposed to itraconazole. Antimicrob Agents Chemother 56:869–874. https://doi.org/10.1128/AAC.05077-11

Bustamante B, Illescas LR, Posadas A, Campos PE (2020) Azole resistance among clinical isolates of *Aspergillus fumigatus* in Lima-Peru. Med Mycol 58:54–60. https://doi.org/10.1093/mmy/myz032

Čadková E, Komárek M, Kaliszová R, Vaněk A, Balíková M (2013) Tebuconazole sorption in contrasting soil types. Soil Sediment Contam 22:404–414. https://doi.org/10.1080/15320383.2013.733448

Camps SM, van der Linden JW, Li Y, Kuijper EJ, van Dissel JT, Verweij PE, Melchers WJ (2012) Rapid induction of multiple resistance mechanisms in *Aspergillus fumigatus* during azole therapy: a case study and review of the literature. Antimicrob Agents Chemother 56:10–16. https://doi.org/10.1128/AAC.05088-11

Chen J, Li H, Li R, Bu D, Wan Z (2005) Mutations in the cyp51A gene and susceptibility to itraconazole in *Aspergillus fumigatus* serially isolated from a patient with lung aspergilloma. J Antimicrob Chemother 55:31–37. https://doi.org/10.1093/jac/dkh507

Chen Y, Lu Z, Zhao J, Zou Z, Gong Y, Qu F, Bao Z, Qiu G, Song M, Zhang Q, Liu L, Hu M, Han X, Tian S, Chen F, Zhang C, Sun Y, Verweij PE, Huang L, Han L (2016) Epidemiology and molecular characterizations of azole resistance in clinical and environmental *Aspergillus fumigatus* isolates from China. Antimicrob Agents Chemother 60:5878–5884. https://doi.org/10.1128/AAC.01005-16

Chen Y, Dong F, Zhao J, Fan H, Qin C, Li R, Verweij PE, Zheng Y, Han L (2020) High azole resistance in *Aspergillus fumigatus* isolates from strawberry fields, China, 2018. Emerg Infect Dis 26:81–89. https://doi.org/10.3201/eid2601.190885

Chowdhary A, Kathuria S, Randhawa HS, Gaur SN, Klaassen CH, Meis JF (2012a) Isolation of multiple-triazole-resistant *Aspergillus fumigatus* strains carrying the TR/L98H mutations in the cyp51A gene in India. J Antimicrob Chemother 67:362–366. https://doi.org/10.1093/jac/dkr443

Chowdhary A, Kathuria S, Xu J, Sharma C, Sundar G, Singh PK, Gaur SN, Hagen F, Klaassen CH, Meis JF (2012b) Clonal expansion and emergence of environmental multiple-triazole-resistant *Aspergillus fumigatus* strains carrying the TR34/L98H mutations in the *cyp51A* gene in India. PLoS One 7:e52871. https://doi.org/10.1371/journal.pone.0052871

Chowdhary A, Kathuria S, Xu J, Meis JF (2013) Emergence of azole-resistant *Aspergillus fumigatus* strains due to agricultural azole use creates an increasing threat to human health. PLoS Pathog 9:e1003633. https://doi.org/10.1371/journal.ppat.1003633

Chowdhary A, Sharma C, Kathuria S, Hagen F, Meis JF (2014a) Azole-resistant *Aspergillus fumigatus* with the environmental TR46/Y121F/T289A mutation in India. J Antimicrob Chemother 69:555–557. https://doi.org/10.1093/jac/dkt397

Chowdhary A, Sharma C, van den Boom M, Yntema JB, Hagen F, Verweij PE, Meis JF (2014b) Multi-azole-resistant *Aspergillus fumigatus* in the environment in Tanzania. J Antimicrob Chemother 69:2979–2983. https://doi.org/10.1093/jac/dku259

Chowdhary A, Sharma C, Kathuria S, Hagen F, Meis JF (2015) Prevalence and mechanism of triazole resistance in *Aspergillus fumigatus* in a referral chest hospital in Delhi, India and an update of the situation in Asia. Front Microbiol 6:428. https://doi.org/10.3389/fmicb.2015.00428

Crini G, Fourmentin S, Fenyvesi É, Torri G, Fourmentin M, Morin-Crini N (2018) Cyclodextrins, from molecules to applications. Environ Chem Lett 16:1361–1375. https://doi.org/10.1007/s10311-018-0763-2

Dauchy C, Bautin N, Nseir S, Reboux G, Wintjens R, Le Rouzic O, Sendid B, Viscogliosi E, Le Pape P, Arendrup MC, Gosset P, Fry S, Fréalle E (2018) Emergence of *Aspergillus fumigatus* azole resistance in azole-naive patients with chronic obstructive pulmonary disease and their homes. Indoor Air 28:298–306. https://doi.org/10.1111/ina.12436

de Ramón-Carbonell M, Sánchez-Torres P (2020) Significance of 195 bp-enhancer of PdCYP51B in the acquisition of *Penicillium digitatum* DMI resistance and increase of fungal virulence. Pestic Biochem Physiol 165:104522. https://doi.org/10.1016/j.pestbp.2020.01.003

Del Valle EMM (2004) Cyclodextrins and their uses: a review. Process Biochem 39:1033–1046. https://doi.org/10.1016/S0032-9592(03)00258-9

Diaz-Guerra TM, Mellado E, Cuenca-Estrella M, Rodriguez-Tudela JL (2003) A point mutation in the 14alpha-sterol demethylase gene cyp51A contributes to itraconazole resistance in *Aspergillus fumigatus*. Antimicrob Agents Chemother 47:1120–1124. https://doi.org/10.1128/aac.47.3.1120-1124.2003

Dousset S, Thevenot M, Schrack D, Gouy V, Carluer N (2010) Effect of grass cover on water and pesticide transport through undisturbed soil columns, comparison with field study (Morcille watershed, Beaujolais). Environ Pollut 158:2446–2453. https://doi.org/10.1016/j.envpol.2010.03.028

Dunne K, Hagen F, Pomeroy N, Meis JF, Rogers TR (2017) Intercountry transfer of triazole-resistant *Aspergillus fumigatus* on plant bulbs. Clin Infect Dis 65:147–149. https://doi.org/10.1093/cid/cix257

El Azhari N, Dermou E, Barnard RL, Storck V, Tourna M, Beguet J, Karas PA, Lucini L, Rouard N, Botteri L, Ferrari F, Trevisan M, Karpouzas DG, Martin-Laurent F (2018) The dissipation and microbial ecotoxicity of tebuconazole and its transformation products in soil under standard laboratory and simulated winter conditions. Sci Total Environ 637–638:892–906. https://doi.org/10.1016/j.scitotenv.2018.05.088

Fischer J, van Koningsbruggen-Rietschel S, Rietschel E, Vehreschild MJ, Wisplinghoff H, Kronke M, Hamprecht A (2014) Prevalence and molecular characterization of azole resistance in *Aspergillus* spp. isolates from German cystic fibrosis patients. J Antimicrob Chemother 69:1533–1536. https://doi.org/10.1093/jac/dku009

Fisher MC, Hawkins NJ, Sanglard D, Gurr SJ (2018) Worldwide emergence of resistance to antifungal drugs challenges human health and food security. Science 360:739–742. https://doi.org/10.1126/science.aap7999

Garcia-Rubio R, Cuenca-Estrella M, Mellado E (2017) Triazole resistance in *Aspergillus* species: an emerging problem. Drugs 77:599–613. https://doi.org/10.1007/s40265-017-0714-4

Gisi U (2014) Assessment of selection and resistance risk for demethylation inhibitor fungicides in *Aspergillus fumigatus* in agriculture and medicine: a critical review. Pest Manag Sci 70:352–364. https://doi.org/10.1002/ps.3664

GLASS (2016) 2nd Meeting of the global antimicrobial resistance surveillance system collaborative platform with global report available online: https://www.who.int/antimicrobial-resistance/events/2nd-GLASS-collaborative-platform-meeting/en/

Godeau C, Reboux G, Scherer E, Laboissiere A, Lechenault-Bergerot C, Millon L, Rocchi S (2019) Azole-resistant *Aspergillus fumigatus* in the hospital: surveillance from flower beds to corridors. Am J Infect Control 48:702–704. https://doi.org/10.1016/j.ajic.2019.10.003

Gonzalez-Lara MF, Roman-Montes CM, Diaz-Lomeli P, Rangel-Cordero A, Valenzuela MO, Ponce-de-Leon A, Sifuentes-Osornio J, Ostrosky-Zeichner L, Martinez-Gamboa A (2019) Azole resistance and *cyp51A* mutation screening in *Aspergillus fumigatus* in Mexico. J Antimicrob Chemother 74:2047–2050. https://doi.org/10.1093/jac/dkz121

Gsaller F, Hortschansky P, Furukawa T, Carr PD, Rash B, Capilla J, Muller C, Bracher F, Bowyer P, Haas H, Brakhage AA, Bromley MJ (2016) Sterol biosynthesis and azole tolerance is governed by the opposing actions of SrbA and the CCAAT binding complex. PLoS Pathog 12:e1005775. https://doi.org/10.1371/journal.ppat.1005775

Guegan H, Chevrier S, Belleguic C, Deneuville E, Robert-Gangneux F, Gangneux JP (2018) Performance of molecular approaches for *Aspergillus* detection and azole resistance surveillance in cystic fibrosis. Front Microbiol 9:531. https://doi.org/10.3389/fmicb.2018.00531

Hagiwara D, Watanabe A, Kamei K, Goldman GH (2016) Epidemiological and genomic landscape of azole resistance mechanisms in *Aspergillus* fungi. Front Microbiol 7:1382. https://doi.org/10.3389/fmicb.2016.01382

Hare RK, Gertsen JB, Astvad KMT, Degn KB, Løkke A, Stegger M, Andersen PS, Kristensen L, Arendrup MC (2019) *In vivo* selection of a unique tandem repeat mediated azole resistance mechanism TR120 in *Aspergillus fumigatus cyp51A*, Denmark. Emerg Infect Dis 25:577–580. https://doi.org/10.3201/eid2503.180297

Hillmann F, Novohradska S, Mattern DJ, Forberger T, Heinekamp T, Westermann M, Winckler T, Brakhage AA (2015) Virulence determinants of the human pathogenic fungus *Aspergillus fumigatus* protect against soil amoeba predation. Environ Microbiol 17:2858–2869. https://doi.org/10.1111/1462-2920.12808

Hodiamont CJ, Dolman KM, Ten Berge IJ, Melchers WJ, Verweij PE, Pajkrt D (2009) Multiple-azole-resistant *Aspergillus fumigatus* osteomyelitis in a patient with chronic granuloma-

tous disease successfully treated with long-term oral posaconazole and surgery. Med Mycol 47:217–220. https://doi.org/10.1080/13693780802545600
Hollomon D (2017) Does agricultural use of azole fungicides contribute to resistance in the human pathogen *Aspergillus fumigatus*? Pest Manag Sci 73:1987–1993. https://doi.org/10.1002/ps.4607
Howard SJ, Webster I, Moore CB, Gardiner RE, Park S, Perlin DS, Denning DW (2006) Multi-azole resistance in *Aspergillus fumigatus*. Int J Antimicrob Agents 28:450–453. https://doi.org/10.1016/j.ijantimicag.2006.08.017
Howard SJ, Cerar D, Anderson MJ, Albarrag A, Fisher MC, Pasqualotto AC, Laverdiere M, Arendrup MC, Perlin DS, Denning DW (2009) Frequency and evolution of azole resistance in *Aspergillus fumigatus* associated with treatment failure. Emerg Infect Dis 5:1068–1076. https://doi.org/10.3201/eid1507.090043
Howard SJ, Pasqualotto AC, Denning DW (2010) Azole resistance in allergic bronchopulmonary aspergillosis and *Aspergillus* bronchitis. Clin Microbiol Infect 16:683–688. https://doi.org/10.1111/j.1469-0691.2009.02911.x
Hurst SF, Berkow EL, Stevenson KL, Litvintseva AP, Lockhart SR (2017) Isolation of azole-resistant *Aspergillus fumigatus* from the environment in the South-Eastern USA. J Antimicrob Chemother 72:2443–2446. https://doi.org/10.1093/jac/dkx168
Ingham CJ, Kalisman O, Finkelshtein A, Ben-Jacob E (2011) Mutually facilitated dispersal between the nonmotile fungus *Aspergillus fumigatus* and the swarming bacterium *Paenibacillus vortex*. Proc Natl Acad Sci U S A 108:19731–19736. https://doi.org/10.1073/pnas.1102097108
Isla G, Leonardelli F, Tiraboschi IN, Refojo N, Hevia A, Vivot W, Szusz W, Cordoba SB, Garcia-Effron G (2018) First clinical isolation of an azole-resistant *Aspergillus fumigatus* isolate harboring a TR46 Y121F T289A mutation in South America. Antimicrob Agents Chemother 62:e00872-18. https://doi.org/10.1128/AAC.00872-18
Jeanvoine A, Rocchi S, Reboux G, Grenouillet F, Benassarou M, Chirouze C, Millon L (2016) Sinus aspergillosis due to an azole-resistant *Aspergillus fumigatus* strain carrying the TR34/L98H mutation in immunocompetent host. Infect Dis (Lond) 48:765–766. https://doi.org/10.1080/23744235.2016.1193791
Jeanvoine A, Rocchi S, Reboux G, Crini N, Crini G, Millon L (2017) Azole-resistant *Aspergillus fumigatus* in sawmills of eastern France. J Appl Microbiol 123:172–184. https://doi.org/10.1111/jam.13488
Jones M (2020) Review of airborne antimicrobial resistance. Report. UK: Environment Agency, Horizon House, Deanery Road, Bristol, BS1 5AH. ISBN: 978-1-84911-424-0. Available online: https://www.gov.uk/government/publications/review-of-airborne-antimicrobial-resistance
Kano R, Kohata E, Tateishi A, Murayama SY, Hirose D, Shibata Y, Kosuge Y, Inoue H, Kamata H, Hasegawa A (2015) Does farm fungicide use induce azole resistance in *Aspergillus fumigatus*? Med Mycol 53:174–177. https://doi.org/10.1093/mmy/myu076
Kemoi EK, Nyerere A, Bii CC (2018) Triazole-resistant *Aspergillus fumigatus* from fungicide-experienced soils in Naivasha Subcounty and Nairobi County, Kenya. Int J Microbiol 2018:7147938. https://doi.org/10.1155/2018/7147938
Kidd SE, Goeman E, Meis JF, Slavin MA, Verweij PE (2015) Multi-triazole-resistant *Aspergillus fumigatus* infections in Australia. Mycoses 58:350–355. https://doi.org/10.1111/myc.12324
Kleinkauf N, Verweij PE, Arendrup MC, Donelly PJ, Cuenca-Estrella M, Fraaije B, Melchers WJG, Adriaenssens N, Kema GHJ, Ullmann A, Bowyer P, Denning DW (2013) European Centre for Disease prevention and Control. Risk assessment on the impact of environmental usage of triazoles on the development and spread of resistance to medical triazoles in Aspergillus species. ECDC, Stockholm. Report available online https://www.ecdc.europa.eu/en/publications-data/risk-assessment-impact-environmental-usage-triazoles-development-and-spread
Komarek M, Cadkova E, Chrastny V, Bordas F, Bollinger JC (2010) Contamination of vineyard soils with fungicides: a review of environmental and toxicological aspects. Environ Int 36:138–151. https://doi.org/10.1016/j.envint.2009.10.005

Korfanty GA, Teng L, Pum N, Xu J (2019) Contemporary gene flow is a major force shaping the *Aspergillus fumigatus* population in Auckland, New Zealand. Mycopathologia 184:479–492. https://doi.org/10.1007/s11046-019-00361-8

Kourkoumpetis TK, Desalermos A, Muhammed M, Mylonakis E (2012) Central nervous system aspergillosis: a series of 14 cases from a general hospital and review of 123 cases from the literature. Medicine (Baltimore) 91:328–336. https://doi.org/10.1097/MD.0b013e318274cd77

Lamoth F (2016) *Aspergillus fumigatus*-related species in clinical practice. Front Microbiol 7:683. https://doi.org/10.3389/fmicb.2016.00683

Latgé JP (1999) *Aspergillus fumigatus* and aspergillosis. Clin Microbiol Rev 12:310–350. https://doi.org/10.1128/CMR.12.2.310

Lavergne RA, Morio F, Favennec L, Dominique S, Meis JF, Gargala G, Verweij PE, Le Pape P (2015) First description of azole-resistant *Aspergillus fumigatus* due to TR46/Y121F/T289A mutation in France. Antimicrob Agents Chemother 59:4331–4335. https://doi.org/10.1128/AAC.00127-15

Lavergne RA, Chouaki T, Hagen F, Toublanc B, Dupont H, Jounieaux V, Meis JF, Morio F, Le Pape P (2017) Home environment as a source of life-threatening azole-resistant *Aspergillus fumigatus* in immunocompromised patients. Clin Infect Dis 64:76–78. https://doi.org/10.1093/cid/ciw664

Lavergne RA, Morio F, Danner-Boucher I, Horeau-Langlard D, David V, Hagen F, Meis JF, Le Pape P (2019) One year prospective survey of azole resistance in *Aspergillus fumigatus* at a French cystic fibrosis reference centre: prevalence and mechanisms of resistance. J Antimicrob Chemother 74:1884–1889. https://doi.org/10.1093/jac/dkz144

Lazzarini C, Esposto MC, Prigitano A, Cogliati M, De Lorenzis G, Tortorano AM (2016) Azole resistance in *Aspergillus fumigatus* clinical isolates from an Italian culture collection. Antimicrob Agents Chemother 60:682–685. https://doi.org/10.1128/AAC.02234-15

Le Pape P, Lavergne RA, Morio F, Alvarez-Moreno C (2016) Multiple fungicide-driven alterations in azole-resistant *Aspergillus fumigatus*, Colombia, 2015. Emerg Infect Dis 22:156–157. https://doi.org/10.3201/eid2201.150978

Léchenault-Bergerot C, Morin-Crini N, Rocchi S, Lichtfouse E, Chanet G, Crini G (2019) Hemp to limit diffusion of difenoconazole in vegetable garden soils. Heliyon 5:e02392. https://doi.org/10.1016/j.heliyon.2019.e02392

Lee H-J, Cho S-Y, Lee D-G, Park C, Chun H-S, Park Y-J (2018) TR34/L98H mutation in *cyp51A* gene in *Aspergillus fumigatus* clinical isolates during posaconazole prophylaxis: first case in Korea. Mycopathologia 183:731–736. https://doi.org/10.1007/s11046-018-0271-8

Leonardelli F, Theill L, Nardin ME, Macedo D, Dudiuk C, Mendez E, Gamarra S, Garcia-Effron G (2017) First itraconazole resistant *Aspergillus fumigatus* clinical isolate harbouring a G54E substitution in Cyp51Ap in South America. Rev Iberoam Micol 34:46–48. https://doi.org/10.1016/j.riam.2016.05.005

Leroux P, Walker AS (2011) Multiple mechanisms account for resistance to sterol 14alpha-demethylation inhibitors in field isolates of *Mycosphaerella graminicola*. Pest Manag Sci 67:44–59. https://doi.org/10.1002/ps.2028

Lestrade PP, Bentvelsen RG, Schauwvlieghe A, Schalekamp S, van der Velden W, Kuiper EJ, van Paassen J, van der Hoven B, van der Lee HA, Melchers WJG, de Haan AF, van der Hoeven HL, Rijnders BJA, van der Beek MT, Verweij PE (2019) Voriconazole resistance and mortality in invasive aspergillosis: a multicenter retrospective cohort study. Clin Infect Dis 68:1463–1471. https://doi.org/10.1093/cid/ciy859

Lockhart SR, Frade JP, Etienne KA, Pfaller MA, Diekema DJ, Balajee SA (2011) Azole resistance in *Aspergillus fumigatus* isolates from the ARTEMIS global surveillance study is primarily due to the TR/L98H mutation in the *cyp51A* gene. Antimicrob Agents Chemother 55:4465–4468. https://doi.org/10.1128/AAC.00185-11

Lockhart SR, Beer K, Toda M (2020) Azole-resistant *Aspergillus fumigatus*: what you need to know. Clin Microbiol Newsl 42:1–6. https://doi.org/10.1016/j.clinmicnews.2019.12.003

Mann PA, Parmegiani RM, Wei SQ, Mendrick CA, Li X, Loebenberg D, DiDomenico B, Hare RS, Walker SS, McNicholas PM (2003) Mutations in *Aspergillus fumigatus* resulting in reduced susceptibility to posaconazole appear to be restricted to a single amino acid in the cytochrome P450 14alpha-demethylase. Antimicrob Agents Chemother 47:577–581. https://doi.org/10.1128/aac.47.2.577-581.2003

Meis JF, Chowdhary A, Rhodes JL, Fisher MC, Verweij PE (2016) Clinical implications of globally emerging azole resistance in *Aspergillus fumigatus*. Philos Trans R Soc Lond Ser B Biol Sci 371:1. https://doi.org/10.1098/rstb.2015.0460

Mellado E, Diaz-Guerra TM, Cuenca-Estrella M, Rodriguez-Tudela JL (2001) Identification of two different 14-alpha sterol demethylase-related genes (*cyp51A* and *cyp51B*) in *Aspergillus fumigatus* and other *Aspergillus* species. J Clin Microbiol 39:2431–2438. https://doi.org/10.1128/JCM.39.7.2431-2438.2001

Mellado E, Garcia-Effron G, Alcazar-Fuoli L, Cuenca-Estrella M, Rodriguez-Tudela JL (2004) Substitutions at methionine 220 in the 14alpha-sterol demethylase (*cyp51A*) of *Aspergillus fumigatus* are responsible for resistance in vitro to azole antifungal drugs. Antimicrob Agents Chemother 48:2747–2750. https://doi.org/10.1128/AAC.48.7.2747-2750.2004

Meneau I, Coste AT, Sanglard D (2016) Identification of *Aspergillus fumigatus* multidrug transporter genes and their potential involvement in antifungal resistance. Med Mycol 54:616–627. https://doi.org/10.1093/mmy/myw005

Montesinos E (2003) Plant-associated microorganisms: a view from the scope of microbiology. Int Microbiol 6:221–223. https://doi.org/10.1007/s10123-003-0141-0

Moore CB, Novak-Frazer L, Muldoon E, Dunn KW, Masania R, Richardson MD, Rautemaa-Richardson R (2017) First isolation of the pan-azole-resistant *Aspergillus fumigatus* cyp51A TR46/Y121F/T289A mutant in a UK patient. Int J Antimicrob Agents 49:512–514. https://doi.org/10.1016/j.ijantimicag.2017.01.004

Morio F, Aubin GG, Danner-Boucher I, Haloun A, Sacchetto E, Garcia-Hermoso D, Bretagne S, Miegeville M, Le Pape P (2012) High prevalence of triazole resistance in *Aspergillus fumigatus*, especially mediated by TR/L98H, in a French cohort of patients with cystic fibrosis. J Antimicrob Chemother 67:1870–1873. https://doi.org/10.1093/jac/dks160

Mortensen KL, Mellado E, Lass-Flörl C, Rodriguez-Tudela JL, Johansen HK, Arendrup MC (2010) Environmental study of azole-resistant *Aspergillus fumigatus* and other aspergilli in Austria, Denmark, and Spain. Antimicrob Agents Chemother 54:4545–4549. https://doi.org/10.1128/AAC.00692-10

Mortensen KL, Jensen RH, Johansen HK, Skov M, Pressler T, Howard SJ, Leatherbarrow H, Mellado E, Arendrup MC (2011) *Aspergillus* species and other molds in respiratory samples from patients with cystic fibrosis: a laboratory-based study with focus on *Aspergillus fumigatus* azole resistance. J Clin Microbiol 49:2243–2251. https://doi.org/10.1128/JCM.00213-11

Nawrot U, Kurzyk E, Arendrup MC, Mroczynska M, Wlodarczyk K, Sulik-Tyszka B, Wroblewska M, Ussowicz M, Zdziarski P, Niewinska K, Brillowska-Dabrowska A (2018) Detection of Polish clinical *Aspergillus fumigatus* isolates resistant to triazoles. Med Mycol 56:121–124. https://doi.org/10.1093/mmy/myx012

O'Gorman CM, Fuller H, Dyer PS (2009) Discovery of a sexual cycle in the opportunistic fungal pathogen *Aspergillus fumigatus*. Nature 457:471–474. https://doi.org/10.1038/nature07528

Onishi K, Muhammad Sarumoh B, Hagiwara D, Watanabe A, Kamei K, Toyotome T (2017) Azole-resistant *Aspergillus fumigatus* containing a 34-bp tandem repeat in *cyp51A* promoter is isolated from the environment in Japan. Med Mycol J 23:579–581. https://doi.org/10.3314/mmj.17-00002

Ozmerdiven GE, Ak S, Ener B, Agca H, Cilo BD, Tunca B, Akalin H (2015) First determination of azole resistance in *Aspergillus fumigatus* strains carrying the TR34/L98H mutations in Turkey. J Infect Chemother 8:581–586. https://doi.org/10.1016/j.jiac.2015.04.012

Pagano L, Akova M, Dimopoulos G, Herbrecht R, Drgona L, Blijlevens N (2011) Risk assessment and prognostic factors for mould-related diseases in immunocompromised patients. J Antimicrob Chemother 66:5–14. https://doi.org/10.1093/jac/dkq437

Pelaez T, Gijon P, Bunsow E, Bouza E, Sanchez-Yebra W, Valerio M, Gama B, Cuenca-Estrella M, Mellado E (2012) Resistance to voriconazole due to a G448S substitution in *Aspergillus fumigatus* in a patient with cerebral aspergillosis. J Clin Microbiol 50:2531–2534. https://doi.org/10.1128/JCM.00329-12

Pelaez T, Monteiro MC, Garcia-Rubio R, Bouza E, Gomez-Lopez A, Mellado E (2015) First detection of *Aspergillus fumigatus* azole-resistant strain due to *cyp51A* TR46/Y121F/T289A in an azole-naive patient in Spain. New Microbes New Infect 6:33–34. https://doi.org/10.1016/j.nmni.2015.04.005

Peyton LR, Gallagher S, Hashemzadeh M (2015) Triazole antifungals: a review. Drugs Today (Barc) 51:705–718. https://doi.org/10.1358/dot.2015.51.12.2421058

Pinto E, Monteiro C, Maia M, Faria MA, Lopes V, Lameiras C, Pinheiro D (2018) *Aspergillus* species and antifungals susceptibility in clinical setting in the north of Portugal: cryptic species and emerging azoles resistance in *A. fumigatus*. Front Microbiol 9:1656. https://doi.org/10.3389/fmicb.2018.01656

Pontes L, Beraquet CAG, Arai T, Pigolli GL, Lyra L, Watanabe A, Moretti ML, Schreiber AZ (2020) *Aspergillus fumigatus* clinical isolates carrying *cyp51A* with TR34/L98H/S297T/F495I substitutions detected after four-year retrospective azole resistance screening in Brazil. Antimicrob Agents Chemother 64:e02059-19. https://doi.org/10.1128/AAC.02059-19

Prigitano A, Venier V, Cogliati M, De Lorenzis G, Esposto MC, Tortorano AM (2014) Azole-resistant *Aspergillus fumigatus* in the environment of northern Italy, May 2011 to June 2012. Euro Surveill 19:20747. https://doi.org/10.2807/1560-7917.es2014.19.12.20747

Prigitano A, Esposto MC, Biffi A, De Lorenzis G, Favuzzi V, Koncan R, Lo Cascio G, Barao Ocampo M, Colombo C, Pizzamiglio G, Romano L, Tortorano AM (2017) Triazole resistance in *Aspergillus fumigatus* isolates from patients with cystic fibrosis in Italy. J Cyst Fibros 16:64–69. https://doi.org/10.1016/j.jcf.2016.06.006

Prigitano A, Esposto MC, Romano L, Auxilia F, Tortorano AM (2019) Azole-resistant *Aspergillus fumigatus* in the Italian environment. J Glob Antimicrob Resist 16:220–224. https://doi.org/10.1016/j.jgar.2018.10.017

Reboux G, Gbaguidi-Haore H, Bellanger AP, Demonmerot F, Houdrouge K, Deconinck E, Bertrand X, Millon L (2014) A 10-year survey of fungal aerocontamination in hospital corridors: a reliable sentinel to predict fungal exposure risk? J Hosp Infect 87:34–40. https://doi.org/10.1016/j.jhin.2014.02.008

Ren J, Jin X, Zhang Q, Zheng Y, Lin D, Yu Y (2017) Fungicides induced triazole-resistance in *Aspergillus fumigatus* associated with mutations of TR46/Y121F/T289A and its appearance in agricultural fields. J Hazard Mater 326:54–60. https://doi.org/10.1016/j.jhazmat.2016.12.013

Resendiz-Sharpe A, Lagrou K, Meis JF, Chowdhary A, Lockhart SR, Verweij PE (2018) Triazole resistance surveillance in *Aspergillus fumigatus*. Med Mycol 56:83–92. https://doi.org/10.1093/mmy/myx144

Resendiz-Sharpe A, Mercier T, Lestrade PPA, van der Beek MT, von dem Borne PA, Cornelissen JJ, De Kort E, Rijnders BJA, Schauwvlieghe A, Verweij PE, Maertens J, Lagrou K (2019) Prevalence of voriconazole-resistant invasive aspergillosis and its impact on mortality in haematology patients. J Antimicrob Chemother 74:2759–2766. https://doi.org/10.1093/jac/dkz258

Riat A, Plojoux J, Gindro K, Schrenzel J, Sanglard D (2018) Azole resistance of environmental and clinical *Aspergillus fumigatus* isolates from Switzerland. Antimicrob Agents Chemother 62:e02088-17. https://doi.org/10.1128/AAC.02088-17

Rocchi S, Daguindau E, Grenouillet F, Deconinck E, Bellanger AP, Garcia-Hermoso D, Bretagne S, Reboux G, Millon L (2014a) Azole-resistant *Aspergillus fumigatus* isolate with the TR34/L98H mutation in both a fungicide-sprayed field and the lung of a hematopoietic stem cell transplant recipient with invasive aspergillosis. J Clin Microbiol 52:1724–1726. https://doi.org/10.1128/JCM.03182-13

Rocchi S, Reboux G, Larosa F, Scherer E, Daguindeau E, Berceanu A, Deconinck E, Millon L, Bellanger AP (2014b) Evaluation of invasive aspergillosis risk of immunocompromised patients alternatively hospitalized in hematology intensive care unit and at home. Indoor Air 24:652–661. https://doi.org/10.1111/ina.12108

Rocchi S, Poncot M, Morin-Crini N, Laboissiere A, Valot B, Godeau C, Lechenault-Bergerot C, Reboux G, Crini G, Millon L (2018) Determination of azole fungal residues in soils and detection of *Aspergillus fumigatus*-resistant strains in market gardens of Eastern France. Environ Sci Pollut Res 25:32015–32023. https://doi.org/10.1007/s11356-018-3177-6

Rocchi S, Morin-Crini N, Léchenault-Bergerot C, Godeau C, Fourmentin M, Millon L, Crini G (2019) Effet de l'interaction cyclodextrine-difénoconazole sur la croissance d'une moisissure responsable d'infections fongiques graves, Environnement, Risques & Santé.

Rohr JR, Barrett CB, Civitello DJ, Craft ME, Delius B, DeLeo GA, Hudson PJ, Jouanard N, Nguyen KH, Ostfeld RS, Remais JV, Riveau G, Sokolow SH, Tilman D (2019) Emerging human infectious diseases and the links to global food production. Nat Sustain 2:445–456. https://doi.org/10.1038/s41893-019-0293-3

Schoustra SE, Debets AJM, Rijs A, Zhang J, Snelders E, Leendertse PC, Melchers WJG, Rietveld AG, Zwaan BJ, Verweij PE (2019a) Environmental hotspots for azole resistance selection of *Aspergillus fumigatus*, the Netherlands. Emerg Infect Dis 25:1347–1353. https://doi.org/10.3201/eid2507.181625

Schoustra SE, Zhang J, Zwaan BJ, Debets AJF, Verweij P, Buijtenhuijs D, Rietveld AG (2019b) New insights in the development of azole-resistance in *Aspergillus fumigatus*. https://doi.org/10.21945/RIVM-2018-0131

Schwartz S, Ruhnke M, Ribaud P, Corey L, Driscoll T, Cornely OA, Schuler U, Lutsar I, Troke P, Thiel E (2005) Improved outcome in central nervous system aspergillosis, using voriconazole treatment. Blood 106:2641–2645. https://doi.org/10.1182/blood-2005-02-0733

Seufert R, Sedlacek L, Kahl B, Hogardt M, Hamprecht A, Haase G, Gunzer F, Haas A, Grauling-Halama S, MacKenzie CR, Essig A, Stehling F, Sutharsan S, Dittmer S, Killengray D, Schmidt D, Eskandarian N, Steinmann E, Buer J, Hagen F, Meis JF, Rath PM, Steinmann J (2018) Prevalence and characterization of azole-resistant *Aspergillus fumigatus* in patients with cystic fibrosis: a prospective multicentre study in Germany. J Antimicrob Chemother 73:2047–2053. https://doi.org/10.1093/jac/dky147

Sewell TR, Zhang Y, Brackin AP, Shelton JMG, Rhodes J, Fisher MC (2019a) Elevated prevalence of azole resistant *Aspergillus fumigatus* in urban versus rural environments in the United Kingdom. Antimicrob Agents Chemother 63:e00548–e00519. https://doi.org/10.1128/AAC.00548-19

Sewell TR, Zhu J, Rhodes J, Hagen F, Meis JF, Fisher MC, Jombart T (2019b) Nonrandom distribution of azole resistance across the global population of *Aspergillus fumigatus*. MBio 10:e00392-19. https://doi.org/10.1128/mBio.00392-19

Seyedmousavi S, Hashemi SJ, Zibafar E, Zoll J, Hedayati MT, Mouton JW, Melchers WJ, Verweij PE (2013) Azole-resistant *Aspergillus fumigatus*, Iran. Emerg Infect Dis 19:832–834. https://doi.org/10.3201/eid1905.130075

Sharma C, Hagen F, Moroti R, Meis JF, Chowdhary A (2015) Triazole-resistant *Aspergillus fumigatus* harbouring G54 mutation: is it de novo or environmentally acquired? J Glob Antimicrob Res 3:69–73. https://doi.org/10.1016/j.jgar.2015.01.005

Snelders E, van der Lee HAL, Kuijpers J, Rijs AJMM, Varga J, Samson RA, Mellado E, Donders ART, Melchers WJG, Verweij PE (2008) Emergence of azole resistance in *Aspergillus fumigatus* and spread of a single resistance mechanism. PLoS Med 5:e219. https://doi.org/10.1371/journal.pmed.0050219

Snelders E, Veld RAGH i't, Rijs AJMM, Kema GHJ, Melchers WJG, Verweij PE (2009) Possible environmental origin of resistance of *Aspergillus fumigatus* to medical triazoles. Appl Environ Microbiol 75:4053–4057. https://doi.org/10.1128/AEM.00231-09

Snelders E, Karawajczyk A, Schaftenaar G, Verweij PE, Melchers WJ (2010) Azole resistance profile of amino acid changes in *Aspergillus fumigatus cyp51A* based on protein homology modeling. Antimicrob Agents Chemother 54:2425–2430. https://doi.org/10.1128/AAC.01599-09

Snelders E, Camps SMT, Karawajczyk A, Schaftenaar G, Kema GHJ, van der Lee HA, Klaassen CH, Melchers WJG, Verweij PE (2012) Triazole fungicides can induce cross-resistance to medical triazoles in *Aspergillus fumigatus*. PLoS One 7:e31801. https://doi.org/10.1371/journal.pone.0031801

Sułowicz S, Cycoń M, Piotrowska-Seget Z (2016) Non-target impact of fungicide tetraconazole on microbial communities in soils with different agricultural management. Ecotoxicology 25:1047–1060. https://doi.org/10.1007/s10646-016-1661-7

Tangwattanachuleeporn M, Minarin N, Saichan S, Sermsri P, Mitkornburee R, Gross U, Chindamporn A, Bader O (2017) Prevalence of azole-resistant *Aspergillus fumigatus* in the environment of Thailand. Med Mycol 55:429–435. https://doi.org/10.1093/mmy/myw090

Tashiro M, Izumikawa K, Minematsu A, Hirano K, Iwanaga N, Ide S, Mihara T, Hosogaya N, Takazono T, Morinaga Y, Nakamura S, Kurihara S, Imamura Y, Miyazaki T, Nishino T, Tsukamoto M, Kakeya H, Yamamoto Y, Yanagihara K, Yasuoka A, Tashiro T, Kohno S (2012) Antifungal susceptibilities of *Aspergillus fumigatus* clinical isolates obtained in Nagasaki, Japan. Antimicrob Agents Chemother 56:584–587. https://doi.org/10.1128/AAC.05394-11

Tilman D, Cassman KG, Matson PA, Naylor R, Polasky S (2002) Agricultural sustainability and intensive production practices. Nature 418:671–677. https://doi.org/10.1038/nature01014

Tornimbene B, Eremin S, Escher M, Griskeviciene J, Manglani S, Pessoa-Silva CL (2018) WHO global antimicrobial resistance surveillance system early implementation 2016–17. Lancet Infect Dis 18:241–242. https://doi.org/10.1016/S1473-3099(18)30060-4

Toyotome T, Fujiwara T, Kida H, Matsumoto M, Wada T, Komatsu R (2016) Azole susceptibility in clinical and environmental isolates of *Aspergillus fumigatus* from eastern Hokkaido, Japan. J Infect Chemother 22:648–650. https://doi.org/10.1016/j.jiac.2016.03.002

Toyotome T, Hagiwara D, Kida H, Ogi T, Watanabe A, Wada T, Komatsu R, Kamei K (2017) First clinical isolation report of azole-resistant *Aspergillus fumigatus* with TR34/L98H-type mutation in Japan. J Infect Chemother 23:579–581. https://doi.org/10.1016/j.jiac.2016.12.004

Ullmann AJ, Aguado JM, Arikan-Akdagli S, Denning DW, Groll AH, Lagrou K, Lass-Florl C, Lewis RE, Munoz P, Verweij PE, Warris A, Ader F, Akova M, Arendrup MC, Barnes RA, Beigelman-Aubry C, Blot S, Bouza E, Bruggemann RJM, Buchheidt D, Cadranel J, Castagnola E, Chakrabarti A, Cuenca-Estrella M, Dimopoulos G, Fortun J, Gangneux JP, Garbino J, Heinz WJ, Herbrecht R, Heussel CP, Kibbler CC, Klimko N, Kullberg BJ, Lange C, Lehrnbecher T, Loffler J, Lortholary O, Maertens J, Marchetti O, Meis JF, Pagano L, Ribaud P, Richardson M, Roilides E, Ruhnke M, Sanguinetti M, Sheppard DC, Sinko J, Skiada A, Vehreschild M, Viscoli C, Cornely OA (2018) Diagnosis and management of *Aspergillus* diseases: executive summary of the 2017 ESCMID-ECMM-ERS guideline. Clin Microbiol Infect 24:e1–e38. https://doi.org/10.1016/j.cmi.2018.01.002

van de Veerdonk FL, Gresnigt MS, Romani L, Netea MG, Latge JP (2017) *Aspergillus fumigatus* morphology and dynamic host interactions. Nat Rev Microbiol 15:661–674. https://doi.org/10.1038/nrmicro.2017.90

Van der Linden J, Snelders E, Kampinga GA, Rijnders BJ, Mattsson E, Debets-Ossenkopp YJ, Kuijper J, Van Tiel FH, Melchers WJ, Verweij PE (2011) Clinical implications of azole resistance in *Aspergillus fumigatus*, the Netherlands, 2007–2009. Emerg Infect Dis 17:1846–1854. https://doi.org/10.3201/eid1710.110226

van der Linden JW, Camps SM, Kampinga GA, Arends JP, Debets-Ossenkopp YJ, Haas PJ, Rijnders BJ, Kuijper EJ, van Tiel FH, Varga J, Karawajczyk A, Zoll J, Melchers WJ, Verweij PE (2013) Aspergillosis due to voriconazole highly resistant *Aspergillus fumigatus* and recovery of genetically related resistant isolates from domiciles. Clin Infect Dis 57:513–520. https://doi.org/10.1093/cid/cit320

van Paassen J, Russcher A, Wingerden AW I't V-v, Verweij PE, Kuijper EJ (2016) Emerging aspergillosis by azole-resistant *Aspergillus fumigatus* at an intensive care unit in the Netherlands, 2010–2013. Euro Surveill 21:30300. https://doi.org/10.2807/1560-7917.ES.2016.21.30.30300

Vermeulen E, Maertens J, De Bel A, Nulens E, Boelens J, Surmont I, Mertens A, Boel A, Lagrou K (2015) Nationwide surveillance of azole resistance in *Aspergillus* diseases. Antimicrob Agents Chemother 59:4569–4576. https://doi.org/10.1128/AAC.00233-15

Verweij PE (2017) Azole-resistance selection in *Aspergillus fumigatus*, final report. Annex to 27858, No. 411. The Centre for Agriculture and Environment, Laboratory of Genetics of Wageningen University, Department of Medical Microbiology of Radboud University Medical Center. Report available online: https://www.parlementairemonitor.nl/9353000/1/j9vvij5epmj1ey0/vkjwoy3asfzz

Verweij PE, Mellado E, Melchers WJ (2007) Multiple-triazole-resistant aspergillosis. In N Engl J Med 1481–1483. United States. https://doi.org/10.1056/NEJMc061720

Verweij PE, Snelders E, Kema GH, Mellado E, Melchers WJ (2009) Azole resistance in *Aspergillus fumigatus*: a side-effect of environmental fungicide use? Lancet Infect Dis 9:789–795. https://doi.org/10.1016/S1473-3099(09)70265-8

Verweij PE, Mellado E, Melchers WJ (2017) Multiple-triazole-resistant aspergillosis. N Engl J Med 356:1481–1483. https://doi.org/10.1056/NEJMc061720

Wang HC, Huang JC, Lin YH, Chen YH, Hsieh MI, Choi PC, Lo HJ, Liu WL, Hsu CS, Shih HI, Wu CJ, Chen YC (2018) Prevalence, mechanisms and genetic relatedness of the human pathogenic fungus *Aspergillus fumigatus* exhibiting resistance to medical azoles in the environment of Taiwan. Environ Microbiol 20:270–280. https://doi.org/10.1111/1462-2920.13988

Wiederhold NP, Patterson TF (2015) Emergence of azole resistance in *Aspergillus*. Semin Respir Crit Care Med 36:673–680. https://doi.org/10.1055/s-0035-1562894

Wiederhold NP, Garcia Gil V, Gutierrez F, Lindner JR, Albataineh MT, McCarthy DI, Sanders C, Fan H, Fothergill AW, Sutton DA (2015) First detection of TR34/L98H and TR46/Y121F/T289A *cyp51* mutations in *Aspergillus fumigatus* isolates in the United States. J Clin Microbiol 54:168–171. https://doi.org/10.1128/JCM.02478-15

Wu CJ, Wang HC, Lee JC, Lo HJ, Dai CT, Chou PH, Ko WC, Chen YC (2015) Azole-resistant *Aspergillus fumigatus* isolates carrying TR34/L98H mutations in Taiwan. Mycoses 58:544–549. https://doi.org/10.1111/myc.12354

Zhang J, Snelders E, Zwaan BJ, Schoustra SE, Meis JE, van Dijk K, Hagen F, van der Beek MT, Kampinga GA, Zoll J, Melchers WJG, Verweij PE, Debets AJM (2017) A novel environmental azole resistance mutation in *Aspergillus fumigatus* and a possible role of sexual reproduction in its emergence. MBio 8:e00791–e00717. https://doi.org/10.1128/mBio.00791-17

Chapter 4
The Fate of Antibiotic-Resistant Bacteria in the Environment

Charles P. Henriot, Daniel Martak, Christophe Dagot, Fabienne Petit, Edward Topp, Xavier Bertrand, Gudrun Bornette, and Didier Hocquet

Abstract The discovery of antibiotics in the twentieth century largely improved health conditions. However, the increasing and disproportionate use of these drugs has led to the emergence of antibiotic-resistant bacteria. Currently, these antibiotic-resistant bacteria are spreading within the human community. Our environment is polluted by these emerging contaminants, mostly through the effluents of wastewater treatment plants and soil amendments with residual organic waste. Farming, intensive pasturing, and aquacultures are also sources of antibiotic-resistant bacteria in the environment. Once released into the environment, antibiotic-resistant bacteria spread mostly in aquatic ecosystems where they can persist in culturable or non-culturable states. Together with antibiotic-resistant bacteria, antimicrobials (i.e.

C. P. Henriot (✉) · G. Bornette
UMR 6249 Chrono-environnement, CNRS, Université Bourgogne Franche-Comté, Besançon, France
e-mail: charles.henriot@univ-fcomte.fr; gudrun.bornette@univ-fcomte.fr

D. Martak · X. Bertrand · D. Hocquet
UMR 6249 Chrono-environnement, CNRS, Université Bourgogne Franche-Comté, Besançon, France

Hygiène Hospitalière, Centre Hospitalier Universitaire de Besançon, Besançon, France
e-mail: dmartak@chu-besancon.fr; xbertrand@chu-besancon.fr; dhocquet@chu-besancon.fr

C. Dagot
UMR Inserm 1092, Agents anti-microbiens, CBRS – Université de Limoges, Limoges, France
e-mail: christophe.dagot@unilim.fr

F. Petit
Normandie Université, UniRouen, UniCaen, CNRS UMR M2C, Rouen, France

Sorbonne Universités, UPMC, CNRS, EPHE, UMR 7619 METIS, Paris, France
e-mail: fabienne.petit@univ-rouen.fr

E. Topp
Agriculture and Agri-Food Canada, University of Western Ontario, London, ON, Canada
e-mail: Ed.topp@canada.ca

N. Morin-Crini et al. (eds.), *Emerging Contaminants Vol. 1*, Environmental Chemistry for a Sustainable World 65,
https://doi.org/10.1007/978-3-030-69079-3_4

antibiotics, metals and biocides) are released most often from the same sources and can exert selective pressure on bacterial communities. To what extent are environments contaminated with antibiotic-resistant bacteria a new risk for human populations? Humans in low and middle income countries are more frequently exposed than humans in high income countries, benefiting from the efficient collection and treatment of wastewater. Indeed, various management measures can be implemented to reduce the likelihood of human exposure to environmental antibiotic resistance and the subsequent risk to human health; such measures include efficient wastewater treatments, the implementation of local policies and knowledge exchange. Moreover, as the spread of antibiotic resistance is both a public and an environmental health concern, this issue can only be understood through a multidisciplinary approach that involves all stakeholders.

Keywords Antibiotic resistance · Antibiotic-resistant bacteria · Wastewater treatment plant · Environmental exposure · Water environment

Abbreviations

ARB	antibiotic-resistant bacteria
ARG	antibiotic resistance gene
CFU	colony-forming unit
ESBL	extended-spectrum β-lactamase
VBNC	viable but non-culturable
WWTP	wastewater treatment plant

4.1 Introduction

The discovery of antibiotics in the twentieth century largely improved health conditions. Antibiotics are natural, semi-synthetic or synthetic molecules capable of destroying bacteria (bactericidal antibiotics) or inhibiting their growth (bacteriostatic antibiotics). A microorganism is said to be resistant when it multiplies in the presence of a given antibiotic. Antibiotic-resistant bacteria are thus bacteria able to grow in the presence of a given antibiotic. This ability to resist can be intrinsic and specific to all the isolates of a given bacterial species or genus (natural resistance). Metagenomic approaches have shown that the soil microbiota, the source of many antibiotic-producing microorganisms, is a reservoir of antibiotic-resistant genes, with nucleotidic sequences close or identical to those found in clinically important pathogenic bacteria (Forsberg et al. 2012; Nesme and Simonet 2015; Jiang et al. 2017). In contrast, acquired resistance differs from natural resistance since it is only observed in a subpopulation of a species that is susceptible to the considered antibiotic. Acquired resistance relies either on the mutation of chromosomal genes (modification of the genetic capital) or the acquisition of genes by horizontal gene transfer. Among human pathogens, some bacterial species are intrinsically resistant to certain antibiotics (e.g. *Pseudomonas aeruginosa*, *Klebsiella pneumoniae*, and *Enterobacter cloacae*) and can acquire resistance to other antibiotics. However, the majority of human pathogenic bacteria are not intrinsically resistant but can acquire resistance to different antibiotics (e.g. *Escherichia coli* and *Staphylococcus aureus*).

The misuse and increasing use of antibiotics has led to the emergence of antibiotic-resistant bacteria (ARB), mainly within bacterial strains of clinical relevance. ARB are therefore emerging contaminants of concern. Antibiotics give ARB a selective advantage over susceptible bacteria (Fig. 4.1). Antibiotics have become less effective at treating bacterial infections, and the proportion of bacteria resistant to a specific antibiotic has increased (Cecchini et al. 2015). The pace at which we are discovering new antibiotics has slowed, and bacteria have invariably become resistant to new antibiotics (Review on Antimicrobial Resistance 2016). Cecchini et al. (2015) highlighted three major threats related to antibiotic-resistant microorganisms: (i) antibiotic-resistant microorganisms prolong morbidity and increase the time spent ill with the infection, (ii) antibiotic-resistant microorganisms increase the likelihood of developing other comorbidities or complications, and (iii) by requiring more intensive treatments, patients are more likely to experience adverse reactions or secondary effects. The authors estimated that patients infected with antibiotic-resistant microorganisms are up to three times more likely to die than are other patients (Cecchini et al. 2015). More than 25,000 European citizens are estimated to die annually due to ARB infections (European Centre for Disease Prevention and Control, European Medicines Agency 2009). Globally, antimicrobial-resistant infections are estimated to cause over 0.7 million deaths per year (Cecchini et al. 2015), although reliable estimates of the true number are scarce. Between 2007 and 2015 in the European Union, cases attributed to ARB infections increased from ~240,000 to 603,000 cases, and the associated deaths increased from ~11,000 to

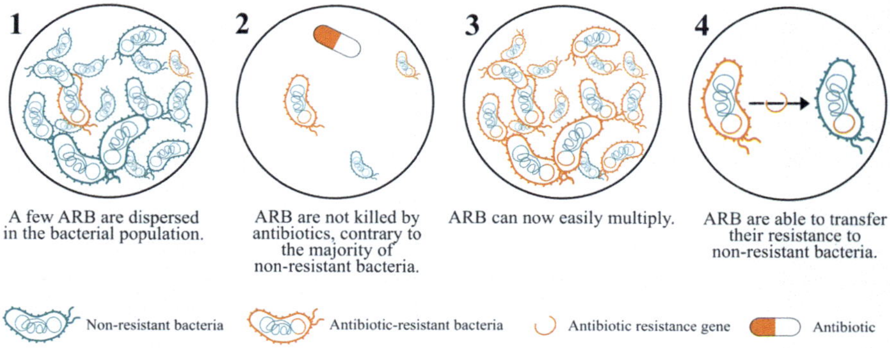

Fig. 4.1 A schematic representation illustrating the selective advantage of antibiotic-resistant bacteria and explaining their spread in the presence of antibiotics compared with that of susceptible bacteria of the same species (*ARB* antibiotic-resistant bacteria; adapted from the World Health Organization)

27,000 (Cassini et al. 2018). A continued increase in resistance by 2050 could result in the death of ten million people worldwide each year (Review on Antimicrobial Resistance 2016). Africa and Asia are expected to be particularly impacted by antimicrobial resistance. Hence, by 2050, antimicrobial-resistant infections could cause 22,000 deaths yearly in Oceania, 317,000 deaths in North America, 390,000 deaths in Europe, 392,000 deaths in Latin America, 4,150,000 deaths in Africa, and 4,730,000 deaths in Asia (Review on Antimicrobial Resistance 2016).

Among bacteria, the class of *γ-proteobacteria*, especially the order *Enterobacteriaceae*, contains a particularly large number of human pathogens that may acquire antibiotic resistance and therefore pose a major threat to human populations (Cassini et al. 2018). The authors showed that four ARB populations were particularly harmful to humans in Europe: third-generation cephalosporin-resistant *E. coli*, methicillin-resistant *S. aureus*, carbapenem-resistant *P. aeruginosa* and third-generation cephalosporin-resistant *K. pneumoniae*. In the same geographical area, the authors estimated that these four species annually cause approximately 9100, 7000, 4200, and 3700 deaths, respectively (Cassini et al. 2018). Bacterial resistance to antibiotics has rapidly evolved over time. Hence, the most prevalent extended-spectrum β-lactamase (ESBL)-producing *Enterobacteriaceae*, which are able to hydrolyse third-generation cephalosporins, were *Enterobacter aerogenes* (34%) and *E. coli* (25%) in healthcare facilities in 2003, whereas *E. coli* (60%) and *K. pneumoniae* (18%) were dominant only seven years later in 2010 (Carbonne et al. 2012). In the G7 countries (Canada, France, Germany, Italy, Japan, the United Kingdom and the USA), Cecchini et al. (2015) estimated that in clinical settings, 12.1% of *E. coli* were resistant to third-generation cephalosporin and 27.7% were resistant to fluoroquinolones; 17.3% of *K. pneumoniae* were resistant to third-generation cephalosporin and 5.5% were resistant to carbapenems; and 30.5% of *S. aureus* were resistant to methicillin. In low and middle income countries the data are scarce (World Health Organization 2014). Some global ARB prevalence data (World Health Organization 2014) are reported in Table 4.1. Currently, these ARB are spreading within the human community. In 2006, 0.6% of healthy subjects living in the Paris area were carrying ESBL *E. coli* in their gut, while in 2011, this rate was 6%, reflecting a ten-fold increase in 5 years (Nicolas-Chanoine et al. 2012).

The intestinal microbiome is the primary reservoir of human-associated bacteria engaged in mutual symbiotic or commensal symbiotic relationships with their host. Thus, the intestinal microbiota of antibiotic-treated humans or livestock become ecological niches where ARB are selected (Casals-Pascual et al. 2018). For these reasons, hospitals (Hocquet et al. 2016) and intensive farms (McKinney et al. 2018; Topp et al. 2018) are the two main reservoirs with the highest prevalence of ARB. Bacteria from the intestinal microbiota, among which ARB, are dispersed into the environment *via* faeces. Therefore, the main sources of ARB emissions into the environment are human and animal excretions. Human excretions mostly end up in sewage systems in high income countries. Animal manures may in some instances be treated prior to land application. Antibiotics and other selective agents (e.g. biocides and trace metals) are also released in wastewater and may exert co-selective pressure on ARB. Bacterial contaminants then reach wastewater treatment plants

Table 4.1 Summary table of the antibiotic-resistant bacteria prevalence in the six regions of the world

	Third-generation cephalosporin-resistant *E. coli* (%)	Fluoroquinolone-resistant *E. coli* (%)	Third-generation cephalosporin-resistant *K. pneumoniae* (%)	Carbapenem-resistant *K. pneumoniae* (%)	β-lactam-resistant *S. aureus* (%)
African Region National data	2–70 (n = 13)	14–71 (n = 14)	8–77 (n = 13)	0–4 (n = 4)	12–80 (n = 9)
Region of the Americas National data or report to ReLAVRA	0–48 (n = 14)	8–58 (n = 16)	4–71 (n = 17)	0–11 (n = 17)	21–90 (n = 15)
Eastern Mediterranean Region National data	22–63 (n = 4)	21–62 (n = 4)	22–50 (n = 4)	0–54 (n = 4)	10–53 (n = 4)
European Region National data or report to EARS-Net	3–82 (n = 35)	8–48 (n = 35)	2–82 (n = 33)	0–68 (n = 33)	0.3–60 (n = 36)
South-East Asia Region National reports	16–68 (n = 5)	32–64 (n = 5)	34–81 (n = 4)	0–8 (n = 4)	10–26 (n = 3)
Western Pacific Region National data	0–77 (n = 13)	3–96 (n = 16)	1–72 (n = 14)	0–8 (n = 9)	4–84 (n = 16)

Source: World Health Organization (2014)
ReLAVRA Latin American Antimicrobial Resistance Surveillance Network, *EARS-Net* European Antimicrobial Resistance Surveillance Network
% range, n = number of countries considered

(WWTPs), which historically were not designed to mitigate contamination by ARB. A non-negligible proportion of the bacteria entering the WWTP are ultimately released into the environment (rivers, coastal waters, lakes) (Marti et al. 2014a). In high income countries, treated wastewater is thus the main source of ARB contamination in aquatic environments (Marti et al. 2014a; Guo et al. 2017). In other regions of the world, where basic hygiene infrastructures may be poorly developed and/or accessible, the situation is quite different. Indeed, approximately 2.5 billion people lack improved sanitation (World Health Organization and United Nations Children's Fund 2013). A sizeable part of the population practices open defecation (e.g. Niger: 68%, Chad: 67%, South Soudan: 63% and India: 26%; World Health Organization 2019), thus releasing ARB directly in the environment. The issue of ARB dispersal and the associated environmental contamination is a contemporary issue that has been of growing interest for only a few decades. The processes of dispersion, survival and selection in sewage systems – from the sewer entrance to the WWTP effluent – and in the environment still raise considerable public health concerns.

This chapter focuses on the release and dispersal of antibiotic-resistant bacteria from wastewater into the environment and the corresponding human-associated likelihood of exposure to environmental antibiotic resistance. Here, we focus on the four antibiotic-resistant bacteria that account for the major part of the resistance burden: *E. coli*, *S. aureus*, *P. aeruginosa* and *K. pneumoniae* (Cassini et al. 2018). To illustrate the release and spread of antibiotic resistance in the environment, we will also describe the spread of other bacterial species and antibiotic resistance genes. This chapter will first focus on antibiotic-resistant bacteria in hospital wastewater and urban wastewater before considering their fate in wastewater treatment plants and their release in the environment. Sources of environmental contamination other than wastewater treatment plants will also be briefly discussed. The dispersion and survival of antibiotic-resistant bacteria in several aquatic and terrestrial environments will then be studied. The resulting environmental risk associated with these antibiotic-resistant bacteria will be investigated. Finally, we will briefly discuss the mitigation processes that could reduce the environmental exposure to antibiotic-resistant bacteria.

4.2 Contamination of Sewage Systems by Antibiotic-Resistant Bacteria

Human faeces are a major source of antibiotic-resistant bacteria. Faeces end up in the sewage system before being treated in WWTPs. Two types of wastewaters will be of concern in this section: hospital wastewater and urban wastewater (Fig. 4.2). Together, hospital and urban wastewater combine to form the raw wastewater flow that enters WWTPs, as described in the following section. Urban wastewater contains antibiotics and antibiotic-resistant bacteria excreted by the human community. Antibiotics are widely used in hospitals, where the antibiotic-resistant bacteria prevalence is *de facto* higher than that in the community. Hospital wastewater is therefore expected to be more contaminated by antibiotic-resistant bacteria than is urban wastewater. Hospital wastewater is generally not treated before reaching a WWTP and is mixed with urban wastewater in the sewage system.

In high income countries, most houses are connected to a sewage system (e.g. ~85% in France (Ministère de la transition écologique et solidaire et Ministère des solidarités et de la santé 2019)). Other houses use non-collective sewage systems to treat their wastewater (mostly septic tanks). Their environmental impact in terms of ARB dispersion is very heterogeneous in view of the different contexts and treatment systems and such variations have been poorly studied. For these reasons, this chapter will not address this issue.

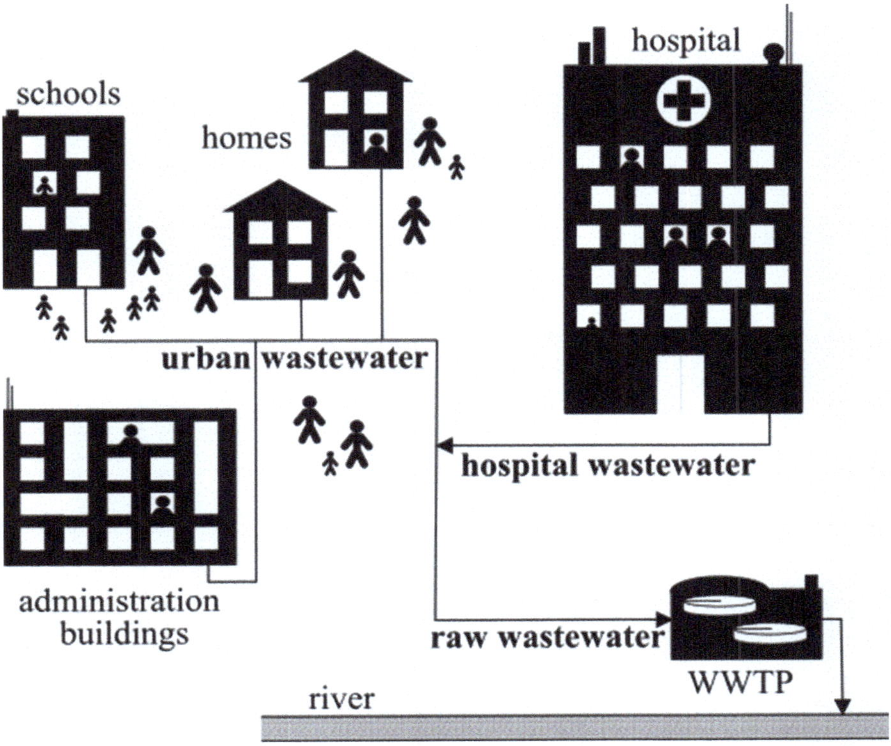

Fig. 4.2 Simplified scheme of a sewage system. Hospital wastewater reaches urban wastewater to form the raw wastewater. The raw wastewater is treated at a wastewater treatment plant, where it releases the treated outflow into the environment (*WWTP* wastewater treatment plant)

4.2.1 Hospital Wastewater

Servais and Passerat (2009) found a higher prevalence of *E. coli* strains resistant to at least one antibiotic of clinical relevance in hospital wastewater (71%) than in urban wastewater effluent (44%). Some authors found no difference in total *E. coli* concentrations in hospital and urban wastewater but found that the proportion of ESBL-producing *E. coli* was higher in hospital wastewater than in urban wastewater (7.55% *vs.* 0.10% of the total *E. coli* strains, respectively, in Bréchet et al. (2014), or 3.8–39% *vs.* 0.1–19% of the total *E. coli* strains (10^5–10^6 in both urban and hospital wastewaters), respectively, in Hocquet et al. (2016)). *P. aeruginosa* concentrations have been found to be higher in hospital wastewater than in urban wastewater, and *P. aeruginosa* found in hospital wastewater was more frequently resistant to one or more antibiotics (Slekovec et al. 2012). ARB are therefore more concentrated in hospital wastewater than in urban wastewater (Fig. 4.3). In German clinical/urban wastewater, 28.2% of all isolates harboured a carbapenemase-encoding gene, whereas this proportion was 1.4% in rural systems, identifying clinical/urban

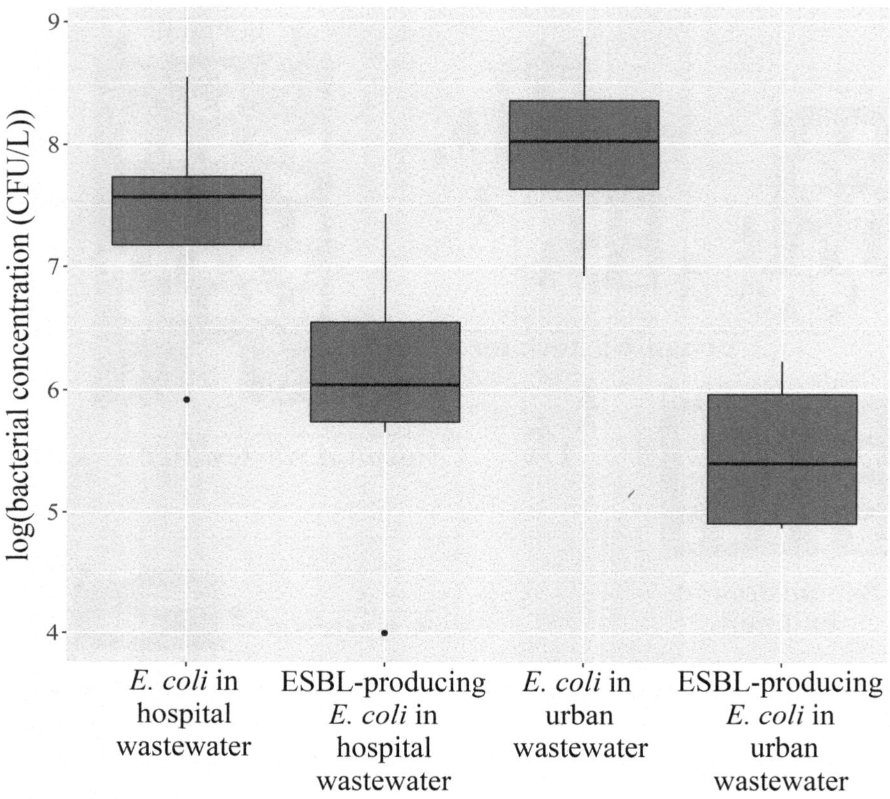

Fig. 4.3 Comparison of *E. coli* and ESBL-producing *E. coli* in hospital and urban wastewater (*CFU* colony forming units, *ESBL* extended-spectrum β-lactamase; data collected from Servais and Passerat 2009; Galvin et al. 2010; Oberlé et al. 2012; Bréchet et al. 2014; Kwak et al. 2015; Drieux et al. 2016)

wastewater as a source of dissemination of bacteria resistant to last-resort antibiotics in the environment (Müller et al. 2018). Similarly, Cahill et al. (2019) reported that 59% and 14% of *Enterobacterales* were resistant to ertapenem and/or meropenem in hospital wastewater and prehospital wastewater, respectively. Furthermore, Paulus et al. (2019) detected the presence of some antibiotics (namely, fluconazole, sulfaclozine and trimethoprim) only in hospital wastewater. These examples show high and specific contamination levels in hospital wastewater by antibiotics and ARB compared to those in urban wastewater.

The minimal selective concentration is the lowest antibiotic concentration at which the bacteria resistant to an antibiotic are favoured compared to the susceptible subpopulation. In other words, the minimal selective concentration is the minimum antibiotic concentration at which the biological cost of resistance is compensated. In addition, Gullberg et al. (2011) reported that subinhibitory concentrations of antibiotics can also select ARB. Hence, antibiotic concentrations found in hospital wastewater could be sufficient for selecting new ARB and to favour their growth (Rodriguez-Mozaz et al. 2015; Buelow et al. 2018). For example, Paulus et al. (2019) reported ciprofloxacin concentrations of 2.7–3.7μg/L in two hospital wastewaters in the Netherlands. These concentrations are much higher than the minimum inhibitory concentration of ciprofloxacin for wild-type *E. coli* reported by Gullberg et al. (2011), which was 0.023μg/L.

4.2.2 Urban Wastewater

Urban wastewater is relatively more contaminated by faecal bacteria but is less contaminated by ARB than is hospital wastewater, as described in the previous section (Servais and Passerat 2009; Slekovec et al. 2012; Bréchet et al. 2014; Hocquet et al. 2016; Fig. 4.3). However, urban wastewater has been described as more diverse than hospital wastewater in terms of antibiotic resistance (Stalder et al. 2013). Furthermore, urban wastewater includes much larger volumes of wastewater than hospital wastewater. For example, hospital wastewater was reported to represent only 2.2% of the volume of raw wastewater in France (Kümmerer 2004; Buelow et al. 2018), reflecting a much higher absolute number of ARB in urban wastewater. Therefore, the majority of ARB that reach WWTPs come from urban wastewater.

4.2.3 Would It Be Preferable to Treat Hospital Wastewater Separately from Urban Wastewater?

In the influents of three Austrian WWTPs, the highest *E. coli* resistance rates were found at the only WWTP treating both hospital and urban wastewater (Reinthaler et al. 2003). However, Buelow et al. (2018) showed that the relative abundance of human-associated ARB at the influent of WWTPs was comparable with or without hospital wastewater. The authors reported that even if they release high loads of ARB, hospitals contributed little to the quantity and diversity of ARB in the investigated sewage systems. Buelow et al. (2020) monitored the microbiota and resistome dynamics of untreated and treated hospital and urban wastewaters over three years. Each wastewater presented a robust, distinct and stable antibiotic resistance signature over time. The different signatures mirrored the differences in antibiotic exposure for each wastewater type. The authors also studied the resistome of mixed hospital and urban wastewaters for one year: the resistome of the mixed wastewater closely resembled that of urban wastewater. Although Buelow et al. (2020) confirmed high rates of antibiotic resistance in hospital wastewater, they also highlighted that separate treatments of hospital and urban wastewater would not reduce the release of ARB in the environment. Since the volume ratio between each type of wastewater used in this study is much higher than that reported in most sewage systems, the contribution of hospital-specific WWTPs would be limited. Overall, the following conclusions can be summarized as follows: (i) hospital wastewater contains specific antibiotic-resistant bacteria due to the use of hospital-specific antibiotics; (ii) the load of ARB is higher in hospital wastewater than in urban wastewater; and (iii) the high volume of urban wastewater (compared to that of hospital wastewater) makes it the major contributor to antibiotic resistance in the raw wastewater entering WWTPs.

4.3 Wastewater Treatment Plants Effluents as the Main Source of Antibiotic-Resistant Bacteria in the Environment

The raw wastewater is treated by wastewater treatment plants (Fig. 4.2) before the discharge of the treated effluents into the receiving environment. This process consists of a multitude of different treatment systems depending on several parameters (e.g. the specific contamination of the raw wastewater, the localization and the financial cost). In general, a WWTP implements the following five treatment steps (Crini and Lichtfouse 2019) (Fig. 4.4): (1) pre-treatment, which consists of grit removal, flow equalization, and fat and grease removal; (2) primary treatment, which includes a primary clarifier: simple decantation (sedimentation of a part of the solids present in the raw wastewater) with the collection of sludge; (3) secondary treatment, which usually consists of attached or suspended growth systems (including activated sludge) in which microorganisms (mainly bacteria and protozoa) consume biodegradable soluble organic contaminants; a secondary clarifier

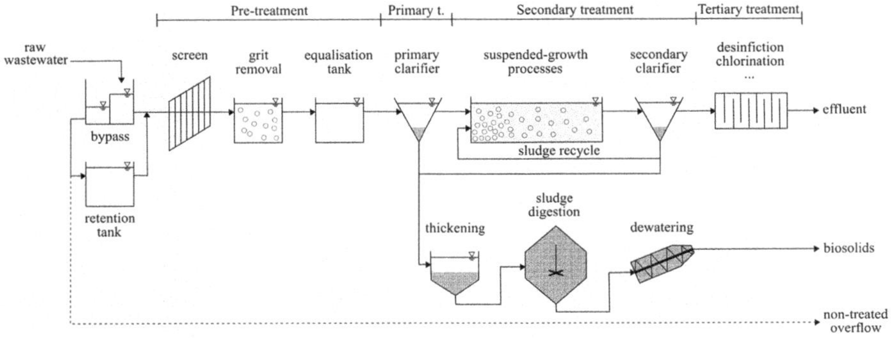

Fig. 4.4 General scheme of a conventional wastewater treatment plant (the untreated raw wastewater enters the WWTP; in combined sewage systems, a part of the entering raw wastewater can directly be rejected in the environment without any treatment, see the end of Sect. 4.3.2; once in a WWTP, the influent wastewater undergoes several treatment steps and is separated between the aqueous and solid phases, thus producing two types of output: aqueous effluents and biosolids)

usually follows and collects the sludge formed during the previous biological process; (4) in some cases, a third and even a fourth treatment is used to further improve the effluent quality before the effluent is discharged into the receiving environment; these treatments can consist of a variety of physical, chemical and biological advanced wastewater processes; and (5) sludge treatment, which includes thickening, dewatering, and eventual drying before agricultural (spreading) or thermal (incineration) recovery. In this chapter, sewage sludge refers to the solids separated during treatment at WWTPs, while biosolids refer to treated sewage sludge.

The main role of WWTPs is to remove most of the organic matter and nutrients present in raw wastewater prior to discharge into the receiving environment, usually a stream or river. The performance of a WWTP is generally monitored using global parameters such as dissolved oxygen, the oxidation-reduction potential, the pH and nutrients (mostly ammonia and nitrogen) (Martín de la Vega and Jaramillo-Morán 2018). The levels of human-associated bacteria are usually not monitored, simply because WWTPs were not designed to treat such bacteria and because clearance and monitoring would add a significant additional cost.

Not all bacteria survive in raw wastewater before reaching WWTPs. A Swedish study showed that the proportion of human-associated bacteria was reduced by at least 40% in the influent of WWTPs compared to the level in human faeces (Bengtsson-Palme et al. 2016). This decrease in the faecal bacteria abundance is due to a shift from anaerobic to aerobic conditions in raw wastewater, which leads to a shift from obligate anaerobic bacteria to facultative anaerobes (Bengtsson-Palme et al. 2016).

4.3.1 Are Antibiotic-Resistant Bacteria Enriched in Wastewater Treatment Plants?

This section first focuses on the aqueous phase (wastewater), from the influent to the effluent of the WWTP, before examining the solid phase (sludge) during the sludge treatment.

In Wastewater Most studies on the removal of ARB by WWTPs have shown good absolute clearance rates. Reinthaler et al. (2003) reported a 2.3-log reduction in the total *E. coli* concentration in the effluents of three studied WWTPs and a decrease in the resistance rates of *E. coli*. In a WWTP in Tokyo (Japan), Iwane et al. (2001) assessed coliform and *E. coli* resistance to a panel of antibiotics and found a decrease in the ARB loads during the treatment process. The percentage of *E. coli* isolates resistant to one or more antibiotics was less than two-thirds in the effluent compared with that in the influent. Schwartz et al. (2003) found that the WWTP reduced the release of vancomycin-resistant enterococci and β-lactam-hydrolysing *Enterobacteriaceae* into the aquatic environment. Overall, ARB loads are reduced by WWTPs (Fig. 4.5).

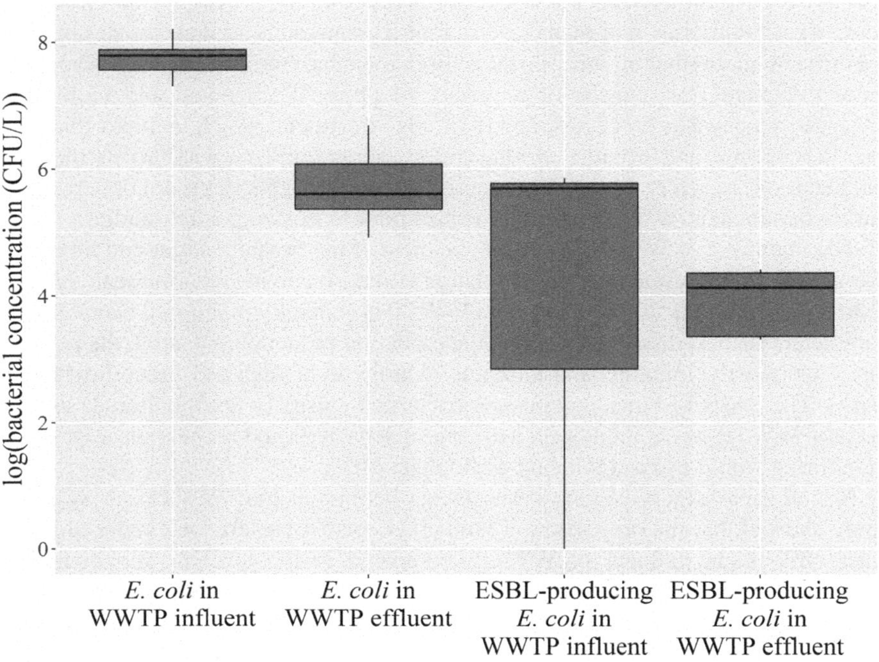

Fig. 4.5 Comparison of *E. coli* and ESBL-producing *E. coli* in wastewater treatment plant influent and effluent (*ESBL* extended-spectrum β-lactamase, *WWTP* wastewater treatment plant; data collected from Reinthaler et al. 2003; Servais and Passerat 2009; Galvin et al. 2010; Oberlé et al. 2012; Blaak et al. 2014a; Bréchet et al. 2014; Ojer-Usoz et al. 2014; Kwak et al. 2015)

Many studies have shown a good absolute clearance rate for ARB in WWTP processes, but other studies have shown a relative enrichment in ARB throughout WWTP processes. Silva et al. (2006) investigated the clearance of antibiotic-resistant coliforms at an activated sludge WWTP in Antofagasta (Chile) and noted a significant increase in the proportion of antibiotic-resistant coliforms in the effluent. Wastewater treatment plants can therefore reduce the total number of coliforms but may increase the relative abundance of antibiotic-resistant coliforms in effluent. The same conclusions were drawn by Bréchet et al. (2014), who showed that absolute total and ESBL-producing *E. coli* concentrations both decreased within a WWTP, whereas the proportion of ESBL-producing *E. coli* increased during the WWTP treatment process, changing from 0.3% of total *E. coli* in the raw wastewater entering the WWTP to 0.6% of total *E. coli* in the effluent. Börjesson et al. (2010) confirmed that although WWTPs reduced the diversity and number of meticillin-resistant *S. aureus* strains, they simultaneously selected antibiotic-resistant strains. For other bacteria, Zhang et al. (2009b) reported that the prevalence of resistance in *Acinetobacter* isolates to amoxicillin/clavulanic acid, chloramphenicol, rifampin and ≥ 3 antibiotics increased from 8.7%, 25.2%, 63.1%

and 33.0%, respectively, in raw wastewater influents, to 37.9%, 69.0%, 84.5% and 72.4% respectively, in WWTP effluents. These conflicting results indicate that while WWTPs may sometimes favor ARB, these most often undergo the same clearance as their susceptible parents.

Some studies have observed a correlation between antibiotic-resistant bacteria (Gao et al. 2012) or antibiotic resistance genes (Gao et al. 2012; Mao et al. 2015; Rodriguez-Mozaz et al. 2015) and the associated antibiotics throughout the WWTP. These correlations suggest the possible selection of antibiotic resistance determinants (i.e. ARB and ARGs) by antibiotics throughout WWTP processes. Even if some studies reported high reductions in antibiotic concentrations in WWTPs (e.g. 50–98% for sulfonamides, 58–6% for tetracyclines, 56–86% for quinolones, and 58–87% for macrolides; Mao et al. 2015), the suggested selection of ARB by antibiotics in WWTPs is reinforced by the fact that Gullberg et al. (2011) revealed that ARB can be selected by subinhibitory concentrations of antibiotics. Nevertheless, we have to be careful: correlation does not imply causation. The ARB and antibiotic concentrations could be correlated because they have the same source (WWTP outflow). This enrichment is not consensually agreed upon in the scientific community. For example, Bengtsson-Palme et al. (2016) observed that despite the relatively high concentrations of tetracycline and ciprofloxacin in raw wastewater influent (above the predicted concentrations for resistance selection), there was no consistent enrichment in antibiotic resistance in the studied WWTPs. Antibiotics may poorly influence antibiotic resistance throughout the treatment process compared to the effect of other selective factors, such as oxygen availability (Bengtsson-Palme et al. 2016). Similarly, Karkman et al. (2019) reported that ARB were correlated with faecal pollution rather than with antibiotics. The previous correlation-based speculations therefore require further studies. General conclusions remain difficult to obtain because of the discrepancies in the type of bacteria and resistance investigated, type of influent wastewater, WWTP processes reported and abundance metrics used (i.e. relative/proportion or absolute value).

In addition to the question of the selection of antibiotic-resistant bacteria by antibiotics in WWTPs, the question of co-selection, in particular by metals and biocides, also arises. Trace metals can induce a similar bacterial resistance mechanism as that of antibiotics (e.g. efflux pumps), which can explain the co-selection of some ARB and ARGs by both antibiotics and metals (Mao et al. 2015; Di Cesare et al. 2016a). However, this question, which involves the selective pressure of antibiotics, is not a subject of consensus among the scientific community. For example, using metagenomics, Bengtsson-Palme et al. (2016) found limited support for co-selection between antibiotic resistance and both biocides and metals in WWTPs.

Some authors have reported that WWTPs are hotspots of horizontal gene transfer and may facilitate the dissemination of antibiotic resistance. Hultman et al. (2018) confirmed the decrease in antibiotic resistance genes during the WWTP treatment process, but found clues related to horizontal gene transfer for antibiotic resistance genes by some bacterial groups. Müller et al. (2018) also suggested that WWTPs act as "melting pots" for the horizontal gene transfer of ARGs. The proportion of ARGs associated with mobile genetic elements has been observed to increase within

WWTPs (Petrovich et al. 2018). Additionally, some studies have indicated that in WWTPs, bacteria harbour various plasmids associated with antibiotic resistance genes that can be transferred to other bacteria, favouring genetic exchange and the dissemination of antibiotic resistance among bacterial communities (Szczepanowski et al. 2009; Ju et al. 2019). Other studies revealed that the environmental conditions in WWTPs may favour horizontal gene transfer (Silva et al. 2006; Rizzo et al. 2013). All these studies have suggested that WWTPs may provide a favourable environment for the horizontal gene transfer of antibiotic resistance genes. However, studies supporting active horizontal gene transfer in WWTPs are rather scarce, and some studies assumed a limited role for horizontal gene transfer in WWTPs (Karkman et al. 2019).

In Sludge As briefly described before, WWTP treatment processes usually separate aqueous and solid phases twice. Primary sludge consists of the settled solids removed during the primary treatment (i.e. in the primary clarifier). Secondary sludge that is collected in the secondary clarifier includes sludge from the secondary treatment bioreactors. The sludge follows a treatment circuit parallel to that of the wastewater. The sludge is usually first thickened before being dewatered (through centrifugation, filtration and/or evaporation). The dewatered sludge is then digested (through anaerobic digestion, aerobic digestion, and/or composting) and sometimes stabilized (e.g. lime and heat stabilization). Finally, the resulting treated sludge is either incinerated (in a minority of cases, due to the cost of the process) or spread as fertilizer in fields.

Gao et al. (2012) reported an enrichment of antibiotic-resistant bacteria in WWTP sludge, as they observed that the tetracycline- and sulfamethoxazole-resistant bacteria concentrations were higher in dewatered sludge than in primary sludge (2-log and 1-log, respectively). When normalized by the total count of bacteria, the concentrations of tetracycline-resistant bacteria were also higher in dewatered sludge than in surplus sludge (from the activated sludge tank) and in primary sludge, suggesting the potential selection of antibiotic-resistant bacteria during the WWTP sludge treatment process. Anaerobic digestion and lime or heat stabilization can reduce the level of antibiotic resistance of sewage sludge compared to that of conventional dewatering and gravity-thickening methods (Munir et al. 2011; Murray et al. 2019). Previous studies suggested that the persistence of antibiotic-resistant bacteria depends on the bacterial species considered and the type of treatment applied to the WWTP sewage sludge. Before being spread on agricultural fields, biosolids are often composted. Composting is commonly reported to reduce the abundance of antibiotic-resistant bacteria (Sadikalay 2018). However, some studies have shown that the antibiotic resistance reduction is not systematic during composting (Chen et al. 2019), or that the abundance and diversity of some ARGs are enriched (Gao et al. 2012; Bengtsson-Palme et al. 2016).

Antibiotic-resistant bacteria (Brooks et al. 2007; Munir et al. 2011; Gao et al. 2012) and antibiotic resistance genes (Auerbach et al. 2007; Zhang et al. 2009a; Munir et al. 2011; Gao et al. 2012; Bengtsson-Palme et al. 2016) have been reported to be more concentrated in biosolids than in WWTP aqueous effluent (Auerbach et al. 2007; Munir et al. 2011), supporting the sorption of antibiotic resistance determinants by sludge. Yang et al. (2014) reported a lower clearance of ARGs in sludge (20.70%, between activated sludge and anaerobic digestion sludge) than in wastewater (99.82%, between influent and effluent). Mao et al. (2015) drew the same conclusion by studying 30 ARGs (conferring resistance to tetracyclines, sulfonamides, quinolones and macrolides) in two WWTPs (North China), with higher ARG concentrations in dewatered than in aqueous WWTP effluent. Munir et al. (2011) reported that tetracycline- and sulfonamide-resistant bacteria had comparable concentrations in biosolids and in raw wastewater, reflecting the low efficiency of WWTP sludge processes in reducing ARB levels. Biosolids may therefore contribute to the spread of antibiotic resistance in the environment.

4.3.2 Environmental Contamination by Wastewater Treatment Plant Effluents

Wastewater treatment plant effluents are highly concentrated with antibiotic-resistant bacteria (e.g. in the range of 5×10^2–6×10^5 CFU/100 mL; Munir et al. 2011). In their review, Hocquet et al. (2016) estimated that WWTPs, depending on their size, release between 10^{10} and 10^{12} ESBL-producing *E. coli* each day in receiving environments, depending on the released discharge. Similarly, Bréchet et al. (2014) estimated that 6×10^{11} ESBL-producing *E. coli* are released daily in the Doubs River (France), and Blaak et al. (2014a) reported that approximately 10^{10}–10^{11} ESBL-producing *E. coli* are discharged daily in four surface freshwaters bodies in the Netherlands from four corresponding WWTPs. LaPara et al. (2011) demonstrated that even with a tertiary treatment, WWTP effluents still constitute a source of dissemination of ARB and ARGs into the environment. As expected and demonstrated previously, WWTP effluents highly contribute to the dissemination of ARB in the environment (Table 4.2).

The proportion of antibiotic-resistant bacteria has been reported to be considerably higher in WWTP effluents compared to that in the receiving river upstream from the WWTP effluent (Iwane et al. 2001; Galvin et al. 2010). Blaak et al. (2014a) reported that the concentrations of ESBL-producing *E. coli* were similar in the investigated WWTP effluents and receiving rivers just downstream of WWTP discharges and 1- to 2-log levels higher than concentrations upstream of the WWTPs. Iwane et al. (2001) showed that the proportion of ARB increased in the receiving river downstream of an effluent discharge; notably, 5.6% *vs.* 29.4% of coliforms were ampicillin-resistant upstream and downstream of the studied WWTP, respectively. In the Doubs River (France), Bréchet et al. (2014) reported 20 *vs.* 1300 ESBL-producing *E. coli* per litre upstream and downstream of a WWTP effluent, respectively; additionally, 0.7% and 0.9% of total *E. coli* produced ESBL upstream and downstream of the WWTP, respectively. In the same case study, Slekovec et al. (2012) showed that the studied WWTP did not decrease the proportion of

Table 4.2 Summary table of environmental contamination with antibiotic-resistant bacteria according to the different sources mentioned in this chapter

	WWTPs	Agriculture	Aquaculture
ARB- or ARG-contaminated media	Effluents and biosolids	Manure, animals, and biosolids from WWTPs	Ponds and fishes
ARB- or ARG-contaminated environments	Mostly aquatic (e.g. rivers, lakes, and coastal waters)	Agricultural lands and the surrounding aquatic environments.	Mostly aquatic (e.g. rivers, lakes, and coastal waters)
Selection of ARB and ARG	Generally low (excepted downstream effluents of pharmaceutical industries)	Low	High
Dispersion of ARB and ARGs	Water fluxes	Spreading, runoff and leaching	Water fluxes
Survival of ARB and ARGs	Generally low, but higher in biofilms	Low for ARB and higher for ARGs	High
Associated health hazard	Water reuse, downstream water consumption, and downstream recreational activities	Food consumption, especially poultry, and workers	Fish consumption, downstream water consumption, downstream recreational activities, and workers

ARB antibiotic-resistant bacteria, *ARG* antibiotic resistance gene, *WWTP* wastewater treatment plant

antibiotic-resistant strains of *P. aeruginosa* and that the concentration of *P. aeruginosa* was higher in the river downstream of the WWTP effluent, demonstrating environmental contamination by hospital-associated antibiotic-resistant strains *via* WWTPs. The greater the contribution of the WWTP effluent to the river discharge is, the greater the concentration of antibiotic resistance determinants in the receiving stream (Proia et al. 2016). In addition to receiving waters, the sediment near WWTP discharge points is also directly contaminated and harbours higher levels of antibiotic resistance determinants compared with the upstream sediment (Marti et al. 2014a; Chu et al. 2018). Marti et al. (2014a) also highlighted significant differences in the community structure and composition of biofilms between upstream and downstream sediment samples, suggesting that WWTP effluents not only disseminate antibiotic resistance determinants in the environment, but may also impact the bacterial communities of the receiving bodies.

Separate *Versus* Combined Sewage Systems WWTPs can become an even more important source of antibiotic resistance during high-precipitation periods. Indeed, there are two types of sewage systems used to collect wastewater before reaching a WWTP: a separate sewage system collects wastewater separately from runoff water (storm water), whereas a combined sewage system collects wastewater and runoff water *via* the same network (Fig. 4.6). Combined sewage systems are more widespread than separate systems because they are older, cheaper and easier to implement.

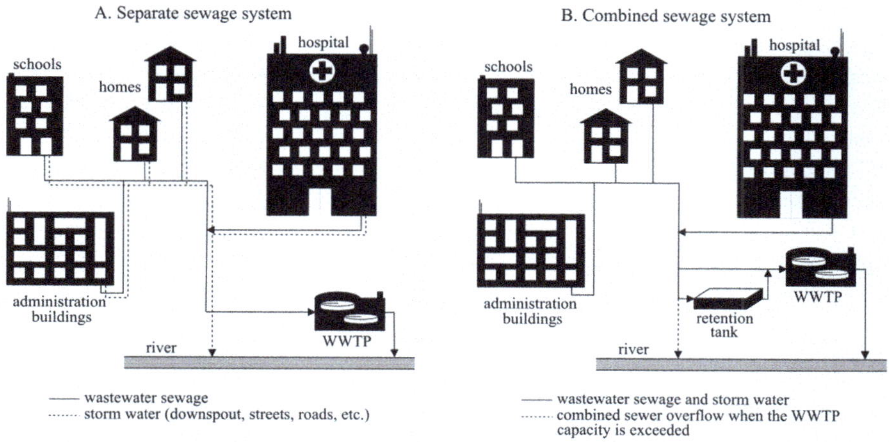

Fig. 4.6 Schematic representation of (**a**) separate and (**b**) combined sewage systems (a separate sewage system collects wastewater separately from runoff water – storm water; a combined sewage system collects wastewater and runoff water *via* the same network; combined sewer overflows, i.e. the release of untreated wastewater into the receiving environment, are frequent in combined sewage systems during rainy periods; *WWTP* wastewater treatment plant)

In a combined sewage system, runoff water, together with raw wastewater, enters a WWTP to be treated. This system therefore encounters two main problems during rainy periods or storm events. First, the dilution of wastewater reduces the WWTP efficiency and the subsequent clearance of pollutants, including antibiotic-resistant bacteria (Capodaglio 2004). Second, an excessive influent flux can saturate the WWTPs, which are designed and optimized to treat a given inflow generally estimated during non-rainy periods. This scenario results in combined sewer overflows (i.e. untreated raw wastewater is deviated from the WWTP by a bypass to the receiving environment; Capodaglio 2004; Jalliffier-Verne et al. 2015; Honda et al. 2020). The discharge of ARB from combined sewer overflows occurs occasionally, depending on precipitations. Honda et al. (2020) studied the combined sewage system in Kanazawa, Japan (29,900 inhabitants, continental climate) over a one-year period. The authors found that 23% of the volume of combined sewage was discharged untreated *via* combined sewer overflows, estimating that the yearly total discharge of antibiotic-resistant *E. coli* from combined sewer overflows in the environment was 3.7-log larger than that from WWTP effluent. Combined sewer overflows therefore appeared to be the primary source of faecal and antibiotic-resistance contamination in the receiving river (Jalliffier-Verne et al. 2015; Honda et al. 2020). The impact of combined sewer overflows obviously depends on climate, and Honda et al. (2020) reported that even if combined sewer overflows are intense in tropical regions, combined sewer overflows in temperate regions could release more pollutant loads, including ARB, due to limited dilution and long-duration rainfall events. To avoid this catastrophic situation, bypasses are generally built upstream of WWTPs (Fig. 4.4). These bypasses hold the raw wastewater in retention tanks that are later discharged in the WWTP. In France and the United States, retention tanks effectively reduce pollutant loads and faecal coliform levels in combined sewer overflows (Jacopin et al. 1999; U.S. Environmental Protection Agency 1999). If a WWTP is not equipped with retention tanks or if the retention tanks are full, the raw wastewater is discharged untreated directly into the receiving environment (Figs. 4.4 and 4.6), as described by Honda et al. (2020; WWTP without a retention tank). The receiving environment is therefore directly exposed to antibiotic-resistant bacteria contamination. The spread of antibiotic resistance can therefore be considerably higher during high-precipitation periods due to combined sewer overflows.

4.3.3 Environmental Contamination by the Spreading of Wastewater Treatment Plant Sewage Sludge and Biosolids

In most countries, the majority of biosolids produced by WWTPs are spread on agricultural lands, raising the question of the dispersal of antibiotic resistance.

Studies have reported the persistence of antibiotic-resistant bacteria in the biosolids produced by WWTPs. Briefly, Brooks et al. (2007) reported that the ARB concentration varied from 7.8×10^5 to 4.5×10^8 CFU/g in the studied biosolids, and Munir et al. (2011) found ARB concentrations ranking between 3.2×10^4 and 1.9×10^9 CFU/g. Compared to the ARB levels found in WWTP effluents (5.0×10^0 – 6.1×10^3 CFU/mL; Munir et al. 2011), biosolids are more concentrated in ARB. Specifically, biosolids can contain 2.6×10^5 ESBL-producing *E. coli*/g (Bréchet et al. 2014) and 5×10^2 antibiotic-resistant *P. aeruginosa*/g (Slekovec et al. 2012). Thus, the spread of biosolids on agricultural lands could play a role in the dispersion of antibiotic resistance in the environment (Table 4.2).

Brooks et al. (2007) determined the prevalence of culturable ARB in soil after the application of biosolids and reported that the application of biosolids did not change the proportion of ARB in agricultural soils. In contrast, Yang et al. (2018) reported an increase in ARG concentrations in soils after the application of biosolids. Similarly, Murray et al. (2019) measured the increase in several ARG abundances in the week following biosolid application. Chen et al. (2016) reported that soils that have received biosolids for multiple years were enriched in ARGs. The application of sewage sludge or biosolids therefore implies an increase in antibiotic resistance in agricultural soils by increasing the load of ARGs rather than viable ARB (Table 4.2).

The risk of using biosolids as a fertilizer is widely debated. Their application on agricultural land is limited or even prohibited in favour of incineration and energy production in some countries, such as Switzerland and Sweden.

4.3.4 Other Sources of Antibiotic Resistance in the Environment

Wastewater treatment plants are hotspot sources for antibiotic-resistance in the environment. However, the contamination of water bodies away from WWTPs indicates that other sources of antibiotic-resistant bacteria exist (Laroche et al. 2010; Blaak et al. 2014a; Henriot et al. 2019).

Aquaculture Farms Between 1986 and 2006, the number of industrial aquaculture farms increased by a factor of 20 (Cabello 2006), especially in Asia and South America (Cabello et al. 2013). The high densities of fish increase the risk of infection, and antibiotics are widely used prohylactically (Cabello 2006). In the early 2000s, high income countries generally restricted the use of antibiotics at aquaculture farms, but this was not necessarily the case globally (Cabello 2006; Bravo et al. 2005). For example, in Chile, 10–12 t of quinolones were used for humans, compared to 100–110 t for animals annually, mainly for aquaculture (Cabello 2006; Bravo et al. 2005). Antibiotics are involved in the selection of ARB in the water column and in sediment (Baquero et al. 2008). Food containing antibiotics, faeces contaminated by antibiotics and antibiotic-resistant bacteria are released in the water column. Ng et al. (2018) retrieved multidrug-resistant *E. coli* from fish farms in Singapore, among which 81.8% were ESBL producers. Fish farms are generally connected with the local aquatic environment, or fish are directly farmed in natural environments (e.g. coastal water), thus facilitating the dissemination of ARB by water fluxes (Table 4.2). Gordon et al. (2007) observed an increase in bacteria resistant to oxytetracyline, quinolones, trimethoprim-sulfamethoxazole, florfenicol and amoxicillin in the vicinity of a fish farm. Cabello (2006) warned about the potential alteration of the composition of sediment bacterial communities (Tamminen et al. 2011a). Cabello (2006) also warned about the potential transfer of antibiotic resistance to animal (both wild and domestic) and human pathogenic bacteria (Rhodes et al. 2000). Tamminen et al. (2011b) demonstrated the long-term impact of aquaculture farms on the surrounding environment by showing the persistence of ARGs after the ban of ARG-specific antibiotics.

Agriculture In 2017, in 31 European countries, 6703 t of antibiotics were consumed in agriculture (European Medicines Agency 2019). Antibiotic residues and ARB are excreted in animal faeces that are directly released on pasture or are collected in manure/slurry. Manure application generally increases the abundance of antibiotic resistance in agricultural soils (McKinney et al. 2018). Once in agricultural soils, ARB can spread in the environment through water fluxes during rainy periods *via* runoff (surface drainage) and leaching (underground drainage). For example, Laroche et al. (2010) investigated a rural watershed with only agriculture and livestock breeding activities and away from any WWTP, and they found that the transport of resistant *E. coli* in the study area was dominated by runoff and leaching phenomena. During a wet period without grazing or during a dry period with grazing, less than 7% of the isolated *E. coli* were antibiotic-resistant. In contrast, during rainfall events with grazing, 30–55% of the *E. coli* were antibiotic-resistant (Laroche et al. 2010). ARB of agricultural origin can further spread in natural aquatic environments (Table 4.2). Beattie et al. (2018) observed that riverbed sediment can temporally accumulate antibiotic resistance determinants during manure fertilization periods.

In addition to the spread of manure and slurry, agricultural areas are also sources of antibiotic resistance because of (i) the application of sewage sludge and biosolids from WWTPs and (ii) the reuse of treated water (WWTP effluents) to irrigate agricultural lands in remote regions (Silva et al. 2006) (Table 4.2).

Slaughterhouses Very few studies have been conducted on antibiotic resistance at slaughterhouses. Slaughterhouse wastewater is also contaminated by antibiotic-resistant bacteria (Um et al. 2016), as the majority of slaughtered farm animals have been treated with antibiotics and can therefore carry ARB. The wastewater at slaughterhouses is either collected in an urban wastewater network or treated in an independent WWTP for larger slaughterhouses. Slaughterhouse WWTPs are estimated to release approximately 10^5–10^6 CFU/L of *E. coli* into the environment (i.e. on the same order as the concentration for urban WWTPs). With respect to ARB, Um et al. (2016) showed that treatment of slaughterhouse wastewater does not alter the proportion of resistant *E. coli* strains. Sludge for land application produced by slaughterhouse WWTPs contains up to 10^5 CFU/g of *E. coli*, less than that produced by urban WWTPs (10^7–10^8 CFU/g; Bréchet et al. 2014). Slaughterhouses are therefore not important sources of environmental contamination by ARB compared to WWTPs.

Pharmaceutical Industries Pharmaceutical industries can release large amounts of antibiotics into the environment and therefore favour the persistence and dispersion of antibiotic-resistant bacteria. The contribution of pharmaceutical industries to environmental antibiotic resistance will be further discussed in Sect. 4.4.2.

Overall, the following conclusions can be summarized: (i) although WWTPs reduce the absolute ARB load in raw wastewater, aqueous effluents and biosolids release ARB into the environment; (ii) in terms of relative abundance, it is difficult to draw clear conclusions; ARB are neither advantaged nor disadvantaged over their susceptible counterparts during treatment at WWTPs; (iii) WWTP products, including aqueous effluents and biosolids, are the main sources of contamination by antibiotic resistance in the environment; (iv) fish farms and agriculture also contribute to the dissemination of antibiotic-resistance in the environment.

4.4 Dispersion and Persistence of Antibiotic Resistance in the Environment

4.4.1 Environmental Contamination Driven by the Level of Anthropization

In addition to suggesting that WWTPs have a strong impact on the receiving environments in terms of antibiotic-resistant bacteria contamination, the literature shows that this impact depends on the degree of anthropization of the watershed considered. Servais and Passerat (2009) investigated the source of faecal ARB in the Seine watershed (France). They observed that the abundance of antibiotic-resistant *E. coli* increases from forest non-point sources (2% of total E. coli) to agricultural non-point sources (16% of total E. coli), municipal wastewater (44% of total E. coli) and hospital wastewater (71% of total E. coli). Along a river continuum characterized by two opposite density gradients for animals (cattle and wild animals) and human populations, Petit et al. (2017) showed that in the upstream part, the low abundance of resistant *E. coli* (5% of total E. coli) could be explained by the low number of cattle receiving antibiotic therapy; in contrast, the higher frequency of resistant *E. coli* (27% of total E. coli) reflected the human contamination of water at downstream urban sites. Along a medical centre-retirement home-wastewater treatment plant-river continuum in France, the relative proportion of *Enterococcus faecium* that was resistant to antibiotics decreased, and subpopulations of *E. faecium* less resistant to antibiotics progressively increased (Leclercq et al. 2013). Guyomard-Rabenirina et al. (2017) investigated the antimicrobial resistance of *Enterobacteriaceae* to fluoroquinolones and third-generation cephalosporins in 70 surface water samples obtained primarily from two rivers in Guadeloupe (France). The authors confirmed that resistant *Enterobacteriaceae* were absent or rare near the river sources but frequent near the surrounding human activities dominated by housing and farming. Similarly, this increase in antibiotic resistance in anthropized environments has also been demonstrated in lake ecosystems, with a decrease in antibiotic resistance as the distance from WWTP discharge points increased (LaPara et al. 2011; Czekalski et al. 2014; Czekalski et al. 2015). These increased antibiotic resistance rates in the vicinity of WWTP discharges and the associated anthropization-dependent levels of ARB in natural environments are intuitive but also suggest that the dissemination of ARB seems limited, as their survival is rather poor in nonfavourable environments such as freshwaters or coastal waters.

4.4.2 Exogenous Parameters that Influence the Dispersion and Survival of Antibiotic-Resistant Bacteria in Natural Environments

In the absence of antibiotic pressure, Berthe et al. (2013) demonstrated in a microcosm study that antibiotic-resistant bacteria survived less (< 4 days) than susceptible bacteria in water (> 14 days), suggesting a cost of resistance to bacterial fitness. However, antibiotic-resistant bacteria are released into the environment along with other pollutants, some of which potentially promote the survival of ARB.

Antibiotics These substances are released by several sources into the environment (e.g. WWTPs, agriculture, fish farms and pharmaceutical industry effluents). Multiple studies/monitoring networks have measured the concentrations of antibiotics in surface water. Three of them will be detailed by way of example. Valcárcel et al. (2011) determined the concentration of several antibiotics in five Spanish rivers and reported ciprofloxacin, clarithromycin, erythromycin, metronidazole, norfloxacin, ofloxacin, sulfamethoxazole, tetracycline and trimethoprim median concentrations of 0.003, 0.235, 0.320, 1.195, 0.010, 0.179, 0.326, 0.023 and 0.424 μg/L, respectively. In China, norfloxacin, ofloxacin, ciprofloxacin and oxytetracycline have been frequently detected at concentrations up to 5.770, 1.290, 0.653 and 0.652 μg/g in sediments from the Pearl River, Yellow River, Hai River and Liao River (Zhou et al. 2011). Finally, Johnson et al. (2015) reported expected means of 0–0.04, 0–0.04, 0–0.06 and 0–0.06μg/L for ciprofloxacin, sulfamethoxazole, trimethoprim and erythromycin, respectively, in European rivers. The concentrations of antibiotics generally increase from upstream to downstream, with a maximum observed downstream of the main WWTP discharge outfalls (Dinh et al. 2017).

Some of the previous antibiotic concentrations measured in rivers are higher than the "environmental predicted no-effect concentrations" reported by Tell et al. (2019), suggesting a possible selection of ARB in natural environments. Moreover, Gullberg et al. (2011) reported that the minimal selective concentrations of streptomycin, ciprofloxacin and tetracycline are approximately 1/4, 1/10 and 1/100 of their minimal inhibitory concentrations respectively. Subinhibitory concentrations of antibiotics can therefore favour ARB. For example, Gullberg et al. (2011) determined that the minimal inhibitory concentration of ciprofloxacin for wild type-*E. coli* was 0.023μg/ml, and the corresponding minimal selective concentration could be estimated at 0.0023μg/ml. The concentrations of ciprofloxacin previously reported in rivers can exceed this minimal selective concentration and therefore exert a selective pressure on resistant *E. coli*. On the other hand, some studies did not find any relation between antibiotic concentrations and ARB abundances in rivers (Oberlé et al. 2012). Karkman et al. (2019) observed that antibiotic resistance was correlated with faecal contamination downstream of WWTP effluents rather than with antibiotic concentrations. However, there is no simple relationship between the presence of ARB and the concentrations of antibiotics in water because ARB have been previously selected in humans and animals treated with numerous antibiotics, whereas only persistent compounds can be detected in water (e.g. sulfonamides, fluoroquinolones, macrolides and diaminopyrimidine).

Antibiotic-resistant bacteria have been reported to be selected in rivers downstream of pharmaceutical industries. Milaković et al. (2019) showed that the discharge of an azithromycin manufacturing factory shifted the bacterial community composition in the downstream sediment of the Sava River (Croatia), and this community displayed higher antibiotic resistance than the bacterial communities located upstream from the WWTP. The five studied ARGs persisted in sediment until 11 km downstream from the effluent discharge point. China and India produce the majority of the world's antibiotics (Changing Markets, Ecostorm 2016). Two studies that investigated the role of pharmaceutical industries producing penicillin G and oxytetracycline in China reported higher multiresistance frequencies in bacteria isolated in the rivers downstream of industrial discharges than in other areas (Li et al. 2009, 2010). Li et al. (2010) found that almost all bacteria (97%) from the wastewater and the river water downstream of the effluent discharge points were multiresistant to antibiotics, while they were less frequent (28%) in upstream river samples. In Patancheru (India), the antibiotic concentration was extremely high in the effluent of the WWTP that served approximately 90 bulk drug manufacturers (Larsson et al. 2007). For example, the concentration of ciprofloxacin, which reached 31,000μg/L, exceeded the level of toxicity to some bacteria by over 1000-fold. Kristiansson et al. (2011) investigated the impact of these effluents on the sediment of the receiving river. Antibiotics can persist for up to 30–40 years in the sediment (Tamtam et al. 2011). Kristiansson et al. (2011) confirmed the selection of antibiotic resistance determinants downstream of pharmaceutical industries in the sediment and also suggested the mobilization of antimicrobial resistance within bacterial communities, potentially from environmental bacteria to human-associated bacteria, including human pathogens.

In the natural environment, antibiotic pollution could favour antibiotic-resistant bacteria. However, environmental matrices are complex, polyspecific and contaminated by multiple pollutants, unlike most experimental matrices. Bacteria survive in communities in the liquid phase or biofilm and are not systematically culturable. Moreover, the bioavailability of antibiotics is poorly known. The resulting selection of antibiotic resistance by antibiotics occurs under specific conditions in natural environments (e.g. downstream pharmaceutical industry effluents, WWTP effluents particularly concentrated in antibiotics and areas near fish farms). Low and middle income countries appear to be more exposed than high income countries to the selection of antibiotic resistance in natural environments. This difference is mostly due to the lower removal of ARB and antibiotics in wastewaters, as well as the many pharmaceutical industries in these countries and their poorly treated effluents (Changing Markets, Ecostorm 2016).

Metals and Biocides Contamination with metals could favour the maintenance of ARB in the environment (Baker-Austin et al. 2006; Wales and Davies 2015). These associated resistances can be explained either by the presence of genes resistant to metals and antibiotics on the same genetic carrier (co-resistance) or by the existence of common mechanisms conferring co-resistance (cross-resistance; e.g. changes in membrane permeability) (Pal et al. 2017). In the environment, many studies have shown a correlation between metal contamination and bacterial populations co-resistant to metals and antibiotics (Dickinson et al. 2019; Milaković et al. 2019). Wales and Davies (2015) and Poole (2017) suggested that some metals, particularly copper and zinc, are capable of complexing with antibiotics, thereby enhancing their antimicrobial activity depending on the molecule and promoting the emergence of co-resistant strains. However, this was not the case reported by Vos et al. (2020). Recently, the contribution of metagenomics has made it possible to gain a comprehensive understanding of this issue. An in-depth genomic analysis of a collection of 5436 genomes from 39 phyla from differing habitats (human, water, soil) showed that the co-occurrence of genes encoding resistance to antibiotics or to metals was very high in the genomes of bacteria isolated from humans and moderate in bacterial genomes from water or soil (Li et al. 2017). Maximum values were found in pathogenic bacteria belonging to *Enterobacteriaceae*.

The relationship between bacterial resistance to antimicrobial biocides and the use of antibiotics has long been known (Chaplin 1952). Like metals, biocides can promote the spread of antimicrobial resistance through co- and cross-resistance mechanisms (Davin-Regli and Pagès 2012; Wales and Davies 2015). Metal and biocide resistance genes were found in 86% of bacterial genomes and coexisted with ARGs in 17% of cases (Pal et al. 2015). Environmental contamination by biocides and metals could therefore favour the maintenance of ARB in natural environments. Agricultural environments are particularly exposed to this risk, especially because of the combined use of antibiotics, biocides and metals (mainly copper and zinc).

However, the data currently available in the literature do not allow to clearly understand the effect of the presence of metals or biocides on the dissemination of antimicrobial resistance in the environment. Thus, in the conclusion of their review, Pal et al. (2017) proposed further studies to assess the relationship between the level of exposure to metals, including their bioavailability and the selection/persistence of antibiotic-resistant bacteria in differing environments.

Microplastics These new emerging contaminants are other exogenous products released by WWTPs into natural environments, which can favour the dissemination of antibiotic resistance determinants. Microplastics have been reported to constitute a reservoir/niche for the bacterial community and more specifically for ARB and ARGs (Arias-Andres et al. 2018; Yang et al. 2019), thus increasing the horizontal gene transfer of ARGs between human-associated or environmental bacteria and the dispersion of ARB and ARGs over long distances (Arias-Andres et al. 2018). Using metagenomics, Yang et al. (2019) studied the ARGs present in microplastics in the Northern Pacific. They showed that both the diversity and richness indices of ARGs were higher in microbiota on plastics compared to those for seawater microbiota. Multidrug and multi-metal resistance genes were the main classes of ARGs detected in the studied microbiota on plastics, suggesting the high horizontal gene transfer of antibiotic resistance determinants (Yang et al. 2019). Recent studies reported high concentrations of microplastics in natural – mainly aquatic – environments (Ivleva et al. 2017), highlighting the risk of antibiotic resistance dissemination by microplastics. Further investigations should focus on the following questions: are microplastics more favourable to the spread of antibiotic resistance than other suspended matter in aquatic environments? In any case, microplastics offer an additional dispersion route for antibiotic resistance.

4.4.3 Dispersal and Survival of Antibiotic-Resistant Bacteria in Receiving Environments

4.4.3.1 Rivers

These aquatic water bodies, i.e. rivers, are the main receivers of wastewater treatment plant effluents and are also the most studied and exposed in terms of antibiotic-resistant bacteria contamination. Antibiotic-resistant bacteria have been studied in the water column and in sediments.

Case of the Water Column In the upstream part of the Seine watershed, Servais and Passerat (2009) found that 83% of enterococci and 42% of *E. coli* were resistant to at least one antibiotic of clinical interest. Forty-nine percent of enterococci and 35% of *E. coli* were multiresistant to antibiotics. In another study, 30–57% of *E. coli* isolated from estuarine water in the Seine watershed were resistant to at least one antibiotic, and of these, 60.5–80% were resistant to ≥ 2 antibiotics (Laroche et al. 2009). The authors determined that 3.5×10^9 antibiotic-resistant *E. coli* circulated in the Seine estuary each second. The estuary of the Seine River contained a proportion of ARB similar to that at a WWTP situated ~50 km upstream (48.3%), suggesting the dispersion of ARB over long distances. Estuary environments, since they are the final destination for ARB, might serve as reservoirs for antibiotic resistance (Laroche et al. 2009). Temporarily, high loads of ARB are often measured in rivers during high precipitation periods (Laroche et al. 2009; Jalliffier-Verne et al. 2015; Henriot et al. 2019), mostly due to combined sewer overflows but also to the runoff and leaching of contaminated water from the surrounding agricultural lands. However, high ARB concentrations are also reported in summer due to the low flow of receiving streams and thus the decreased dilution of pollutants (Zhang et al. 2009b).

Garcia-Armisen and Servais (2004) studied the viable but non-culturable (VBNC) bacteria in microcosms and natural environments by comparing the concentrations of viable *E. coli* assessed by a culture-based method with that of the VBNC *E. coli* with a culture-independent method. On the one hand, 200 hours after inoculation, only 1% of the bacteria were still culturable, whereas 21% were VBNC. On the other hand, along the Seine River downstream of a large WWTP, the ratio of the *E. coli* concentrations VBNC/viable varied from 11 (directly downstream of the WWTP discharge) to 71 (120 km downstream of the WWTP). A metagenomic approach confirmed the high proportion of VBNC bacteria in the effluents of WWTP release in surface water (Jäger et al. 2018a). This suggests that harsh conditions for bacteria (in WWTP or in natural environments) favour the VBNC state. Faecal bacteria, including ARB, are therefore capable of surviving over long distances downstream of WWTP discharges in the environment. The role of VBNC in ARB as a vehicle for the spread of antibiotic resistance in the environment still needs to be explored.

Once in natural environments, antibiotic resistance determinants can be spread by water fluxes and disrupt environmental bacterial communities, as shown above. The ARB can also be disseminated through fauna (Bollache et al. 2018). ESBL-producing *E. coli* can be transferred from WWTPs to receiving rivers and then to biofilms grazed by omnivorous fishes that they are able to colonize (Bollache et al. 2018). Most compartments of the aquatic food networks can thus become vectors in the environmental dispersion of antibiotic resistance. However, the survival of such human-associated bacteria in the gut of fish and other aquatic animals is poorly understood.

Case of the Sediment Few studies have investigated ARB in riverbed sediment, but several have done so for total faecal bacteria. Some studies have reported the rapid decay of bacteria in sediment, mostly due to the absence of water column mixing and thus the lack of dissolved oxygen (Anderson et al. 2005). In contrast, many authors considered sediment to be a favourable habitat for bacteria, as it provides nutrients (Knox et al. 2008) and protection against predators, parasites, desiccation, UV, toxins and antibiotics (Brennan et al. 2014). Other studies have shown that the concentration of bacteria in the river water column is influenced by the concentration in the sediment because of sediment remobilization by water movements (Smith et al. 2008). However, Berthe et al. (2008) observed that in the Seine estuary mudflats, abundance of faecal bacteria decreased with the sediment depth. This suggests that the sediment does not favour the survival of a culturable form of ARB. The authors described that bacteria associated with easily settleable particles were deposited on mudflats, which can temporally constitute a storage site, and lost their culturability. They also suggested that bacteria may be favoured at low temperatures on mudflats. Although the sediment can contain higher bacterial loads than the water column, resuspension moderately impacts the ARB concentration in the water column (Heß et al. 2018).

4.4.3.2 Riverine Wetlands

River annexes, mainly wetlands, are supplied by different types of water, mostly river water, groundwater and runoff water (Amoros and Bornette 2002), that can potentially contaminate wetlands with ARB. The physico-chemical characteristics of these ecosystems greatly vary according to the geomorphological, anthropic and hydrogeological contexts (Amoros and Bornette 2002; Bornette et al. 2008).

Very few studies have been conducted on the antibiotic resistance contamination of riverine wetlands. Henriot et al. (2019) studied 16 isolated wetlands in three floodplains in eastern France. ESBL-producing *E. coli* was able to reach the riverine wetlands, but only during high-flow periods, meaning that the main stream of the river was the leading source of antibiotic resistance in these low-anthropized ecosystems (Henriot et al. 2019). This seasonal presence of ESBL-producing *E. coli* in wetlands suggests that although ARB are transferred by rivers over long distances, their culturability quickly decreases in natural environments. Moreover, some physico-chemical parameters (i.e. phosphate, pH and nitrate levels) were positively related to the abundance of total *E. coli* in wetlands (Henriot et al. 2019). Nevertheless, ESBL-producing *E. coli* loads were independent of the studied physico-chemical parameters and only depended on the connectivity of wetlands with a river. This result demonstrates that although ESBL-producing *E. coli* are able to be dispersed from rivers to wetlands, they struggle to survive there. The higher abundance of ESBL-producing *E. coli* during high–flow periods than in other periods suggests that the retrieved isolates are coming from untreated discharges associated with combined sewer overflows and/or that both runoff and leaching from the surrounding areas contribute to the contamination of wetlands by ARB.

Due to lower flow and favourable physico-chemical conditions, riverine wetlands contain more abundant and diversified vegetation than the main stream of most rivers. Aquatic vegetation possibly affects ARB. Several studies have explored the sheltering role of bacteria by aquatic plants, thus promoting the bacterial fixation, the formation of biofilm on the surface of submerged parts and the presence in vegetal tissues (Knox et al. 2008; Kleinheinz et al. 2009). Furthermore, secondary compounds produced by aquatic plant species can reduce the survival of bacteria (Özbay and Alim 2009; Chiapponi et al. 2020). Finally, due to its high nutrient content (Berg et al. 2005), the rhizosphere could favour bacterial concentration and interactions, including horizontal gene transfer. However, these different processes have been poorly studied, especially for ARB. Further investigations are needed to better understand the role of vegetation in the spread of antibiotic resistance in aquatic environments.

4.4.3.3 Lakes

To the best of our knowledge, no studies have been conducted on the survival of the four major bacterial species considered in this chapter in lakes. However, several studies have investigated ARGs in lakes. Chu et al. (2018) showed the direct influence of WWTP discharges in the sediment of Lake Michigan (USA). Their results suggest that WWTP effluents modify indigenous microbial communities in the sediment through the simple deposition of ARGs and horizontal gene transfer. Low concentrations of ARGs were found in the centre of Lake Michigan (100–200 km from a WWTP effluent), reflecting the dilution and poor dispersion of antibiotic resistance in lakes. In Lake Geneva (Switzerland), Czekalski et al. (2014) showed that ARG abundances near a WWTP effluent were up to 200-fold higher than those measured at the centre of the lake (~7 km from the effluent source). However, some ARGs were transported over long distances in the lake. Czekalski et al. (2015)

measured the ARG concentrations in 21 Swiss lakes and found that some ARGs were directly related to the anthropization level of the catchment area of the lakes. In contrast, ARG concentrations have been correlated with several ecological parameters in lakes (i.e. positively with dissolved oxygen and negatively with chlorophyll-a (Di Cesare et al. 2015), as well as positively with water residence times and the eutrophication status (Czekalski et al. 2015)). This finding suggests that freshwater lakes may constitute a niche for antibiotic resistance determinants. However, this role of reservoirs is very limited, as antibiotic resistance determinants are generally found near WWTP effluents (Czekalski et al. 2014; Chu et al. 2018).

4.4.3.4 Coastal Water

These environments can be contaminated with antibiotic resistance determinants transported by river estuaries, WWTP effluents, surface runoff and contaminated groundwater (Laroche et al. 2009). WWTP outfalls in the open sea and cruise ship discharges cannot be neglected. Entering seawater, faecal bacteria face several abiotic (light radiation and the associated oxidative stress, salinity and nutrient deprivation) and biotic stresses (predation protozoa, competition and bacteriophage) (Rozen and Belkin 2001). However, antibiotic-resistant *E. coli* have been detected in several coastal waters (Blaak et al. 2014a; Leonard et al. 2018a, b). Leonard et al. (2018a) estimated that each *E. coli* retrieved in 13 studied coastal waters (England) harboured an average of 1.24 ARGs. Fifteen percent of 97 bathing water samples obtained off the coast of England and Wales, contained third-generation cephalosporin-resistant *E. coli,* with 0.12% of all collected *E. coli* being third-generation cephalosporin-resistant *E. coli* (Leonard et al. 2018a). Contrary to the findings of Blaak et al. (2014a), Leonard et al. (2018a, b) studied sites that were not under the direct influence of a WWTP, and the results suggested the possible dispersion of antibiotic resistance in coastal water. Coastal water is generally less polluted than rivers due to the dilution from the mouth of a river to the sea (Guyomard-Rabenirina et al. 2017).

In addition to direct contamination sources, fauna can be a vector for the spread of antibiotic resistance to coastal waters. Vittecoq et al. (2017) investigated the presence of carbapenem-resistant strains of *E. coli* in two species of gulls in southern France: one living in close association with humans (*Larius michahellis*) and the other feeding at sea (*Chroicocephalus genei*). The authors found that 19% and 0% of 'anthropized' and 'non-anthropized' gulls carried the gene $bla_{\text{VIM-1}}$ (encoding a carbapenemase) in their gut, respectively. This result confirmed that fauna can be contaminated with ARB directly by the ingestion of water or of particulate matter as plastic previously contaminated by ARB and thus could spatially disseminate the antibiotic resistance determinants.

4.4.3.5 Agricultural soils

Agricultural soils can be contaminated by antibiotic-resistant determinants through the application of biosolids from WWTPs and manure from agricultural farms.

Case of Sewage Sludge or Biosolid Application As previously described, Brooks et al. (2007) reported that the application of biosolids did not change the proportion of culturable ARB in soils. In contrast, Chen et al. (2016) demonstrated that the long-term application of sewage sludge significantly modified the composition and structure of the bacterial soil community by increasing ARG abundance and diversity in the studied soils. In addition, the authors found that mobile genetic elements were positively correlated with total ARGs in soils. This result suggests that the dissemination of ARGs in agricultural soils is partly related to mobile genetic elements and the associated horizontal gene transfer, which in turn could favour the accumulation and persistence of ARGs in amended soils. Similarly, Yang et al. (2018) reported an increase in ARG abundances in soils after the application of biosolids. They also reported that the application of biosolids could drive a shift in ARGs through horizontal gene transfer.

Case of Manure and Slurry Application Five percent of cattle faeces sampled in eastern France carried ESBL-producing *E. coli* (Hartmann et al. 2012). Crops harvested one year after manure application carried the same clones detected in manure, suggesting their persistence for at least one year in the field environment. In a Canadian study performed in an area with a humid continental climate, viable *E. coli* and *Enterococcus* spp. populations quantified by plate counts fell below the detection limit within six months of a swine or dairy manure application (Marti et al. 2014b). In contrast, ARG concentrations determined by qPCR remained higher in manured plots than in non-manured controls throughout two growing seasons, a period of approximately a year and a half. The long-term application of chicken manure and short-term application (one week) of composted and liquid swine manures increased the abundance and diversity of ARGs in soil, with the abundance returning to background levels six months after application (Chen et al. 2016, 2019). In contrast, analyses of archived soils from field plots treating with composted swine manure indicated that some gene targets persisted for at least five years post application (Scott et al. 2018). Overall, ARG targets persist in soils far longer than typical viable indicator bacteria. However, it is unknown whether the gene targets quantified using culture-independent methods are carried in viable bacteria.

The persistence of antibiotic-resistant bacteria is variable in amended soils and is condition dependent (type of fertilizer applied, type of soil). Combined, these results demonstrate that the ARG concentration increases after manure application and suggest that ARGs can persist in soils (several weeks to more than one year). Soils are not very favourable habitats for enteric bacteria, and a non-negligible part of them can become VBNC.

4.4.3.6 Groundwater

This environment can be contaminated with antibiotic-resistant bacteria by the leaching of contaminated soils depending on the land use in the watershed and the hydrological conditions. In general, groundwater is minimally contaminated due to the poor survival conditions of human-associated bacteria within aquifers (John and Rose 2005). This section will not describe riverine aquifers that can be contaminated by water fluxes coming from contaminated rivers and streams. We will rather focus on karstic aquifers that are comparatively easily contaminated due to their high porosity and the resulting rapid infiltration. As for rivers in rural areas, the presence of ARB in aquifers can be explained by soil leaching during rainy events (Laroche et al. 2010; Ribeiro et al. 2012). Laroche et al. (2010) investigated a rural karstic hydrosystem not influenced by any WWTP but only by livestock breeding activities. They found that despite the permanent circulation of *E. coli* in the studied karstic hydrosystem, antibiotic-resistant *E. coli* infiltrated the groundwater during rainy events, indicating leaching from agricultural surfaces. Petit et al. (2018) reported the same presence of antibiotic-resistant *E. coli* in a karstic hydrosystem during high-rainfall events. Laroche et al. (2010) and Petit et al. (2018) observed a decrease in culturable bacteria in a spring (downstream of a karstic hydrosystem system) compared to the bacteria level in a creek (upstream of the karstic hydrosystem) of 1–2 log. This decrease may be due to the dilution of ARB within the aquifer, the trapping/decay in the karst and/or the loss of cultivability.

Groundwater has been described as a permanent reservoir of VBNC *E. coli* (Petit et al. 2018). Further studies are needed to better investigate the ARGs in groundwater and improve our knowledge of antibiotic resistance dispersion in these particular systems. Aquifers – even karstic ones – seem less exposed to antibiotic resistance than surface waters.

Overall, the following conclusions can be summarized: (i) the high dispersion of antibiotic resistance is reported in aquatic environments; the level of anthropization of an ecosystem is a determinant, and ARB are quickly diluted with the distance from a source; (ii) the dissemination of antibiotic resistance depends on precipitation and land use in a watershed; (iii) in a natural environment, the selective pressure is generally low, with the exception of environments that receive effluents from pharmaceutical industries; (iv) the survival of ARB is difficult to quantify due to VBNC forms; natural environments do not allow the long-term culturability/survival of ARB, but it could be interesting to assess the persistence of genetic support in such environments; (v) ARB survival can depend on the specific conditions of the receiving environment; (vi) the environment could therefore be a secondary reservoir of antibiotic resistance, but the ARB present in natural environments are rather the result of spatial dissemination than of proliferation; (vii) fauna and flora could favour ARB spatial dispersion and possibly survival in the environment, but their role needs further investigation; and (viii) the persistence of VBNC ARB and the corresponding genetic information have been poorly investigated and require special attention.

4.5 What Are the Risks for Humans ?

The presence of antibiotic-resistant bacteria in natural environments raises the question of the risk for human population health. The concept of risk involves aspects of both hazard and exposure. The risk is the exposure to a hazard, here antibiotic resistance, and more specifically, human contamination by ARB. In human health, risk assessment is generally used to estimate the probability of illness or death for individuals in a given population exposed to a given hazard (Bürgmann et al. 2018). However, the risk is comparatively complex to assess considering antibiotic resistance. Indeed, several parameters must be taken into account, such as the pathogenicity of the resistant bacterial strains, the importance of the antibiotic to which resistance is confirmed, the ease of establishment, the dissemination of resistant strains in individuals and the possibility of transfer of resistance within the microbiota of the population studied (Bürgmann et al. 2018). Because of the difficulty of assessing all of the above parameters, very few studies have assessed the risk for human population conferred by antibiotic resistance in the environment. The studies referred to in this section estimated the likelihood of exposure to antibiotic-resistant bacteria, rather than the actual risk associated with antibiotic resistance.

4.5.1 Exposure Through Water Consumption

Groundwater is one of the main sources of water consumption (41% of the population of the USA depends on groundwater for its drinking water supply, National Groundwater Association 2020). Laroche et al. (2010) retrieved antibiotic-resistant *E. coli* in boreholes, a water resource intended for human consumption, demonstrating that drinking water resources withdrawn from karst groundwater are vulnerable to contamination by ARB (Laroche et al. 2010; Ribeiro et al. 2012). Admittedly, such contamination is uneven and temporary, but after pumping and distribution systems are contaminated, antibiotic-resistant bacteria could be trapped and grow in biofilms; therefore, if so, the entire system could be vulnerable, and not only during rainy periods.

Surface freshwater is also a direct or indirect source of water consumption for humans and animals. Antibiotic pollution and the spread of ARB discussed above may pose serious risks for human and animal health. In Lake Geneva (Switzerland), Czekalski et al. (2014) retrieved ARGs near the location of a drinking water pumping station. Schwartz et al. (2003) investigated antibiotic resistance in biofilms in a water distribution system in Mainz (Germany). This water distribution system is supplied with bank-filtered water (from the Rhine River) followed by UV irradiation at waterworks. Antibiotic-resistant bacteria (i.e. enterococci and *Enterobacteriaceae*) were absent from the drinking water distribution system, but ARGs (i.e. *vanA* and *ampC*) were retrieved from the same samples. This discrepancy was either due to the presence of dead bacteria that released their DNA or contamination with VBNC forms of ARB. For example, Jeanvoine et al. (2019) found VBNC cells of the high-risk clone *P. aeruginosa* ST308 in the plumbing system of a healthcare facility. The health hazard associated with the presence of VBNC bacteria has generally been poorly evaluated (Schwartz et al. 2003), but such bacteria may pose a health risk.

In New Delhi (India), Tanner et al. (2015) investigated 19 drinking water samples from the distribution system. The drinking water in New Delhi mostly comes from rivers (Safe Water Network 2016). The authors found isolates resistant to meropenem at five sampling sites. The presence of carbapenem-resistant bacteria retrieved in this drinking water is particularly alarming since it directly exposes 18 million people to bacteria resistant to an antibiotic of last resort (Safe Water Network 2016).

Environmental water contaminated with antibiotic-resistant bacteria or antibiotic resistance genes can therefore constitute a risk for the human population through drinking water. There is an urgent need to find a low-cost and easy-implemented method to control the presence of antibiotic resistance determinants in drinking water networks, especially those conferring resistance to antibiotics of last resort and for highly exposed populations in low and middle income countries.

4.5.2 Exposure Through Food Consumption

Vegetable Contamination Blaak et al. (2014b) found that third-generation cephalosporin-resistant *Enterobacteriaceae* were present on raw consumed fresh produce (isolated from 2.7%, 1.3% and 1.1% of vegetables from supermarkets, iceberg lettuce from farms and agricultural soil, respectively). By demonstrating the link between pig manure spreading (1–2 weeks before planting) and the contamination of vegetables, the authors also showed that the ARB contamination of fresh vegetables mirrored that of the agricultural soil in which they were grown. However, the percentages of contaminated vegetables were rather low, suggesting a low hazard for contamination through vegetable consumption for human populations. Moreover, the authors found that the ESBL-producing *Enterobacteriaceae* isolated from vegetables and the environment were all environmental species, therefore limiting the hazard for human populations (Blaak et al. 2014b). Randall et al. (2017) reported that none of the fruits and vegetables sampled from five retail stores in the UK were positive for ESBL-producing *E. coli*. The likelihoof of exposure to ARB through the consumption of vegetables is limited and dependent on agricultural conditions (type of soil, type of amendment, type of vegetable).

Meat and Fish Meat Contamination Cabello (2006) reported the presence of antibiotic residues and ARB in fish grew in fish farms. In the same way, Abraham (2011) reported that more than 50% of the strains of human-associated bacteria retrieved in farm-raised freshwater fishes (India) exhibited resistance to multiple antibiotics, again suggesting the potential hazard associated with the consumption of farm-raised fishes. Finally, Obaidat et al. (2015) observed that half of the sampled traded fish (from Egypt, India, and Yemen) were contaminated with *S. aureus*, and most were resistant to at least one antibiotic. The authors also found that the presence of an enterotoxin-encoding gene (a toxin causing gastrointestinal symptoms) was positively correlated with antibiotic resistance, suggesting a pathogenic hazard – in addition to the antibiotic resistance hazard – linked to the consumption of fish meats.

Concerning meat, Schwaiger et al. (2012) showed that *E. coli* strains retrieved in chicken and pork meat sampled in a slaughterhouse and a retail setting (Germany) were often resistant to penicillins, streptomycin, spectinomycin, doxycycline and sulfamethoxazole/trimethoprim. Among several coliforms of interest, *E. coli* presented the highest resistance rates. The prevalence of ARB was higher in meat from a retail store than in meat from a slaughterhouse, and resistant and multiresistant isolates were more abundant in chicken than in pork meat (Schwaiger et al. 2012). Randall et al. (2017) reported that approximately 2% of beef and pork samples and 65% of chicken samples were positive for ESBL-producing *E. coli*. These two studies suggest that poultry meat provides a higher level of resistance than pork and beef meats. Tansawai et al. (2018) showed that the prevalence of ESBL-producing *E. coli* reached 70% in poultry meat from supermarkets in Thailand. Moreover, the authors found that pathogenic *E. coli* ST131 isolates harboured ARGs, thus raising serious concerns for food safety. Monte et al. (2017) retrieved colistin-resistant *E. coli*-carrying plasmids previously identified in animals and human isolates in commercial chicken meat (Brazil), suggesting the potential dispersion of ARB between farm animals and humans. Similarly, Berg et al. (2017) found that clones of cephalosporin-resistant *E. coli* were shared by poultry and patients, thus supporting their transfer from chicken meat to humans. All these studies revealed that human contamination with ARB through meat consumption may lead to difficult-to-treat infections. However, the likelihood of exposure appears to be low except for poultry meat.

Several studies have been conducted on the contamination of vegetables and meats by antibiotic-resistant bacteria and assumed an existing but limited ARB dispersion from food to humans. Poultry meat is relatively more contaminated by ARB than are other meats, but the associated risk is very limited if the meat is well cooked. Some studies have suggested that ARB present in the food chain could constitute a reservoir of transferable ARGs among pathogens (Koo and Woo 2012). Losasso et al. (2018) measured a lower total load of ARGs in vegan diets compared to those in omnivorous and vegetarian diets, suggesting a possible role of animal-derived food consumption in the risk of contamination by antibiotic resistance determinants. However, considering each studied ARG separately, the authors found no differences among the three diet groups, thereby reflecting the results of Meijs et al. (2020) who conlude that eating meat is not an important risk factor for ESBL E. coli and K. pneumoniae carriage. Moreover, Königer et al. (2014) and Milanović et al. (2017) found that the diet did not impact ARG diversity in the gut and the load of ESBL-producing *Enterobacteriaceae* in faeces, respectively. More generally, the likelihood of exposure is also linked to the cooking conditions (cooking, kitchen environment, hygiene measures). Based on these findings, food is associated with a moderate risk of transmission of antibiotic resistance in high income countries.

4.5.3 Exposure through Recreational Activities

Coastal water and freshwater, potentially contaminated by antibiotic-resistant bacteria of faecal origin, are the sites of multiple recreational activities, thus raising the question of the hazard to which human populations are exposed. The environmental health hazard associated with antibiotic resistance through recreational activities is poorly known. ESBL-producing *E. coli* have been found in coastal waters popular for swimming and other recreational activities (Blaak et al. 2014a). However, the associated risk for human populations has not been assessed. Nevertheless, Leonard et al. (2018a) investigated antibiotic-resistant *E. coli* in 13 coastal bathing waters in England. As already reported, each *E. coli* isolate retrieved from these sites harboured an average of 1.24 ARGs. Nearly 2.5 million water sports sessions occurred in England in 2016, exposing people to the ingestion of at least 100 ARGs (Leonard et al. 2018a). This estimation only took into account one species (*E. coli*) and did not report the additional resistance hazard associated with other species of bacteria. Moreover, this study was carried out on viable *E. coli*, which are able to directly colonize the human gut. This result demonstrates that humans can be exposed to ARGs borne by bacterial species able to colonize the human intestinal microbiota during recreational aquatic activities. In a second work, Leonard et al. (2018b) found that 15% of the water samples from 97 coastal bathing waters (England and Wales) contained third-generation cephalosporin-resistant *E. coli* and that 0.12% of all retrieved *E. coli* were resistant to third-generation cephalosporin (and 0.07% harboured $bla_{\mathrm{CTX\text{-}M}}$ genes). The authors noted a low hazard for such a high level of exposure but concluded that the risk of environmental contamination does exist. Furthermore, Leonard et al. (2018b) estimated the risk of contamination for surfers by third-generation cephalosporin-resistant *E. coli*. To do so, 143 surfers were recruited together with 130 non-surfer controls. The authors found that 9.1% of the surfers were contaminated by third-generation cephalosporin-resistant *E. coli vs.* 3.1% of the non-surfers. Additionally, 6.3% of the surfers carried a $bla_{\mathrm{CTX\text{-}M}}$ gene *vs.* 3.1% for the controls. The authors determined that surfers have an increased risk (3–4 times) of being exposed to antibiotic resistance through the environment than non-surfers.

Fishing is a recreational activity that includes a lower environmental contamination exposure for human populations. Bollache et al. (2018) demonstrated that fish could be contaminated by ARB (cefotaxime-resistant *E. coli* and ESBL-producing *E. coli*), therefore supporting the health hazard associated with the consumption of fish from recreational fishing. However, fishing represents a very low health risk if the fish are cooked before eating.

4.5.4 *One Health or Not One Health?*

The One Health approach (initiated in the early 2000s) promotes an integrated, systemic and unified approach to public, animal and environmental health; this implies that human and animal health are interdependent and linked to the health of the ecosystems in which they coexist. We have previously seen that animals and the environment can be contaminated with antibiotic-resistant bacteria, becoming a potential source of contamination for humans.

Some studies have found a clonal relatedness between clinical and environmental antibiotic-resistant bacteria. Hence, Seni et al. (2018) showed that part of the human, animal and environmental chains in Tanzania shared some sequence types of ESBL-producing *Enterobacteriaceae*, including *E. coli* ST131. Ojer-Usoz et al. (2017) studied ESBL-producing *E. coli* retrieved in humans, food, WWTPs, water and rivers and reported that there was no sequence-type segregation according to the origin of the isolates, and Martak et al. (2020) found similar results. However, the resolution of the multilocus sequence-typing methods is too low to identify transmissions among compartments. Runcharoen et al. (2017) sequenced the genomes of clinical and environmental isolates of ESBL-producing *K. pneumoniae* collected in Bangkok (Thailand) and found no divergence between the two groups, suggesting the exchange of antibiotic-resistant bacteria between humans and the environment.

However, the majority of genomic studies generally sorted antibiotic-resistant bacteria from clinical and environmental sources into distinct clusters, thereby suggesting they are not exchanged. For example, Ludden et al. (2019) compared the genomes of ESBL-producing and ESBL-non-producing *E. coli* isolated from livestock farms and retail meat with those of *E. coli* isolates responsible for bloodstream infection in the UK. They found that livestock and patient isolates were genetically distinct, and some bacteria isolated from different farms were similar. This finding suggests that *E. coli* originating from livestock do not directly transfer to humans but can spread among animals. In a second study, the same group compared the genomes of *K. pneumoniae* retrieved in a hospital with those isolated from livestock farms, meat products, hospital sewers and WWTPs in England (Ludden et al. 2020). The authors observed a weak genetic link between hospital and extra-hospital isolates. This segregation was also observed at the gene level. Hence, although ARGs with resistance to cephalosporin were widespread, they were borne on different plasmids by hospital and extra-hospital isolates. As for *E. coli*, *K. pneumoniae* originating from livestock or meat are rarely observed in human infections. Previous results reported limited evidence of antibiotic-resistant bacteria transfer between livestock or food and humans, thus contradicting the One Health approach in the context of high income countries.

Mughini-Gras et al. (2019) conducted a population-based modelling study of ESBL-encoding and plasmid-mediated cephalosporinase genes collected from human sources (clinical patients, asymptomatic carriers in the open community, poultry and pig farmers, and people who travelled recently) and non-human sources (companion animals, farm animals, wild birds, food products and surface freshwater). Their model demonstrated that the majority (67.0%) of the ESBL-producing and plasmid-mediated cephalosporinase *E. coli* was acquired by interhuman transmission. In descending order, food accounted for 18.9%, companion animals accounted for 7.9% and farm animals accounted for 3.6% of carriage. The environment (i.e. freshwater and wild birds) only accounted for 2.6% of the community-acquired ESBL-producing and plasmid-mediated cephalosporinase *E. coli* carriage. However, frequent swimming in freshwater increased the environmental-associated carriage of the ESBL-producing and plasmid-mediated cephalosporinase E. coli in summer (from 2.3% to 5.7%) (Mughini-Gras et al. 2019). This study therefore suggests that the likelihood of human contamination with antibiotic resistance determinants from the environment may be low in high income countries. The likelihood of exposure to waterborne antibiotic resistance is much higher in low and middle income countries with poor water sanitation and hygiene infrastructure than it is in high income countries with more robust water sanitation and hygiene infrastructure.

4.5.5 Climate Change and Antibiotic Resistance

Some studies have noted that climate change could increase environmental contamination with antibiotic resistance. For example, MacFadden et al. (2018) demonstrated that a 10 °C increase in the minimum temperature could be associated with increases in antibiotic resistance of 4.2%, 2.2% and 2.7% for *E. coli*, *K pneumoniae* and *S. aureus*, respectively. Similarly, Kaba et al. (2020) confirmed the potential impact of increasing temperatures on antimicrobial resistance by showing that a 0.5 °C increase in the year-wise temperature would result in a 1.02-fold increase in carbapenem-resistant *P. aeruginosa* prevalence. Moreover, Pärnänen et al. (2019) showed that WWTPs in southern European countries (i.e. Portugal, Spain, and Cyprus) generally released more ARGs than those in northern European countries (i.e. Germany, Norway, and Finland). Knowing the optimal growth temperatures of *E. coli* (37°C), *K. pneumoniae* (30–35°C), *P. aeruginosa* (37°C), and *Acinetobacter* spp. (33–35°C), the authors suggested that this higher resistance dispersion in southern European countries may be partly due to the increased survival of antibiotic-resistant bacteria in these warmer environments. Nevertheless, it is important to keep in mind that an increase in temperature can also promote the development of predators (especially protozoa). Moreover, Czekalski et al. (2015) related some ARG abundances with the level of eutrophication of lakes, which is expected to increase with global change. In addition to the global increase in temperature, the general decrease in precipitation, increase in the precipitation intensity (European Academies Science Advisory Council 2018) and subsequent increase in combined sewer overflows could also accelerate the dissemination of antibiotic resistance in the environment downstream of combined sewage systems (Jalliffier-Verne et al. 2015; Olds et al. 2018). Moreover, the future increase in droughts (European academies' science advisory council 2018) will decrease river flows, and thus the

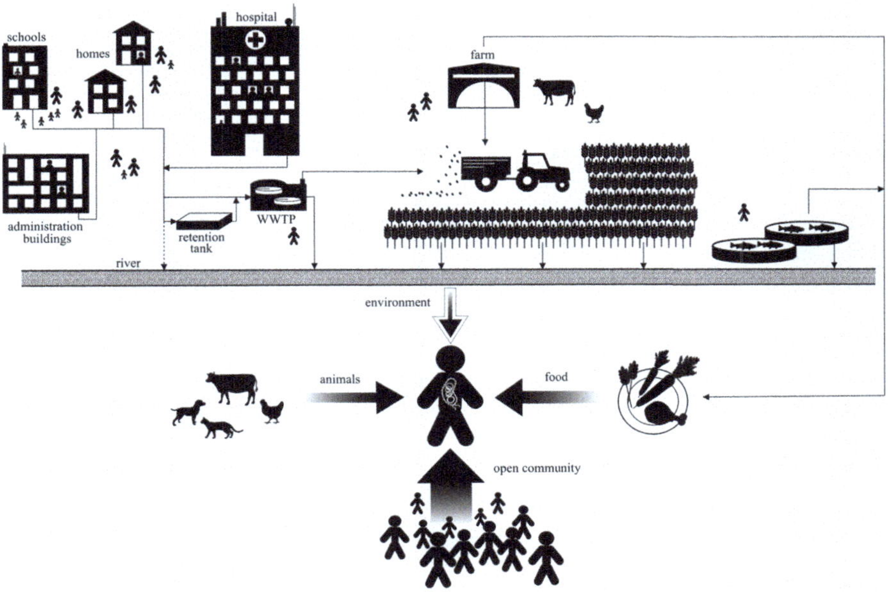

Fig. 4.7 General summary scheme. Wastewater treatment plants, agriculture and aquaculture are the main sources of antibiotic resistance in the environment. The environmental antibiotic resistance hazard is limited in high income countries but is highest in summer (i.e. increased exposure) and in low and middle income countries (i.e. increased hazard and exposure) (*WWTP* wastewater treatment plant)

dilution of ARB released in the receiving water through the treated effluent from WWTPs. Finally, the future increase in the world population will lead to a global increase in antibiotic resistance (Jalliffier-Verne et al. 2015) and its associated risk, *via* both hazard and exposure increases, especially in low and middle income countries.

Overall, the following conclusions can be summarized (Fig. 4.7): (i) the antibiotic resistance exposome (i.e. the sum of an individual's lifetime environmental exposure to antibiotic resistance in water, food and other environmental compartments) can pose a health hazard for human populations; (ii) poultry meat is highly contaminated by antibiotic-resistant bacteria, but the associated hazard is low if the meat is well cooked; (iii) the general environmental exposure associated with antibiotic resistance is limited in high income countries; (iv) few studies have estimated the real risk of contamination of humans by antibiotic-resistant bacteria after exposure to the environment; (v) studies are needed to assess this risk, especially in low and middle income countries; and (vi) climate change and the increase of human population could increase antibiotic resistance in the environment.

4.6 What Are the Solutions to Reduce Antibiotic-Resistant Bacteria Dispersion in the Environment and the Associated Likelihood of Exposure?

4.6.1 The Design of Wastewater Sewage Systems

Combined sewage systems need particular attention. First, wastewater treatment plants that collect wastewater from combined sewage systems should be equipped with retention tanks to limit combined sewer overflows. Moreover, separate sewage systems, when appropriate, need to be prioritized in newly built areas. Obviously, separate sewage systems are not the solution in all contexts, especially in densely anthropized areas where runoff is also highly polluted, and effluent needs to be treated before being discharged into the environment. However, separate sewer systems need to be considered and encouraged in moderately and minimally anthropized areas. A risk assessment of environmental contamination by microorganisms needs to be performed to properly design sewage systems.

4.6.2 Wastewater Treatment Plants

Reduction of Antibiotic Resistance in Wastewater by Advanced Wastewater Treatment Methods Although WWTPs reduce the absolute load of antibiotic-resistant bacteria, they are a major source of the spread of antibiotic-resistant bacteria in the environment. Some studies even reported a relative enrichment of antibiotic-resistant bacteria through WWTP processes (i.e. compared to the total bacterial load). Biological treatments are not designed to reduce antibiotic-resistance. Nevertheless, advanced wastewater treatments can effectively reduce bacterial pollution within WWTPs.

Water chlorination adds chlorine or chlorine compounds to the water. Chlorine is highly toxic to microorganisms and can efficiently clear them from wastewater. Disinfection by chlorination is one of the most commonly used methods to kill microorganisms, including ARB, in the tertiary treatment of WWTPs. However, chlorination does not significantly reduce antibiotic resistance in WWTPs (Iwane et al. 2001; Auerbach et al. 2007; Gao et al. 2012; Munir et al. 2011). Moreover, disinfections based on the chemically aggressive destruction of bacterial cell structures, such as chlorination, can promote residual microbial communities that are increasingly resistant to antibiotics and competitive in natural environments (Dodd 2012; Mao et al. 2015; Di Cesare et al. 2016b).

Conventional advanced oxidation processes include UV radiation, ozone treatment and hydrogen peroxide. Gouliouris et al. (2019) showed that UV radiation reduced environmental contamination with vancomycin-resistant *E. faecium*. In contrast, Auerbach et al. (2007) reported that the UV radiation of wastewater effluent showed no reduction in 10 tetracycline antibiotic resistance genes (Auerbach et al. 2007). Jäger et al. (2018b) found that the efficiency of ozone treatment depends on the ARGs considered. Among conventional advanced oxidation processes, the authors reported that ozone treatment was the most efficient in the reduction of ARGs. Generally, advanced oxidation processes have been reported to be efficient in reducing antibiotic-resistant bacteria compared to the use of conventional secondary treatment by WWTPs.

The solar photo-Fenton process is an advanced oxidation process that improves solar disinfection by the addition of H_2O_2 and Fe^{2+} to produce hydroxyl radicals that degrade the external membrane of bacteria and eventually inactivate the cell (Giannakis et al. 2018). Giannakis et al. (2018) tested the effect of this process on a range of antibiotic-resistant *E. coli*, *S. aureus* and *K. pneumoniae*. They found that antibiotic-resistant bacteria reduction depends on the treatment duration. A notable decrease in ARB occurred after 30 min of solar photo-Fenton treatment or after a 3-h exposure to light. *S. aureus* and *E. coli* were totally inactivated by the solar photo-Fenton treatment within 1.5 h, whereas 4 h was required with solar light only (Giannakis et al. 2018).

In low and middle income countries with poor water sanitation infrastructure, the implementation of the costly processes described above is unreasonable. Affordable treatment processes that could limit the contamination of the environment, such as the construction of wetlands or oxidation ponds, could be implemented (Sleytr et al. 2007; Butler et al. 2017).

Reduction of Antibiotic Resistance in Sewage Sludge Anaerobic treatments are generally described as more efficient than aerobic treatments in reducing antibiotic resistance in sewage sludge (Diehl and LaPara 2010; Tian et al. 2016; Bürgmann et al. 2018), even if the majority of studies have focused on antibiotic resistance genes rather than antibiotic-resistant bacteria. Several studies have suggested that thermophilic anaerobic digestion (48°C–55°C) is comparatively more efficient for removing antibiotic resistance genes (Diehl and LaPara 2010; Tian et al. 2016).

Composting is also widely used in the sewage sludge treatment process. The majority of investigations have been conducted on antibiotic resistance genes rather than antibiotic-resistant bacteria. Tien et al. (2017) tested several processes associated with ARG removal before manure spreading (untreated, anaerobically digested, mechanically dewatered or composted) and reported that composted manure displayed the lowest ARG abundances. The ARG loading rates were 10- to 1000-fold lower in soils receiving composted manure than those for raw manure, reflecting the efficiency of composting in lowering the antibiotic resistance dissemination. Nevertheless, composted manure has also been shown to increase the diversity of antibiotic resistance genes (Chen et al. 2019). Tien et al. (2017) recommended a delay of one season between the application of manure and the harvest of fresh produce to minimize the antibiotic resistance hazard.

Aerobic digestion, anaerobic digestion and composting are not considered as stabilization processes; they need to be completed with heat or lime stabilization, for example, to reduce the antibiotic-resistant bacteria load of sewage sludge/manure.

Finally, Lau et al. (2017) observed that with the application of pelletized biosolids, the abundance of antibiotic resistance genes in receiving soils was lowered.

Reducing the Selective Pressure It is also possible to control pollutants that exert a selective pressure on ARB, such as antibiotics, to limit the selection of ARB in sewers, WWTPs and receiving environments. Some antibiotics are stable in wastewater (e.g. quinolones, macrolides and sulfamides) and then for a long time in the receiving environments. Briefly, the major removal pathways of antibiotics include adsorption, biodegradation, disinfection and membrane filtration in wastewater treatment processes (Li and Zhang 2010; Berendonk et al. 2015). UV radiation can efficiently reduce the antibiotic concentration in wastewater, therefore reducing the selective pressure within the WWTP and downstream area (Liu et al. 2017; Bürgmann et al. 2018). Composting and anaerobic digestion have also been demonstrated to be efficient in reducing the levels of some antibiotics in sewage sludge (Ezzariai 2018).

Some advanced wastewater treatments are therefore efficient in reducing the dispersion of antibiotic resistance in the environment; their efficiency depends on the antibiotic resistance determinant and the context of the WWTP. Further investigations are needed to better understand the microbial ecological processes that take place at WWTPs, and to test the ability of new advanced wastewater treatments and sewage sludge processes to reduce the ARB and ARG loads (e.g. activated carbon followed by ultrafiltration; Margot et al. 2013).

4.6.3 Other Sources of Antibiotic-Resistant Bacteria

For hospital, we have previously seen that it is inefficient to separately treat hospital wastewater and urban wastewater (Buelow et al. 2020). Moreover, the cost/benefit ratio of such a pretreatment would not be favourable because it is too costly for a very limited reduction in antibiotic-resistant bacteria when taking into account the entire sewage system. Nevertheless, the appropriate use of antibiotics will reduce the antibiotic concentrations in hospital wastewater, the selective pressure on sewage systems and WWTPs and the abundance of antibiotic-resistant bacteria in WWTP effluents and biosolids.

The reduction of antibiotics in agriculture and fish farms will also limit the contamination of the environment. Furthermore, the spread of biosolids or manure needs to be adapted to consider weather events, such as precipitation and frost (Beattie et al. 2018). Composting can also help in the reduction of antibiotic resistance on agricultural lands. As described above, 'buffer periods' could be established between spreading and cereal/vegetable cultivation (Tien et al. 2017).

Finally, the treatment of the effluents of pharmaceutical industries is crucial for reducing the selective pressure in receiving environments. This need is particularly notable in India, where a poor sanitation system and many important pharmaceutical industries exist (Changing Markets and Ecostorm 2016; World Health Organization 2019).

4.6.4 Education and Practices

Generally, education, knowledge exchange and antibiotic prescription practices are pillars for reducing the dispersion of antibiotic resistance (Topp et al. 2018; Bürgmann et al. 2018). If not already implemented, policies need to be adopted to better manage the use of antibiotics. These policies must not only be implemented but also monitored over long periods so that they can be adapted to changing situations or local contexts. Policies and research are needed at large scales, but even more at low scales, to limit the dispersion of antibiotic resistance, considering local parameters. At the landscape scale, the factors that control the dissemination of antibiotic-resistant bacteria and the associated risk are poorly understood and need further investigation in environments affected by faecal pollution.

Low and middle income countries are particularly concerned with these recommendations because the contamination of the environment by antibiotic resistance is high, and the populations are therefore often exposed due to poor sanitation infrastructure (World Health Organization 2019). Improving sanitation, increasing access to clean water and good governance are necessary to reduce the spread of global antimicrobial resistance (Collignon et al. 2018). Moreover, knowledge and water treatment techniques/technologies for a limited environmental contamination by antibiotic-resistant bacteria must be fully accessible to local decision makers: this is also true for the implementation of uncontaminated drinking water networks. As in high income countries, populations in low and middle income countries need to be made aware of antibiotic resistance and its possible health consequences in the near future.

Overall, the following conclusions can be summarized: (i) particular importance should be placed on the adequate design adequacy of the sewage systems; when a combined sewage system is already operative, the corresponding WWTP should be equipped with retention tanks to reduce combined sewer overflows; (ii) new processes for wastewater or sewage sludge treatment could reduce ARB and ARG loads but require further investigation; (iii) decreasing of antibiotic use for humans and animals can reduce the spread of antibiotics and antibiotic resistance determinants in the environment; (iv) special attention must be paid to the effluents of pharmaceutical industries; (v) managing the spread of antimicrobial resistance in the environment will require increased public awareness, knowledge exchange and the implementation of local policies; and (vi) ultimately, efforts will have to be concentrated in low and middle income countries, where the environment and populations are the most exposed to antibiotic resistance hazards.

4.7 Conclusions

Even if wastewater treatment plants reduce the antibiotic-resistant bacteria load of wastewater, their outputs (treated wastewater and biosolids) are the main sources of antibiotic resistance determinants into the environment in high income countries, especially if wastewater treatment plants are connected to a combined sewage system (Fig. 4.7). Even if hospital wastewater is highly contaminated with antibiotic-resistant bacteria, urban wastewater contributes to most of the antibiotic resistance related to wastewater treatment plants. Agriculture and aquaculture practices also release antibiotic-resistant bacteria in the environment (Table 4.2, Fig. 4.7). Resulting from these multiple sources of release, antibiotic-resistant bacteria are retrieved in many ecosystems, mostly aquatic ones. Logically, contamination with antibiotic-resistant bacteria is related to the level of anthropization. The roles of flora and fauna in the dissemination and survival of human-associated antibiotic resistance determinants are poorly known and require further investigation. The selective pressures exerted by antibiotics, metals and biocides could have little influence on antibiotic-resistant bacteria enrichment in contaminated environments (Table 4.2). However, the situation is different in environments heavily contaminated by antibiotics, especially downstream of pharmaceutical industries. Despite the dispersal of antibiotic resistance in our environment, the associated risk is limited in high income countries (Fig. 4.7). In low and middle income countries with poor water sanitation and hygiene infrastructure, the conclusions need to be nuanced because the majority of human waste goes untreated into the environment. It is clear that antibiotic resistance spread via the environment and throughout the One Health continuum is enormously problematic in these countries.

References

Abraham TJ (2011) Food safety hazards related to emerging antibiotic resistant bacteria in cultured freshwater fishes of Kolkata, India. Adv J Food Sci Technol 3:69–72

Amoros C, Bornette G (2002) Connectivity and biocomplexity in waterbodies of riverine floodplains. Freshw Biol 47:761–776. https://doi.org/10.1046/j.1365-2427.2002.00905.x

Anderson KL, Whitlock JE, Harwood VJ (2005) Persistence and differential survival of fecal indicator bacteria in subtropical waters and sediment. Appl Environ Microbiol 71:3041–3048. https://doi.org/10.1128/AEM.71.6.3041-3048.2005

Arias-Andres M, Klümper U, Rojas-Jimenez K, Grossart HP (2018) Microplastic pollution increases gene exchange in aquatic ecosystems. Environ Pollut 237:253–261. https://doi.org/10.1016/j.envpol.2018.02.058

Auerbach EA, Seyfried EE, McMahon KD (2007) Tetracycline resistance genes in activated sludge wastewater treatment plants. Water Res 41:1143–1151. https://doi.org/10.1016/j.watres.2006.11.045

Baker-Austin C, Wright MS, Stepanauskas R, McArthur JV (2006) Co-selection of antibiotic and metal resistance. Trends Microbiol 14:176–182. https://doi.org/10.1016/j.tim.2006.02.006

Baquero F, Martínez JL, Cantón R (2008) Antibiotics and antibiotic resistance in water environments. Curr Opin Biotechnol 19:260–265. https://doi.org/10.1016/j.copbio.2008.05.006

Beattie RE, Walsh M, Cruz MC, McAliley LR, Dodgen L, Zheng W, Hristova KR (2018) Agricultural contamination impacts antibiotic resistance gene abundances in river bed sediment temporally. FEMS Microbiol Ecol 94. https://doi.org/10.1093/femsec/fiy131

Bengtsson-Palme J, Hammaren R, Pal C, Östman M, Björlenius B, Flach CF, Fick J, Kristiansson E, Tysklind M, Larsson DGJ (2016) Elucidating selection processes for antibiotic resistance in sewage treatment plants using metagenomics. Sci Total Environ 572:697–712. https://doi.org/10.1016/j.scitotenv.2016.06.228

Berendonk TU, Manaia CM, Merlin C, Fatta-Kassinos D, Cytryn E, Walsh F, Bürgmann H, Sørum H, Norström M, Pons M-N (2015) Tackling antibiotic resistance: the environmental framework. Nat Rev Microbiol 13:310–317. https://doi.org/10.1038/nrmicro3439

Berg G, Eberl L, Hartmann A (2005) The rhizosphere as a reservoir for opportunistic human pathogenic bacteria. Environ Microbiol 7:1673–1685. https://doi.org/10.1111/j.1462-2920.2005.00891.x

Berg ES, Wester AL, Ahrenfeldt J, Mo SS, Slettemeås JS, Steinbakk M, Samuelsen Ø, Grude N, Simonsen GS, Løhr IH, Jørgensen SB (2017) Norwegian patients and retail chicken meat share cephalosporin-resistant *Escherichia coli* and IncK/blaCMY-2 resistance plasmids. Clin Microbiol Infect 23:407.e9-407.e15. https://doi.org/10.1016/j.cmi.2016.12.035

Berthe T, Touron A, Leloup J, Deloffre J, Petit F (2008) Faecal-indicator bacteria and sedimentary processes in estuarine mudflats (seine, France). Mar Pollut Bull 57:59–67. https://doi.org/10.1016/j.marpolbul.2007.10.013

Berthe T, Ratajczak M, Clermont O, Denamur O, Petit F (2013) Evidence for coexistence of distinct *Escherichia coli* populations in various aquatic environments and their survival in estuary water. Appl Environ Microbiol 79:4684–4693. https://doi.org/10.1128/AEM.00698-13

Blaak H, de Kruijf P, Hamidjaja RA, van Hoek A, de Roda Husman AM, Schets FM (2014a) Prevalence and characteristics of ESBL-producing *E. coli* in Dutch recreational waters influenced by wastewater treatment plants. Vet Microbiol 171:448–459. https://doi.org/10.1016/j.vetmic.2014.03.007

Blaak H, van Hoek AHAM, Veenman C, Docters van Leeuwen AE, Lynch G, van Overbeek WM, de Roda Husman AM (2014b) Extended spectrum ß-lactamase- and constitutively AmpC-producing *Enterobacteriaceae* on fresh produce and in the agricultural environment. Int J Food Microbiol 168-169:8–16. https://doi.org/10.1016/j.ijfoodmicro.2013.10.006

Bollache L, Bardet E, Depret G, Motreuil S, Neuwirth C, Moreau J, Hartmann A (2018) Dissemination of CTX-M-producing *Escherichia coli* in freshwater fishes from a French watershed (Burgundy). Front Microbiol 9:32–39. https://doi.org/10.3389/fmicb.2018.03239

Börjesson S, Matussek A, Melin S, Löfgren S, Lindgren PE (2010) Methicillin-resistant *Staphylococcus aureus* (MRSA) in municipal wastewater: an uncharted threat? J Appl Microbiol 108:1244–1251. https://doi.org/10.1111/j.1365-2672.2009.04515.x

Bornette G, Tabacchi E, Hupp C, Puijalon S, Rostan JC (2008) A model of plant strategies in fluvial hydrosystems. Freshw Biol 53:1692–1705. https://doi.org/10.1111/j.1365-2427.2008.01994.x

Bravo S, Dolz H, Silva MT, Lagos C, Millanao A, Urbina M (2005) Diagnóstico del uso de fármacos y otros productos químicos en la acuicultura. Universidad Austral de Chile Facultad de Pesquerías y Oceanografía, Instituto de Acuicultura, Puerto Montt, Chile Informe Final Proyecto 2003–28:256

Bréchet C, Plantin J, Sauget M, Thouverez M, Talon D, Cholley P, Guyeux C, Hocquet D, Bertrand X (2014) Wastewater treatment plants release large amounts of extended-spectrum β-lactamase-producing *Escherichia coli* into the environment. Clin Infect Dis 58:1658–1665. https://doi.org/10.1093/cid/ciu190

Brennan FP, Moynihan E, Griffiths BS, Hillier S, Owen J, Pendlowski H, Avery LM (2014) Clay mineral type effect on bacterial enteropathogen survival in soil. Sci Total Environ 468:302–305. https://doi.org/10.1016/j.scitotenv.2013.08.037

Brooks JP, Maxwell SL, Rensing C, Gerba CP, Pepper IL (2007) Occurrence of antibiotic-resistant bacteria and endotoxin associated with the land application of biosolids. Can J Microbiol 53:616–622. https://doi.org/10.1139/W07-021

Buelow E, Bayjanov JR, Majoor E, Willems RJL, Bonten MJM, Schmitt H, van Schaik W (2018) Limited influence of hospital wastewater on the microbiome and resistome of wastewater in a community sewerage system. FEMS Microbiol Ecol 94. https://doi.org/10.1093/femsec/fiy087

Buelow E, Rico A, Gaschet M, Lourenço J, Kennedy SP, Wiest L, Ploy MC, Dagot C (2020) Hospital discharges in urban sanitation systems: long-term monitoring of wastewater resistome and microbiota in relationship to their eco-exposome. Water Res. https://doi.org/10.1016/j.wroa.2020.100045

Bürgmann H, Frigon D, Gaze WH, Manaia CM, Pruden A, Singer AC, Smets BF, Zhang T (2018) Water and sanitation: an essential battlefront in the war on antimicrobial resistance. FEMS Microbiol Ecol 94. https://doi.org/10.1093/femsec/fiy101

Butler E, Hung YT, Al Ahmad MS, Yeh RYL, Liu RLH, Fu YP (2017) Oxidation pond for municipal wastewater treatment. Appl Water Sci 7:31–51. https://doi.org/10.1007/s13201-015-0285-z

Cabello FC (2006) Heavy use of prophylactic antibiotics in aquaculture: a growing problem for human and animal health and for the environment. Environ Microbiol 8:1137–1144. https://doi.org/10.1111/j.1462-2920.2006.01054.x

Cabello FC, Godfrey HP, Tomova A, Ivanova L, Dölz H, Millanao A, Buschmann AH (2013) Antimicrobial use in aquaculture re-examined: its relevance to antimicrobial resistance and to animal and human health. Environ Microbiol 15:1917–1942. https://doi.org/10.1111/1462-2920.12134

Cahill N, O'Connor L, Mahon B, Varley A, McGrath E, Ryan P, Cormican M, Brehony C, Jolley KA, Maiden MC (2019) Hospital effluent: a reservoir for carbapenemase-producing *Enterobacterales*? Sci Total Environ 672:618–624. https://doi.org/10.1016/j.scitotenv.2019.03.428

Capodaglio AG (2004) Improving sewage treatment plant performance in wet weather. In: Enhancing urban environment by environmental upgrading and restoration. Springer, Dordrecht, pp 175–185

Carbonne A, Arnaud I, Maugat S, Marty S, Dumartin C, Bertrand X, Bajolet O, Savey A, Fosse T, Eveillard M (2012) National multidrug-resistant bacteria (MDRB) surveillance in France through the RAISIN network: a 9 year experience. J Antimicrob Chemother 68:954–959. https://doi.org/10.1093/jac/dks464

Casals-Pascual C, Vergara A, Vila J (2018) Intestinal microbiota and antibiotic resistance: perspectives and solutions. Hum Microbiome J 9:11–15. https://doi.org/10.1016/j.humic.2018.05.002

Cassini A, Högberg LD, Plachouras D, Quattrocchi A, Hoxha A, Simonsen GS, Colomb-Cotinat M, Kretzschmar ME, Devleesschauwer B, Cecchini M (2018) Attributable deaths and disability-adjusted life-years caused by infections with antibiotic-resistant bacteria in the EU and the European economic area in 2015: a population-level modelling analysis. Lancet Infect Dis 19:56–66. https://doi.org/10.1016/S1473-3099(18)30605-4

Cecchini M, Langer J, Slawomirski L (2015) Antimicrobial resistance in G7 countries and beyond. Economic issues, policies, and options for action report of organization for economic co-operation and Development (OECD) 1–75

Changing Markets, Ecostorm (2016) Superbugs in the supply chain: how pollution from antibiotics factories in India and China is fuelling the global rise of drug-resistant infections. London, Foundation Changing Markets

Chaplin CE (1952) Bacterial resistance to quaternary ammonium disinfectants. J Bacteriol 63:453–458

Chen Q, An X, Li H, Su J, Ma Y, Zhu YG (2016) Long-term field application of sewage sludge increases the abundance of antibiotic resistance genes in soil. Environ Int 92:1–10. https://doi.org/10.1016/j.envint.2016.03.026

Chen Z, Zhang W, Yang L, Stedtfeld RD, Peng A, Gu C, Boyd SA, Li H (2019) Antibiotic resistance genes and bacterial communities in cornfield and pasture soils receiving swine and dairy manures. Environ Poll 248:947–957. https://doi.org/10.1016/j.envpol.2019.02.093

Chiapponi E, Henriot CP, Bertrand X, Hocquet D, Bornette G (2020) Using GFP-tagged *Escherichia coli* to investigate the persistence of fecal bacteria in vegetated wetlands: an experimental approach. Antibiotics 9:335. https://doi.org/10.3390/antibiotics9060335

Chu BT, Petrovich ML, Chaudhary A, Wright D, Murphy B, Wells G, Poretsky R (2018) Metagenomics reveals the impact of wastewater treatment plants on the dispersal of microorganisms and genes in aquatic sediment. Appl Environ Microbiol 84. https://doi.org/10.1128/AEM.02168-17

Collignon P, Beggs JJ, Walsh TR, Gandra S, Laxminarayan S (2018) Anthropological and socioeconomic factors contributing to global antimicrobial resistance: a univariate and multivariable analysis. Lancet Planet Health 2:398–405. https://doi.org/10.1016/S2542-5196(18)30186-4

Crini G, Lichtfouse E (2019) Advantages and disadvantages of techniques used for wastewater treatment. Environ Chem Lett 17:145–155. https://doi.org/10.1007/s10311-018-0785-9

Czekalski N, Díez EG, Bürgmann H (2014) Wastewater as a point source of antibiotic-resistance genes in the sediment of a freshwater lake. ISME J 8:1381–1390. https://doi.org/10.1038/ismej.2014.8

Czekalski N, Sigdel R, Birtel J, Matthews B, Bürgmann H (2015) Does human activity impact the natural antibiotic resistance background? Abundance of antibiotic resistance genes in 21 Swiss lakes. Environ Int 81:45–55. https://doi.org/10.1016/j.envint.2015.04.005

Davin-Regli A, Pagès JM (2012) Cross-resistance between biocides and antimicrobials: an emerging question. Revue Scientifique et Technique-OIE 31:89

Di Cesare A, Eckert EM, Teruggi A, Fontaneto D, Bertoni R, Callieri C, Corno G (2015) Constitutive presence of antibiotic resistance genes within the bacterial community of a large subalpine lake. Mol Ecol 24:3888–3900. https://doi.org/10.1111/mec.13293

Di Cesare A, Eckert E, Corno G (2016a) Co-selection of antibiotic and heavy metal resistance in freshwater bacteria. J Limnol 75:59–66. https://doi.org/10.4081/jlimnol.2016.1198

Di Cesare A, Fontaneto D, Doppelbauer J, Corno G (2016b) Fitness and recovery of bacterial communities and antibiotic resistance genes in urban wastewaters exposed to classical disinfection treatments. Environ Sci Technol 50:10153–10161. https://doi.org/10.1016/j.ecolmodel.2017.04.006

Dickinson AW, Power A, Hansen MG, Brandt KK, Piliposian G, Appleby P, O'Neill PA, Jones RT, Sierocinski P, Koskella B, Vos M (2019) Heavy metal pollution and co-selection for antibiotic resistance: a microbial palaeontology approach. Environ Int 132:105117. https://doi.org/10.1016/j.envint.2019.105116

Diehl DL, LaPara TM (2010) Effect of temperature on the fate of genes encoding tetracycline resistance and the integrase of class 1 integrons within anaerobic and aerobic digesters treating municipal wastewater solids. Environ Sci Technol 44:9128–9133. https://doi.org/10.1021/es102765a

Dinh QT, Moreau-Guigon E, Labadie P, Alliot F, Teil MJ, Blanchard M, Eurin J, Chevreuil M (2017) Fate of antibiotics from hospital and domestic sources in a sewage network. Sci Total Environ 575:758–766. https://doi.org/10.1016/j.scitotenv.2016.09.118

Dodd MC (2012) Potential impacts of disinfection processes on elimination and deactivation of antibiotic resistance genes during water and wastewater treatment. J Environ Monit 14:1754–1771. https://doi.org/10.1039/C2EM00006G

Drieux L, Haenn S, Moulin L, Jarlier V (2016) Quantitative evaluation of extended-spectrum β-lactamase-producing *Escherichia coli* strains in the wastewater of a French teaching hospital and relation to patient strain. Antimicrob Resist Infect Control 5:9. https://doi.org/10.1186/s13756-016-0108-5

European Academies Science Advisory Council (EASAC) (2018) Extreme weather events in Europe: preparing for climate change adaptation: an update on EASAC's 2013 study. https://easac.eu/publications/details/extreme-weather-events-in-europe

European Centre for Disease Prevention and Control (ECDC), European Medicines Agency (EMEA) (2009) Joint technical report: the bacterial challenge: time to react. https://www.ecdc.europa.eu/en/publications-data/ecdcemea-joint-technical-report-bacterial-challenge-time-react

European Medicines Agency (EMEA) (2019) European surveillance of veterinary antimicrobial consumption. Sales of veterinary antimicrobial agents in 31 European countries in 2017. EMA/294674/2019

Ezzariai A (2018) Devenir des antibiotiques lors du traitement aérobie et anaérobie des boues de STEPs pour une valorisation agronomique. Doctoral dissertation, Institut National Polytechnique de Toulouse, France

Forsberg KJ, Reyes A, Wang B, Selleck EM, Sommer MOA, Dantas G (2012) The shared antibiotic resistome of soil bacteria and human pathogens. Science 337:1107–1111. https://doi.org/10.1126/science.1220761

Galvin S, Boyle F, Hickey P, Vellinga A, Morris D, Cormican M (2010) Enumeration and characterization of antimicrobial-resistant *Escherichia coli* bacteria in effluent from municipal, hospital, and secondary treatment facility sources. Appl Environ Microbiol 76:4772–4779. https://doi.org/10.1128/AEM.02898-09

Gao P, Munir M, Xagoraraki I (2012) Correlation of tetracycline and sulfonamide antibiotics with corresponding resistance genes and resistant bacteria in a conventional municipal wastewater treatment plant. Sci Total Environ 421:173–183. https://doi.org/10.1016/j.scitotenv.2012.01.061

Garcia-Armisen T, Servais P (2004) Enumeration of viable *E. coli* in rivers and wastewaters by fluorescent in situ hybridization. J Microbiol Methods 58:269–279. https://doi.org/10.1016/j.mimet.2004.04.014

Giannakis S, Le TTM, Entenza JM, Pulgarin C (2018) Solar photo-Fenton disinfection of 11 antibiotic-resistant bacteria (ARB) and elimination of representative AR genes. Evidence that antibiotic resistance does not imply resistance to oxidative treatment. Water Res 143:334–345. https://doi.org/10.1016/j.watres.2018.06.062

Gordon L, Giraud E, Ganière JP, Armand F, Bouju-Albert A, De la Cotte N, Mangion C, Le Bris H (2007) Antimicrobial resistance survey in a river receiving effluents from freshwater fish farms. J Appl Microbiol 102:1167–1176. https://doi.org/10.1111/j.1365-2672.2006.03138.x

Gouliouris T, Raven KE, Moradigaravand D, Ludden C, Coll F, Blane B, Naydenova P, Horner C, Brown NM, Corander J (2019) Detection of vancomycin-resistant *Enterococcus faecium* hospital-adapted lineages in municipal wastewater treatment plants indicates widespread distribution and release into the environment. Genome Res 29:626–634. https://doi.org/10.1101/gr.232629.117

Gullberg E, Cao S, Berg OG, Ilbäck C, Sandegren L, Hughes D, Andersson DI (2011) Selection of resistant bacteria at very low antibiotic concentrations. PLoS Pathog 7. https://doi.org/10.1371/journal.ppat.1002158

Guo J, Li J, Chen H, Bond PL, Yuan Z (2017) Metagenomic analysis reveals wastewater treatment plants as hotspots of antibiotic resistance genes and mobile genetic elements. Water Res 123:468–478. https://doi.org/10.1016/j.watres.2017.07.002

Guyomard-Rabenirina S, Dartron C, Falord M, Sadikalay S, Ducat C, Richard V, Breurec S, Gros O, Talarmin A (2017) Resistance to antimicrobial drugs in different surface waters and wastewaters of Guadeloupe. PLoS One 12. https://doi.org/10.1371/journal.pone.0173155

Hartmann A, Amoureux L, Locatelli A, Depret G, Jolivet C, Gueneau E, Neuwirth C (2012) Occurrence of CTX-M producing *Escherichia coli* in soils, cattle, and farm environment in France (Burgundy region). Front Microbiol 3:83. https://doi.org/10.3389/fmicb.2012.00083

Henriot CP, Martak D, Cuenot Q, Loup C, Masclaux H, Gillet F, Bertrand X, Hocquet D, Bornette G (2019) Occurrence and ecological determinants of the contamination of floodplain wetlands with *Klebsiella pneumoniae* and pathogenic or antibiotic-resistant *Escherichia coli*. FEMS Microbiol Ecol 95. https://doi.org/10.1093/femsec/fiz097

Heß S, Berendonk TU, Kneis D (2018) Antibiotic resistant bacteria and resistance genes in the bottom sediment of a small stream and the potential impact of remobilization. FEMS Microbiol Ecol 94. https://doi.org/10.1093/femsec/fiy128

Hocquet D, Muller A, Bertrand X (2016) What happens in hospitals does not stay in hospitals: antibiotic-resistant bacteria in hospital wastewater systems. J Hosp Infect 93:395–402. https://doi.org/10.1016/j.jhin.2016.01.010

Honda R, Tachi C, Yasuda K, Hirata T, Noguchi M, Hara-Yamamura H, Yamamoto-Ikemoto R, Watanabe T (2020) Estimated discharge of antibiotic-resistant bacteria from combined sewer overflows of urban sewage system. Clean Water 3:1–7. https://doi.org/10.1038/s41545-020-0059-5

Hultman J, Tamminen M, Pärnänen K, Cairns J, Karkman A, Virta M (2018) Host range of antibiotic resistance genes in wastewater treatment plant influent and effluent. FEMS Microbiol Ecol 94. https://doi.org/10.1093/femsec/fiy038

Ivleva NP, Wiesheu AC, Niessner R (2017) Microplastic in aquatic ecosystems. Angew Chem Int Ed 56:1720–1739. https://doi.org/10.1002/anie.201606957

Iwane T, Urase T, Yamamoto K (2001) Possible impact of treated wastewater discharge on incidence of antibiotic resistant bacteria in river water. Water Sci Technol 43:91–99. https://doi.org/10.2166/wst.2001.0077

Jacopin C, Bertrand-Krajewski JL, Desbordes M (1999) Characterisation and settling of solids in an open, grassed, stormwater sewer network detention basin. Water Sci Technol 39:135–144. https://doi.org/10.1016/S0273-1223(99)00017-7

Jäger T, Alexander J, Kirchen S, Dötsch A, Wieland A, Hiller C, Schwartz T (2018a) Live-dead discrimination analysis, qPCR assessment for opportunistic pathogens, and population analysis at ozone wastewater treatment plants. Environ Pollut 232:571–579. https://doi.org/10.1016/j.envpol.2017.09.089

Jäger T, Hembach N, Elpers C, Wieland A, Alexander J, Hiller C, Krauter G, Schwartz T (2018b) Reduction of antibiotic resistant bacteria during conventional and advanced wastewater treatment, and the disseminated loads released to the environment. Front Microbiol 9:2599. https://doi.org/10.3389/fmicb.2018.02599

Jalliffier-Verne I, Leconte R, Huaringa-Alvarez U, Madoux-Humery A-S, Galarneau M, Servais P, Prévost M, Dorner S (2015) Impacts of global change on the concentrations and dilution of combined sewer overflows in a drinking water source. Sci Total Environ 508:462–476. https://doi.org/10.1016/j.scitotenv.2014.11.059

Jeanvoine A, Meunier A, Puja H, Bertrand X, Valot B, Hocquet D (2019) Contamination of a hospital plumbing system by persister cells of a copper-tolerant high-risk clone of Pseudomonas aeruginosa. Water Res 157:579–586. https://doi.org/10.1016/j.watres.2019.04.011

Jiang X, Ellabaan MMH, Charusanti P, Munck C, Blin K, Tong Y, Weber T, Sommer MOA, Lee SY (2017) Dissemination of antibiotic resistance genes from antibiotic producers to pathogens. Nat Commun 8:1–7. https://doi.org/10.1038/ncomms15784

John DE, Rose JB (2005) Review of factors affecting microbial survival in groundwater. Environ Sci Technol 39:7345–7356. https://doi.org/10.1021/es047995w

Johnson AC, Keller V, Dumont E, Sumpter JP (2015) Assessing the concentrations and risks of toxicity from the antibiotics ciprofloxacin, sulfamethoxazole, trimethoprim and erythromycin in European rivers. Sci Total Environ 511:747–755. https://doi.org/10.1016/j.scitotenv.2014.12.055

Ju F, Beck K, Yin X, Maccagnan A, McArdell CS, Singer HP, Johnson DR, Zhang T, Bürgmann H (2019) Wastewater treatment plant resistomes are shaped by bacterial composition, genetic exchange, and upregulated expression in the effluent microbiomes. ISME J 13:346–360. https://doi.org/10.1038/s41396-018-0277-8

Kaba HEJ, Kuhlmann E, Scheithauer S (2020) Thinking outside the box: association of antimicrobial resistance with climate warming in Europe – a 30 country observational study. Int J Hyg Environ Health 223:151–158. https://doi.org/10.1016/j.ijheh.2019.09.008

Karkman A, Pärnänen K, Larsson DJ (2019) Fecal pollution can explain antibiotic resistance gene abundances in anthropogenically impacted environments. Nat Commun 10:1–8. https://doi.org/10.1038/s41467-018-07992-3

Kleinheinz G, Coenan A, Zehms T, Preedit J, Leewis MC, Becker D, McDermott C (2009) Effect of aquatic macrophytes on the survival of *Escherichia coli* in a laboratory microcosm. Lake Reservoir Manage 25:149–154. https://doi.org/10.1080/07438140902821413

Knox AK, Dahlgren RA, Tate KW, Atwill ER (2008) Efficacy of natural wetlands to retain nutrient, sediment and microbial pollutants. J Environ Qual 37:1837–1846. https://doi.org/10.2134/jeq2007.0067

Königer D, Gastmeier P, Kola A, Schwab F, Meyer E (2014) Vegetarians are not less colonized with extended-spectrum-beta-lactamase-producing bacteria than meat eaters. J Antimicrob Chemother 69:281–282. https://doi.org/10.1093/jac/dkt335

Koo HJ, Woo GJ (2012) Characterization of antimicrobial resistance of *Escherichia coli* recovered from foods of animal and fish origin in Korea. J Food Prot 75:966–972. https://doi.org/10.4315/0362-028X.JFP-11-003

Kristiansson E, Fick J, Janzon A, Grabic R, Rutgersson C, Weijdegård B, Söderström H, Larsson DGJ (2011) Pyrosequencing of antibiotic-contaminated river sediment reveals high levels of resistance and gene transfer elements. PLoS One 6. https://doi.org/10.1371/journal.pone.0017038

Kümmerer K (2004) Resistance in the environment. J Antimicrob Chemother 54:311–320. https://doi.org/10.1093/jac/dkh325

Kwak YK, Colque P, Byfors S, Giske CG, Möllby R, Kühn I (2015) Surveillance of antimicrobial resistance among *Escherichia coli* in wastewater in Stockholm during 1 year: does it reflect the resistance trends in the society? Int J Antimicrob Agents 45:25–32. https://doi.org/10.1016/j.ijantimicag.2014.09.016

LaPara TM, Burch TR, McNamara PJ, Tan DT, Yan M, Eichmiller JJ (2011) Tertiary-treated municipal wastewater is a significant point source of antibiotic resistance genes into Duluth-Superior Harbor. Environ Sci Technol 45:9543–9549. https://doi.org/10.1021/es202775r

Laroche E, Pawlak B, Berthe T, Skurnik D, Petit F (2009) Occurrence of antibiotic resistance and class 1, 2 and 3 integrons in *Escherichia coli* isolated from a densely populated estuary (seine, France). FEMS Microbiol Ecol 68:118–130. https://doi.org/10.1111/j.1574-6941.2009.00655.x

Laroche E, Petit F, Fournier M, Pawlak B (2010) Transport of antibiotic-resistant *Escherichia coli* in a public rural karst water supply. J Hydrol 392:12–21. https://doi.org/10.1016/j.jhydrol.2010.07.022

Larsson DGJ, de Pedro C, Paxeus N (2007) Effluent from drug manufactures contains extremely high levels of pharmaceuticals. J Hazard Mater 148:751–755. https://doi.org/10.1016/j.jhazmat.2007.07.008

Lau CHF, Li B, Zhang T, Tien YC, Scott A, Murray R, Sabourin L, Lapen DR, Duenk P, Topp E (2017) Impact of pre-application treatment on municipal sludge composition, soil dynamics of antibiotic resistance genes, and abundance of antibiotic-resistance genes on vegetables at harvest. Sci Total Environ 587:214–222. https://doi.org/10.1016/j.scitotenv.2017.02.123

Leclercq R, Oberlé K, Galopin S, Cattoir V, Budzinski H, Petit F (2013) Changes in Enterococcal populations and related antibiotic resistance along a medical center-wastewater treatment plant-river continuum. Appl Environ Microbiol 79:2428–2434. https://doi.org/10.1128/AEM.03586-12

Leonard AFC, Yin XL, Zhang T, Hui M, Gaze WH (2018a) A coliform-targeted metagenomic method facilitating human exposure estimates to *Escherichia coli*-borne antibiotic resistance genes. FEMS Microbiol Ecol 94. https://doi.org/10.1093/femsec/fiy024

Leonard AF, Zhang L, Balfour AJ, Garside R, Hawkey PM, Murray AK, Ukoumunne OC, Gaze WH (2018b) Exposure to and colonisation by antibiotic-resistant *E. coli* in UK coastal water users: environmental surveillance, exposure assessment, and epidemiological study (beach bum survey). Environ Int 114:326–333. https://doi.org/10.1016/j.envint.2017.11.003

Li B, Zhang T (2010) Biodegradation and adsorption of antibiotics in the activated sludge process. Environ Sci Technol 44:3468–3473. https://doi.org/10.1021/es903490h

Li D, Yang M, Hu J, Zhang J, Liu R, Gu X, Zhang Y, Wang Z (2009) Antibiotic-resistance profile in environmental bacteria isolated from penicillin production wastewater treatment plant and the receiving river. Environ Microbiol 11:1506–1517. https://doi.org/10.1111/j.1462-2920.2009.01878.x

Li D, Yu T, Zhang Y, Yang M, Li Z, Liu M, Qi R (2010) Antibiotic resistance characteristics of environmental bacteria from an oxytetracycline production wastewater treatment plant and the receiving river. Appl Environ Microbiol 76:3444–3451. https://doi.org/10.1128/AEM.02964-09

Li LG, Xia Y, Zhang T (2017) Co-occurrence of antibiotic and metal resistance genes revealed in complete genome collection. ISME J 11:651–662. https://doi.org/10.1038/ismej.2016.155

Liu M, Zhang Y, Zhang H, Zhang H, Li K, Tian Z, Yang M (2017) Ozonation as an effective pre-treatment for reducing antibiotic resistance selection potency in oxytetracycline production wastewater. Des Water Treat 74:155–162. https://doi.org/10.5004/dwt.2017.20731

Losasso C, Di Cesare A, Mastrorilli E, Patuzzi I, Cibin V, Eckert EM, Fontaneto D, Vanzo A, Ricci A, Corno G (2018) Assessing antimicrobial resistance gene load in vegan, vegetarian and omnivore human gut microbiota. Int J Antimicrob Agents 52:702–705. https://doi.org/10.1016/j.ijantimicag.2018.07.023

Ludden C, Raven KE, Jamrozy D, Gouliouris T, Blane B, Coll F, de Goffau M, Naydenova P, Horner C, Hernandez-Garcia J (2019) One health genomic surveillance of *Escherichia coli* demonstrates distinct lineages and mobile genetic elements in isolates from humans versus livestock. MBio 10:e02693-18. https://doi.org/10.1128/mBio.02693-18

Ludden C, Moradigaravand D, Jamrozy D, Gouliouris T, Blane B, Naydenova P, Hernandez-Garcia J, Wood P, Hadjirin N, Radakovic M (2020) A one health study of the genetic relatedness of *Klebsiella pneumoniae* and their mobile elements in the east of England. Clin Infect Dis 70:219–226. https://doi.org/10.1093/cid/ciz174

MacFadden DR, McGough SF, Fisman D, Santillana M, Brownstein JS (2018) Antibiotic resistance increases with local temperature. Nat Clim Chang 8:510. https://doi.org/10.1038/s41558-018-0161-6

Mao D, Yu S, Rysz M, Luo Y, Yang F, Li F, Hou J, Mu Q, Alvarez PJJ (2015) Prevalence and proliferation of antibiotic resistance genes in two municipal wastewater treatment plants. Water Res 85:458–466. https://doi.org/10.1016/j.watres.2015.09.010

Margot J, Kienle C, Magnet A, Weil M, Rossi L, De Alencastro LF, Abegglen C, Thonney D, Chèvre N, Schärer M (2013) Treatment of micropollutants in municipal wastewater: ozone or powdered activated carbon? Sci Total Environ 461:480–498. https://doi.org/10.1016/j.scitotenv.2013.05.034

Martak D, Henriot CP, Broussier M, Couchoud C, Valot B, Richard M, Couchot J, Bornette G, Hocquet D, Bertrand X (2020) High prevalence of human-associated *Escherichia coli* in wetlands located in eastern France. Front Microbiol 11:552566. https://doi.org/10.3389/fmicb.2020.552566

Marti E, Variatza E, Balcazar JL (2014a) The role of aquatic ecosystems as reservoirs of antibiotic resistance. Trends Microbiol 22:36–41. https://doi.org/10.1016/j.tim.2013.11.001

Marti R, Tien YC, Murray R, Scott A, Sabourin L, Topp E (2014b) Safely coupling livestock and crop production systems: how rapidly do antibiotic resistance genes dissipate in soil following a commercial application of swine or dairy manure? Appl Environ Microbiol 80:3258–3265. https://doi.org/10.1128/AEM.00231-14

Martín de la Vega PT, Jaramillo-Morán MA (2018) Obtaining key parameters and working conditions of wastewater biological nutrient removal by means of artificial intelligence tools. Water 10:685. https://doi.org/10.3390/w10060685

McKinney CW, Dungan RS, Moore A, Leytem AB (2018) Occurrence and abundance of antibiotic resistance genes in agricultural soil receiving dairy manure. FEMS Microbiol Ecol 94. https://doi.org/10.1093/femsec/fiy010

Meijs AP, Gijsbers EF, Hengeveld PD, Veenman C, van Roon AM, van Hoek AHAM, de Greeff SC, van Duijk E, Dierikx CM (2020) Do vegetarians less frequently carry ESBL/pAmpC-producing Escherichia coli/Klebsiella pneumoniae compared with non-vegetarians? J. Antimicrob. Chemother. 75:550–558. https://doi.org/10.1093/jac/dkz483

Milaković M, Vestergaard G, González-Plaza JJ, Petrić I, Šimatović A, Senta I, Kublik S, Schloter M, Smalla K, Udiković-Kolić N (2019) Pollution from azithromycin-manufacturing promotes macrolide-resistance gene propagation and induces spatial and seasonal bacterial commu-

nity shifts in receiving river sediment. Environ Int 123:501–511. https://doi.org/10.1016/j.envint.2018.12.050
Milanović V, Osimani A, Aquilanti L, Tavoletti S, Garofalo C, Polverigiani S, Litta-Mulondo A, Cocolin L, Ferrocino I, Di Cagno R et al (2017) Occurrence of antibiotic resistance genes in the fecal DNA of healthy omnivores, ovo-lacto vegetarians and vegans. Mol Nutr Food Res 61:1601098. https://doi.org/10.1002/mnfr.201601098
Ministère de la transition écologique et solidaire et Ministère des solidarités et de la santé (2019) Portail sur l'assainissement non collectif. www.assainissement-non-collectif.developpement--durable.gouv.fr/
Monte DF, Mem A, Fernandes MR, Cerdeira L, Esposito F, Galvão JA, Franco BDGM, Lincopan N, Landgraf M (2017) Chicken meat as a reservoir of colistin-resistant *Escherichia coli* strains carrying mcr-1 genes in south america. Antimicrob Agents Chemother 61:e02718–e02716. https://doi.org/10.1128/AAC.02718-16
Mughini-Gras L, Dorado-García A, van Duijkeren E, van den Bunt G, Dierikx CM, Bonten MJM, Bootsma MCJ, Schmitt H, Hald T, Evers EG et al (2019) Attributable sources of community-acquired carriage of *Escherichia coli* containing β-lactam antibiotic resistance genes: a population-based modelling study. Lancet Planet Health 3:357–369. https://doi.org/10.1016/S2542-5196(19)30130-5
Müller H, Sib E, Gajdiss M, Klanke U, Lenz-Plet F, Barabasch V, Albert C, Schallenberg A, Timm C, Zacharias N et al (2018) Dissemination of multi-resistant gram-negative bacteria into German wastewater and surface waters. FEMS Microbiol Ecol 94. https://doi.org/10.1093/femsec/fiy057
Munir M, Wong K, Xagoraraki I (2011) Release of antibiotic resistant bacteria and genes in the effluent and biosolids of five wastewater utilities in Michigan. Water Res 45:681–693. https://doi.org/10.1016/j.watres.2010.08.033
Murray R, Tien YC, Scott A, Topp E (2019) The impact of municipal sewage sludge stabilization processes on the abundance, field persistence, and transmission of antibiotic resistant bacteria and antibiotic resistance genes to vegetables at harvest. Sci Total Environ 651:1680–1687. https://doi.org/10.1016/j.scitotenv.2018.10.030
National Ground Water Association (2020) Groundwater use in the United States of America. In: NGWA The Groundwater Association. https://www.ngwa.org/
Nesme J, Simonet P (2015) The soil resistome: a critical review on antibiotic resistance origins, ecology and dissemination potential in telluric bacteria. Environ Microbiol 17:913–930. https://doi.org/10.1111/1462-2920.12631
Ng C, Chen H, Goh SG, Haller L, Wu Z, Charles FR, Trottet A, Gin K (2018) Microbial water quality and the detection of multidrug resistant *E. coli* and antibiotic resistance genes in aquaculture sites of Singapore. Mar Pollut Bull 135:475–480. https://doi.org/10.1016/j.marpolbul.2018.07.055
Nicolas-Chanoine M-H, Gruson C, Bialek-Davenet S, Bertrand X, Thomas-Jean F, Bert F, Moyat M, Meiller E, Marcon E, Danchin N (2012) 10-fold increase (2006-11) in the rate of healthy subjects with extended-spectrum β-lactamase-producing *Escherichia coli* faecal carriage in a Parisian check-up Centre. J Antimicrob Chemother 68:562–568. https://doi.org/10.1093/jac/dks429
Obaidat MM, Bani Salman AE, Lafi SQ (2015) Prevalence of *Staphylococcus aureus* in imported fish and correlations between antibiotic resistance and enterotoxigenicity. J Food Prot 78:1999–2005. https://doi.org/10.4315/0362-028X.JFP-15-104
Oberlé K, Capdeville MJ, Berthe T, Budzinski H, Petit F (2012) Evidence for a complex relationship between antibiotics and antibiotic-resistant *Escherichia coli*: from medical center patients to a receiving environment. Environ Sci Technol 46:1859–1868. https://doi.org/10.1021/es203399h
Ojer-Usoz E, González D, García-Jalón I, Vitas AI (2014) High dissemination of extended-spectrum β-lactamase-producing *Enterobacteriaceae* in effluents from wastewater treatment plants. Water Res 56:37–47. https://doi.org/10.1016/j.watres.2014.02.041

Ojer-Usoz E, González D, Vitas A (2017) Clonal diversity of ESBL-producing *Escherichia coli* isolated from environmental, human and food samples. Int J Environ Res Public Health 14:676. https://doi.org/10.3390/ijerph14070676

Olds HT, Corsi SR, Dila DK, Halmo KM, Bootsma MJ, McLellan SL (2018) High levels of sewage contamination released from urban areas after storm events: a quantitative survey with sewage specific bacterial indicators. PLoS Med 15. https://doi.org/10.1371/journal.pmed.1002614

Özbay H, Alim A (2009) Antimicrobial activity of some water plants from the northeastern Anatolian region of Turkey. Molecules 14:321–328. https://doi.org/10.3390/molecules14010321

Pal C, Bengtsson-Palme J, Kristiansson E, Larsson DJ (2015) Co-occurrence of resistance genes to antibiotics, biocides and metals reveals novel insights into their co-selection potential. BMC Genomics 16:964. https://doi.org/10.1186/s12864-015-2153-5

Pal C, Asiani K, Arya S, Rensing C, Stekel DJ, Larsson DGJ, Hobman JL (2017) Metal resistance and its association with antibiotic resistance. In: Advances in microbial physiology. Elsevier, San Diego, pp 261–313

Pärnänen KM, Narciso-da-Rocha C, Kneis D, Berendonk TU, Cacace D, Do TT, Elpers C, Fatta-Kassinos D, Henriques I, Jaeger T (2019) Antibiotic resistance in European wastewater treatment plants mirrors the pattern of clinical antibiotic resistance prevalence. Sci Adv 5. https://doi.org/10.1126/sciadv.aau9124

Paulus GK, Hornstra LM, Alygizakis N, Slobodnik J, Thomaidis N, Medema G (2019) The impact of on-site hospital wastewater treatment on the downstream communal wastewater system in terms of antibiotics and antibiotic resistance genes. Int J Hyg Environ Health 222:635–644. https://doi.org/10.1016/j.ijheh.2019.01.004

Petit F, Clermont O, Delannoy S, Servais P, Gourmelon M, Fach P, Oberlé K, Fournier M, Denamur E, Berthe T (2017) Change in the structure of *Escherichia coli* population and the pattern of virulence genes along a rural aquatic continuum. Front Microbiol 8:609. https://doi.org/10.3389/fmicb.2017.00609

Petit F, Berthe T, Chaix G, Denamur E, Clermont O, Massei N, Dupont J-P (2018) Factors influencing the occurrence and the fate of *E. coli* population in karst hydrosystems. In: Karst groundwater contamination and public health. Springer, Cham, pp 219–230. https://doi.org/10.1007/978-3-319-51070-5_25

Petrovich M, Chu B, Wright D, Griffin J, Elfeki M, Murphy BT, Poretsky R, Wells G (2018) Antibiotic resistance genes show enhanced mobilization through suspended growth and biofilm-based wastewater treatment processes. FEMS Microbiol Ecol 94. https://doi.org/10.1093/femsec/fiy041

Poole K (2017) At the nexus of antibiotics and metals: the impact of Cu and Zn on antibiotic activity and resistance. Trends Microbiol 25:820–832. https://doi.org/10.1016/j.tim.2017.04.010

Proia L, von Schiller D, Sànchez-Melsió A, Sabater S, Borrego CM, Rodríguez-Mozaz S, Balcázar JL (2016) Occurrence and persistence of antibiotic resistance genes in river biofilms after wastewater inputs in small rivers. Environ Pollut 210:121–128. https://doi.org/10.1016/j.envpol.2015.11.035

Randall LP, Lodge MP, Elviss NC, Lemma FL, Hopkins KL, Teale CJ, Woodford N (2017) Evaluation of meat, fruit and vegetables from retail stores in five United Kingdom regions as sources of extended-spectrum beta-lactamase (ESBL)-producing and carbapenem-resistant *Escherichia coli*. Int J Food Microbiol 241:283–290. https://doi.org/10.1016/j.ijfoodmicro.2016.10.036

Reinthaler FF, Posch J, Feierl G, Wüst G, Haas D, Ruckenbauer G, Mascher F, Marth E (2003) Antibiotic resistance of *E. coli* in sewage and sludge. Water Res 37:1685–1690. https://doi.org/10.1016/S0043-1354(02)00569-9

Review on antimicrobial resistance (2016) Tackling drug-resistant infections globally: final report and recommendations. http://amr-review.org/sites/default/files/160525_Final%20paper_with%20cover.pdf

Rhodes G, Huys G, Swings J, Mcgann P, Hiney M, Smith P, Pickup RW (2000) Distribution of oxytetracycline resistance plasmids between *aeromonads* in hospital and aquaculture environments:

implication of Tn1721 in dissemination of the tetracycline resistance determinant Tet. Appl Environ Microbiol 66:3883–3890. https://doi.org/10.1128/AEM.66.9.3883-3890.2000

Ribeiro AF, Laroche E, Hanin G, Fournier M, Quillet L, Dupont JP, Pawlak B (2012) Antibiotic-resistant *Escherichia coli* in karstic systems: a biological indicator of the origin of fecal contamination? FEMS Microbiol Ecol 81:267–280. https://doi.org/10.1111/j.1574-6941.2012.01382.x

Rizzo L, Manaia C, Merlin C, Schwartz T, Dagot C, Ploy MC, Michael I, Fatta-Kassinos D (2013) Urban wastewater treatment plants as hotspots for antibiotic resistant bacteria and genes spread into the environment: a review. Sci Total Environ 447:345–360. https://doi.org/10.1016/j.scitotenv.2013.01.032

Rodriguez-Mozaz S, Chamorro S, Marti E, Huerta B, Gros M, Sànchez-Melsió A, Borrego CM, Barceló D, Balcázar JL (2015) Occurrence of antibiotics and antibiotic resistance genes in hospital and urban wastewaters and their impact on the receiving river. Water Res 69:234–242. https://doi.org/10.1016/j.watres.2014.11.021

Rozen Y, Belkin S (2001) Survival of enteric bacteria in seawater. FEMS Microbiol Rev 25:513–529. https://doi.org/10.1111/j.1574-6976.2001.tb00589.x

Runcharoen C, Moradigaravand D, Blane B, Paksanont S, Thammachote J, Anun S, Parkhill J, Chantratita N, Peacock SJ (2017) Whole genome sequencing reveals high-resolution epidemiological links between clinical and environmental *Klebsiella pneumoniae*. Genome Med 9:6. https://doi.org/10.1186/s13073-017-0397-1

Sadikalay S (2018) Influence des rejets humains et animaux sur la diffusion de l'antibiorésistance à l'homme, aux animaux et à l'environnement en Guadeloupe. Doctoral dissertation, Université des Antilles et de la Guyane, France

Safe Water Network (2016) Drinking water supply for urban poor: city of New Delhi. In: Urban small water enterprises: city assessment. https://www.safewaternetwork.org/

Schwaiger K, Huther S, Hölzel C, Kämpf P, Bauer J (2012) Prevalence of antibiotic-resistant *enterobacteriaceae* isolated from chicken and pork meat purchased at the slaughterhouse and at retail in Bavaria, Germany. Int J Food Microbiol 154:206–211. https://doi.org/10.1016/j.ijfoodmicro.2011.12.014

Schwartz T, Kohnen W, Jansen B, Obst U (2003) Detection of antibiotic-resistant bacteria and their resistance genes in wastewater, surface water, and drinking water biofilms. FEMS Microbiol Ecol 43:325–335. https://doi.org/10.1111/j.1574-6941.2003.tb01073.x

Scott A, Tien YC, Drury CF, Reynolds WD, Topp E (2018) Enrichment of antibiotic resistance genes in soil receiving composts derived from swine manure, yard wastes, or food wastes, and evidence for multiyear persistence of swine *Clostridium* spp. Can J Microbiol 64:201–208. https://doi.org/10.1139/cjm-2017-0642

Seni J, Moremi N, Matee M, van der Meer F, DeVinney R, Mshana SE, D Pitout JD (2018) Preliminary insights into the occurrence of similar clones of extended-spectrum beta-lactamase-producing bacteria in humans, animals and the environment in Tanzania: a systematic review and meta-analysis between 2005 and 2016. Zoonoses Public Health 65:1–10. https://doi.org/10.1111/zph.1238

Servais P, Passerat J (2009) Antimicrobial resistance of fecal bacteria in waters of the seine river watershed (France). Sci Total Environ 408:365–372. https://doi.org/10.1016/j.scitotenv.2009.09.042

Silva J, Castillo G, Callejas L, López H, Olmos J (2006) Frequency of transferable multiple antibiotic resistance amongst coliform bacteria isolated from a treated sewage effluent in Antofagasta, Chile. Electron J Biotechnol 9. https://doi.org/10.2225/vol9-issue5-fulltext-7

Slekovec C, Plantin J, Cholley P, Thouverez M, Talon D, Bertrand X, Hocquet D (2012) Tracking down antibiotic-resistant *Pseudomonas aeruginosa* isolates in a wastewater network. PLoS One 7. https://doi.org/10.1371/journal.pone.0049300

Sleytr K, Tietz A, Langergraber G, Haberl R (2007) Investigation of bacterial removal during the filtration process in constructed wetlands. Sci Total Environ 380:173–180. https://doi.org/10.1016/j.scitotenv.2007.03.001

Smith J, Edwards J, Hilger H, Steck TR (2008) Sediment can be a reservoir for coliform bacteria released into streams. J Gen Appl Microbiol 54:173–179. https://doi.org/10.2323/jgam.54.173

Stalder T, Barraud O, Jové T (2013) Quantitative and qualitative impact of hospital effluent on dissemination of the integron pool. ISME J 8:768–777. https://doi.org/10.1038/ismej.2013.189

Szczepanowski R, Linke B, Krahn I, Gartemann K-H, Gützkow T, Eichler W, Pühler A, Schlüter A (2009) Detection of 140 clinically relevant antibiotic-resistance genes in the plasmid metagenome of wastewater treatment plant bacteria showing reduced susceptibility to selected antibiotics. Microbiology 155:2306–2319. https://doi.org/10.1099/mic.0.028233-0

Tamminen M, Karkman A, Corander J, Paulin L, Virta M (2011a) Differences in bacterial community composition in Baltic Sea sediment in response to fish farming. Aquaculture 313:15–23. https://doi.org/10.1016/j.aquaculture.2011.01.020

Tamminen M, Karkman A, Lohmus A, Muziasari WI, Takasu H, Wada S, Suzuki S, Virta M (2011b) Tetracycline resistance genes persist at aquaculture farms in the absence of selection pressure. Environ Sci Technol 45:386–391. https://doi.org/10.1021/es102725n

Tamtam F, Le Bot B, Dinh T, Mompelat S, Eurin J, Chevreuil M, Bonté P, Mouchel JM, Ayrault S (2011) A 50-year record of quinolone and sulphonamide antimicrobial agents in Seine River sediments. J Soils Sediments 11:852–859. https://doi.org/10.1007/s11368-011-0364-1

Tanner WD, VanDerslice JA, Toor D, Benson LS, Porucznik CA, Goel RK, Atkinson RM (2015) Development and field evaluation of a method for detecting carbapenem-resistant bacteria in drinking water. Syst Appl Microbiol 38:351–357. https://doi.org/10.1016/j.syapm.2015.03.010

Tansawai U, Sanguansermsri D, Na-udom A, Walsh TR, Niumsup PR (2018) Occurrence of extended spectrum β-lactamase and AmpC genes among multidrug-resistant *Escherichia coli* and emergence of ST131 from poultry meat in Thailand. Food Control 84:159–164. https://doi.org/10.1016/j.foodcont.2017.07.028

Tell J, Caldwell DJ, Häner A, Hellstern J, Hoeger B, Journel R, Mastrocco F, Ryan JJ, Snape J, Straub JO, Vestel J (2019) Science-based targets for antibiotics in receiving waters from pharmaceutical manufacturing operations. Integr Environ Assess Manag 15:312–319. https://doi.org/10.1002/ieam.4141

Tian Z, Zhang Y, Yu B, Yang M (2016) Changes of resistome, mobilome and potential hosts of antibiotic resistance genes during the transformation of anaerobic digestion from mesophilic to thermophilic. Water Res 98:261–269. https://doi.org/10.1016/j.watres.2016.04.031

Tien YC, Li B, Zhang T, Scott A, Murray R, Sabourin L, Marti R, Topp E (2017) Impact of dairy manure pre-application treatment on manure composition, soil dynamics of antibiotic resistance genes, and abundance of antibiotic-resistance genes on vegetables at harvest. Sci Total Environ 581-582:32–39. https://doi.org/10.1016/j.scitotenv.2016.12.138

Topp E, Larsson DGJ, Miller DN, Van den Eede C, Vitra MPJ (2018) Antimicrobial resistance and the environment: assessment of advances, gaps and recommendations for agriculture, aquaculture and pharmaceutical manufacturing. FEMS Microbiol Ecol 94. https://doi.org/10.1093/femsec/fix185

U.S. Environmental Protection Agency (EPA) (1999) Combined sewer overflow technology fact sheet. EPA 832-F-99-042

Um MM, Barraud O, Kérourédan M, Gaschet M, Stalder T, Osward E, Dagot C, Ploy M-C, Brugere H, Biddal D (2016) Comparison of the incidence of pathogenic and antibiotic-resistant *Escherichia coli* strains in adult cattle and veal calf slaughterhouse effluents highlighted different risks for public health. Water Res 88:30–38. https://doi.org/10.1016/j.watres.2015.09.029

Valcárcel Y, Alonso SG, Rodríguez-Gil JL, Gil A, Catalá M (2011) Detection of pharmaceutically active compounds in the rivers and tap water of the Madrid region (Spain) and potential ecotoxicological risk. Chemosphere 84:1336–1348. https://doi.org/10.1016/j.chemosphere.2011.05.014

Vittecoq M, Laurens C, Brazier L, Durand P, Elguero E, Arnal A, Thomas F, Aberkane S, Renaud N, Prugnolle F (2017) VIM-1 carbapenemase-producing *Escherichia coli* in gulls from southern France. Ecol Evol 7:1224–1232. https://doi.org/10.1002/ece3.2707

Vos M, Sibleyras L, Lo LK, Hesse E, Gaze W, Klümper U (2020) Zinc can counteract selection for ciprofloxacin resistance. FEMS Microbiol Lett 367. https://doi.org/10.1093/femsle/fnaa038

Wales A, Davies R (2015) Co-selection of resistance to antibiotics, biocides and heavy metals, and its relevance to foodborne pathogens. Antibiotics 4:567–604. https://doi.org/10.3390/antibiotics4040567

World Health Organization (WHO) (2014) Antimicrobial resistance: global report on surveillance. WHO, Geneva

World Health Organization (WHO) (2019) Open defecation – data by country. https://apps.who.int/gho/data/node.main.WSHOPENDEFECATION?lang=fr

World Health Organization (WHO), United Nations Children's Fund (UNICEF) (2013) Joint monitoring programme. Progress on drinking water and sanitation: 2013 update

Yang Y, Li B, Zou S, Fang HHP, Zhang T (2014) Fate of antibiotic resistance genes in sewage treatment plant revealed by metagenomic approach. Water Res 62:97–106. https://doi.org/10.1016/j.watres.2014.05.019

Yang L, Liu W, Zhu D, Hou J, Ma T, Wu L, Zhu Y, Christie P (2018) Application of biosolids drives the diversity of antibiotic resistance genes in soil and lettuce at harvest. Soil Biol Biochem 122:131–140. https://doi.org/10.1016/j.soilbio.2018.04.017

Yang Y, Liu G, Song W, Ye C, Lin H, Li Z, Liu W (2019) Plastics in the marine environment are reservoirs for antibiotic and metal resistance genes. Environ Int 123:79–86. https://doi.org/10.1016/j.envint.2018.11.061

Zhang T, Zhang M, Zhang X, Fang HH (2009a) Tetracycline resistance genes and tetracycline resistant lactose-fermenting *Enterobacteriaceae* in activated sludge of sewage treatment plants. Environ Sci Technol 43:3455–3460. https://doi.org/10.1021/es803309m

Zhang Y, Marrs CF, Simon C, Xi C (2009b) Wastewater treatment contributes to selective increase of antibiotic resistance among *Acinetobacter* spp. Sci Total Environ 407:3702–3706. https://doi.org/10.1016/j.scitotenv.2009.02.013

Zhou LJ, Ying GG, Zhao JL, Yang JF, Wang L, Yang B, Liu S (2011) Trends in the occurrence of human and veterinary antibiotics in the sediment of the Yellow River, Hai River and Liao River in northern China. Environ Pollut 159:1877–1885. https://doi.org/10.1016/j.envpol.2011.03.034

Chapter 5
Relationships Between Outdoor Ambient Air Pollution and Cardiovascular Disorders

Ewa Konduracka and Paweł Rostoff

Abstract Most disorders of the circulatory system have a complex etiology, involving genetic and environmental factors, lifestyle, aging, or infections. In recent years, it has been shown that air pollution not only worsens the course of cardiovascular diseases but also contributes to their occurrence. It is also known to increase all-cause and cardiovascular mortality. Although the actual number of deaths caused by air pollution is difficult to determine, the World Health Organization estimates that at least 7 million premature deaths per year may result from the negative impact of air pollution. More than 70% of these deaths are due to cardiovascular complications. The response of the circulatory system to air pollution is modified by the duration of exposure, chemical composition of the pollutants, comorbidities and their treatment, age of an exposed individual, as well as changes of atmospheric conditions such as temperature, humidity, and pressure. The present narrative review discusses the most frequent cardiovascular disorders that are associated with exposure to air pollutants, based on a review of literature. Publications were identified through the search of PubMed, EBSCO, Cochrane, and Science Direct databases. The scope of the chapter was determined by the PECOS search tool, which is generally used to identify the relevant population, exposure, comparison, outcome, and study design. Using the PECOS tool, the following studies were identified and discussed: experimental studies if available as well as epidemiological studies (panel, case-crossover, time-series studies as well as meta-analyses and systematic reviews).

Although the detailed mechanisms underlying the effects of air pollution on the circulatory system are still the subject of research, it is currently assumed that the potential impact is associated with the ability of fine particulate matter, nanoparticles, and some gaseous components to penetrate from the lungs to the blood and vessels, causing inflammation and vasospasm followed by an increase in peripheral resistance and secondary activation of the coagulation system. Although the respira-

E. Konduracka (✉) · P. Rostoff
Department of Coronary Disease and Heart Failure, Jagiellonian University Medical College, John Paul II Hospital, Kraków, Poland
e-mail: ekonduracka@interia.eu

N. Morin-Crini et al. (eds.), *Emerging Contaminants Vol. 1*, Environmental Chemistry for a Sustainable World 65,
https://doi.org/10.1007/978-3-030-69079-3_5

tory system and eye conjunctiva are the major ways by which pollutants enter the circulatory system, the gastrointestinal tract can also be involved, but the exact percentage of pollutants that enter the body via this route has not yet been determined. Some of the chemicals are persistent organic substances that penetrate into the soil, water, crop plants, and animals, thus getting into the body with food, where they accumulate and may affect the circulatory system. The simultaneous effect of various pollutants on the cardiovascular system blurs the fact that particulate matter has a stronger impact than other pollutants according to the available evidence. Numerous studies have reported a relationship between both short- and long-term exposure to air pollution and cardiovascular complications. The strongest evidence exists for an association between air pollutants and global and cardiovascular mortality. Other studies have shown an increased risk of coronary artery disease development and progression. Exposure to air pollution is also currently seen as another cause (in addition to classic cardiovascular risk factors) of a higher incidence of some cardiovascular diseases, such as hypertension and ischemic stroke. A number of studies have indicated a positive association of exposure to particulate matter and selected gaseous pollutants with the risk of hospitalization or death from heart failure. On the other hand, the impact on coagulation disorders, especially venous thrombosis, is less evident and requires further research. It should be emphasized that the occurrence of the above diseases is always multifactorial, and the percentage share of various risk factors is in fact hard to determine. Difficulties in assessing the percentage share of air pollution as a factor contributing to cardiovascular diseases can be clearly seen in large epidemiologic and smaller cohort studies, which often provide significantly different, or sometimes even conflicting, results. This discrepancy is caused by numerous factors. First, there are variations in air pollution and its chemical composition between different parts of the world. Second, air pollution overlaps with different atmospheric conditions, especially heat waves in countries with a mild climate, high humidity and temperature, or sudden temperature drops. Third, potential pollutants are countless and most of them have not been evaluated in terms of their toxicity and health risks. Also epidemiologic studies that contribute to the establishment of dose–response relationships are often subject to errors in the assessment or classification of exposure, errors in the diagnosis or reporting of health outcomes, inadequate sample size, inadequate adjustment for confounding or effect modification by other factors, as well as sampling and statistical bias. There is evidence that air pollution contributes to the risk of cardiovascular diseases and mortality. Therefore, efforts to reduce exposure to air pollution should be urgently intensified and supported by effective legislation. Medical staff plays an important role in educating patients and political decision makers about the harms of air pollution. Further research should explore the optimal methods of air pollution reduction and document its effect on the incidence of cardiovascular disease and mortality.

Keywords Air pollution · Cardiovascular complications · Coronary artery disease · Heart failure · Hypertension · Stroke

Abbreviations

CI	confidence interval
CIMT	carotid intima-media thickness
HR	hazard ratio
IQR	interquartile range
OR	odds ratio
PM	particulate matter
PM_{10}	coarse particulate matter
$PM_{2.5}$	fine particulate matter
$PM_{0.1}$	ultrafine particulate matter
RR	relative risk
STEMI	ST-segment elevation myocardial infarction
Non-STEMI	myocardial infarction without ST-segment elevation

5.1 Introduction

Most disorders of the circulatory system have a multifactorial etiology, involving lifestyle, aging, genetic predisposition, infections, and environmental factors. In recent years, it has been demonstrated that air pollution not only worsens the course of cardiovascular diseases but also contributes to their occurrence. Moreover, it is also known to increase all-cause and cardiovascular mortality. Apart from cardiovascular complications, air pollution has been shown to be related to a broad spectrum of other diseases. Previous studies reported an association of air pollution with an increased incidence of malignancies, respiratory diseases, endocrine disorders, neurodegenerative diseases, immune dysfunction, as well as fetal complications (Mazurek and Badyda 2018). Although the actual number of deaths caused by air pollution is difficult to determine, the World Health Organization estimates that at least 7 million premature deaths per year may result from the negative impact of both household and outdoor air pollution (World Health Organization (2020)). Most of these deaths are due to cardiovascular complications (World Health Organization (2020)).

The response of the circulatory system to air pollution is modified by the duration of exposure, chemical composition of the pollutants and exposure concentration, coexisting comorbidities, individual susceptibility as well as changes of atmospheric conditions such as temperature, humidity, and pressure. In addition, the chemical composition of air pollutants varies significantly depending on the source of emission (Konduracka 2019). Not all pollutants are being monitored. In fact, potential pollutants are countless and most of them have never been evaluated in terms of their toxicity and human health risk (Briggs 2003), especially that new chemicals emerge all the time due to interactions between the existing ones (Briggs 2003; Konduracka 2019).

Air pollutants consist of gaseous components and particulate matter (PM). Among the gaseous components (Mazurek and Badyda 2018), the most commonly monitored chemicals include nitrogen dioxide, nitric oxide, sulfur dioxide, benzene, ozone, and carbon monoxide. On the other hand, particulate matter components consist of elemental and organic carbon particles, transition metals, nitrates, sulfates, polycyclic aromatic hydrocarbons, alkanes, alkenes, dioxins, furans, microorganisms, endotoxins, and other chemicals (Brook et al. 2010; Newby et al. 2015; Samek et al. 2020). Among various air pollutants, a number of different studies have explored the effects of particular matter and have provided powerful scientific evidence for its significant impact on health (Integrated Science Assessment for Particulate Matter 2019. United States Environment Protection Agency website). Recent studies have shown that one of the strongest sources of air pollutant emission are combustion processes and traffic.

The present narrative review discusses the most frequent cardiovascular disorders that are associated with exposure to outdoor air pollutants in adult population. We also describe the potential mechanisms underlying the available clinical findings.

5.2 Methods

We reviewed available literature published in the last 30 years to determine the associations of outdoor air pollution with cardiovascular morbidity and mortality as well as to summarize the proposed mechanisms underlying these associations as reported by experimental and human studies. Publications were identified by searching PubMed, EBSCO, Cochrane, and Science Direct databases, using the following MESH terms: air pollution, particulate matter, and gaseous air pollutants. The following keywords were also used: all-cause mortality, cardiovascular mortality, hypertension, coronary artery disease, hospital admissions, cardiac arrhythmias, myocardial infarction, stroke, and coagulation system. The scope of the review was determined by the PECOS search tool that generally which is generally used to identify the relevant population, exposure, comparison, outcome, and study design. Using the PECOS tool, the following studies were identified and discussed: experimental studies if available (controlled human exposure and/or animal studies) as well as epidemiological studies (panel, case-crossover, time-series studies as well as meta-analyses and systematic reviews). Some studies were excluded due to redundancy. We also excluded studies involving children and those focusing on indoor exposure and exposure to persistent organic chemicals.

5.3 Potential Mechanisms of the Effect of Air Pollution on the Cardiovascular System

The direct way of exposure to air pollution is via the respiratory tract and eye conjunctiva (Briggs 2003; Brook et al. 2010; De Brouwere et al. 2012; Chin 2015). The gastrointestinal tract can also be involved, although the exact percentage of pollutants that enter the body via this route has not yet been determined (Briggs 2003; Brook et al. 2010; De Brouwere et al. 2012; Chin 2015). Some of the air pollutants can penetrate from the air into the soil, water, plants, and animals, thus getting into the body with food, where they accumulate and may affect the circulatory system.

Although the detailed mechanism of the effect of air pollutants on the circulatory system via the respiratory tract is still the subject of research, it is currently assumed that the potential impact is associated with the ability of fine particulate matter ($PM_{2.5}$, aerodynamic diameter <2.5 μm) and ultrafine particulate matter ($PM_{0.1}$, aerodynamic diameter ≤0.1 μm) as well as some gaseous components to penetrate from the lungs into the bloodstream (Brook et al. 2010; Newby et al. 2015; Chin 2015). Ultrafine particulate matter tends to be especially dangerous for health due to its large reactive surface area, greater capacity to bind harmful chemicals, very high deposition in lung alveoli, ability to activate inflammatory cells that can penetrate into the bloodstream, as well as accumulation in peripheral organs (Ohlwein et al. 2019). The concentration of $PM_{0.1}$ is not routinely measured by all air-quality monitoring stations in the world. Moreover, other chemicals such as endotoxins are simultaneously present and may interact with particulate matter and gaseous components, which makes it difficult to assess health effects (Rajagopalan et al. 2018; Miller and Newby 2020).

Based on animal and human experimental studies with exposure or epidemiological studies, the following pathomechanisms have been postulated to underlie the adverse effects of air pollution on the cardiovascular system:

1. activation of local inflammatory responses in the lungs by particulate matter and gaseous pollutants, leading to cytokine-mediated and secondary systemic inflammation (Ghio et al. 2000; Terzano et al. 2010; Brook et al. 2010; Newby et al. 2015; Jiang et al. 2016, Lopez-Campos et al. 2016);
2. activation of systemic inflammation by $PM_{2.5}$ and some gaseous components that can penetrate from the lungs into the bloodstream and promote inflammation, endothelial dysfunction, hypercoagulability, and thrombosis (Brook et al. 2010; Rich et al. 2012; Newby et al. 2015; Hamanaka and Mutlu 2018);
3. stimulation of sensory receptors in the lungs, leading to the imbalance of the autonomic nervous system and sympathetic predominance, resulting in alterations in heart rate and vasoconstriction; however, there are considerable discrepancies between studies in terms of the analyzed parameters and results (Brook et al. 2010; Barbosa et al. 2012; Rich et al. 2012; Newby et al. 2015; Buteau and Goldberg 2016; Hamanaka and Mutlu 2018);
4. promotion of oxidative stress, formation of reactive oxygen species, DNA damage (Sun et al. 2005; Vinzents et al. 2005; Novotna et al. 2007; Rossner Jr et al. 2007; Ha et al. 2007; Araujo et al. 2008; Bai et al. 2011; Min et al. 2011; Lind and Lind 2012; Li et al. 2019);
5. direct interaction of $PM_{2.5}$ with target tissues including vessel wall and cardiac tissues, although this mechanism remains controversial (Hamanaka and Mutlu 2018);

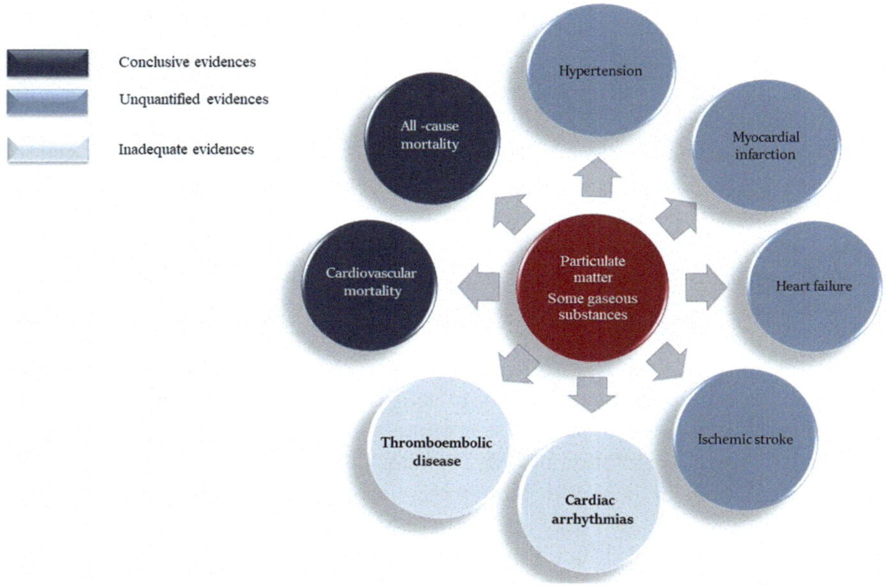

Fig. 5.1 Effect of air pollution on main cardiovascular clinical outcomes (the different color shades reflect the strength of scientific evidence for a given health effect)

6. an increase in the levels of selected stress hormones, such as norepinephrine, epinephrine, and cortisol, as well as enhanced insulin resistance (Li et al. 2017);
7. modification of intestinal microbiota composition, leading to inflammation, atherosclerosis, and host metabolic disorders such as obesity and diabetes (Zhang et al. 2015; Wang et al. 2018).

It is believed that all these mechanisms can potentially contribute to the development and progression of atherosclerosis, hypertension, stroke, cardiac arrhythmia, and death. The analysis of the available studies indicates the main cardiovascular consequences of air pollution. These effects are presented in Fig. 5.1.

5.4 Air Pollution and Mortality

Historically, the relationship between air pollution and cardiovascular morbidity and mortality was first noted in the first half of the 20th century (Hamanaka and Mutlu 2018). The smog incidents in Meuse Valley (Belgium, 1930), Donora (United States, 1948), and London (United Kingdom, 1952) caused a sudden increase in the number of hospitalizations and deaths, especially in elderly patients and those with preexisting cardiovascular or respiratory diseases. The most dramatic lesson for posterity was the so called Great Smog of London, in which about 4000 people died as a direct result of smog and 100,000 suffered health damage (Hamanaka and Mutlu 2018).

There is strong evidence that the majority of premature deaths related to air pollution are due to acute ischemic cardiovascular events, particularly acute coronary syndromes (Hamanaka and Mutlu 2018). Recently, Lelieveld et al. (2019), using a novel method of modelling the effects of various sources of air pollution on death rates, showed that outdoor air pollutants caused an estimated number of 790,000 additional deaths in the whole of Europe in 2015 and 659,000 deaths in the 28 Member States of the European Union. Of these deaths, 40% to 80% were due to cardiovascular diseases. Importantly, air pollution accounted for twice as many deaths from cardiovascular diseases as respiratory diseases (Lelieveld et al. 2019).

5.4.1 Short-Term Effects on Mortality

So far, numerous short-term studies have demonstrated a close relationship between changes in daily (or few-day) averages of particulate matter and gaseous co-pollutants and the number of events, including all-cause and cardiovascular deaths and hospitalizations. The 2 largest multi-city studies were NMMAPS (National Morbidity, Mortality, and Air Pollution Study) in the United States (Dominici et al. 2005) and APHEA (Air Pollution and Health: a European Approach) (Analitis et al. 2006; Samoli et al. 2006, 2008). The findings from those studies were highly consistent and showed that particulate matter levels were significantly related to daily all-cause, cardiovascular, and respiratory mortality (Analitis et al. 2006; Samoli et al. 2006, 2008).

The NMMAPS, a time-series study of 95 large urban centers in the United States, conducted from 1987 to 2000 and involving approximately 50 million American adults, found that the rate of daily cardiopulmonary mortality increased by 0.6% per 20-$\mu g/m^3$ increase in PM_{10} levels (Dominici et al. 2005; Brook et al. 2017). In addition, the relationship between particulate matter exposure and mortality in this study was independent of gaseous pollutants such as nitrogen dioxide, sulfur dioxide, and carbon monoxide (Dominici et al. 2005; Hamanaka and Mutlu 2018).

The APHEA-2 study, which included about 43 million adults from 29 European cities, reported a 1.5% increase in daily cardiovascular mortality per 20-$\mu g/m^3$ increase in PM_{10} levels (Samoli et al. 2006, 2008; Brook 2008). Furthermore, the APHEA study demonstrated a greater effect of PM_{10} on daily mortality in cities with a higher temperature and a higher share of road traffic as the source of particulate matter emissions (Newby et al. 2015). All-cause mortality increased by 0.30% (95% confidence interval CI, 0.22–0.38); cardiovascular mortality, by 0.40% (95% CI, 0.29–0.52); and respiratory mortality, by 0.38% (95% CI, 0.17–0.58) for every 10-$\mu g/m^3$ increase in daily nitrogen dioxide concentrations (Samoli et al. 2006; Newby et al. 2015). Both NMMAPS and APHEA studies showed not only regional but also seasonal variations possibly due to meteorological conditions, different sources of pollutants, and population differences (Dominici et al. 2005; Samoli et al. 2006, 2008; Hamanaka and Mutlu 2018). An important conclusion from these multi-city studies is that there is no safe threshold below which an increase in particulate matter would not be associated with increased mortality (Dominici et al. 2005; Samoli et al. 2006, 2008; Hamanaka and Mutlu 2018).

In the APHENA study (Air Pollution and Health: a Combined European and North American Approach), which was a collaborative analysis of multicity time-series data on the effect of air pollution on population health that brought together findings from the APHEA and NMMAPS projects as well as Canadian data, a 10-μg/m^3 increase in PM_{10} levels was associated with an increase of 0.2% to 0.6% in all-cause mortality rates (Samoli et al. 2008; Katsouyanni et al. 2009; Newby et al. 2015). The APHENA investigators found weak but statistically significant associations between a 10 μg/m^3 increase in PM_{10} levels and cardiovascular mortality in Europe, Canada, and the United States. However the effect of PM_{10} on respiratory mortality was less consistent across regions (Katsouyanni et al. 2009).

According to a report by the Committee on the Medical Effects of Air Pollutants from the United Kingdom, a 10-μg/m^3 increase in $PM_{2.5}$ levels significantly increased the risk of daily cardiovascular mortality by 1.4% (95% CI, 0.7–2.2) (Ayres 2006). On the other hand, a 10-μg/m^3 increase in PM_{10} levels increased the risk of daily cardiovascular mortality by 0.9% (95% CI, 0.7–1.2) (Ayres 2006). The estimated random effect of nitrogen dioxide on daily cardiovascular mortality was 1.0% (95% CI, 0.8–1.3) for every 10-μg/m^3 increase in nitrogen dioxide concentrations. The estimated impact of an 8-hour exposure to ozone on cardiovascular mortality was 0.4% (95% CI, 0.3–0.5) for every 10-μg/m^3 in ozone levels. Finally the estimated effects of carbon monoxide and sulfur dioxide on cardiovascular mortality were 1.1% (95% CI, 0.2–2.1) and 0.8% (95% CI, 0.6–1.0), respectively (Ayres 2006).

Interesting conclusions can be drawn from studies conducted in Asian and developing countries. These studies have generally shown lesser effects of particulate matter on short-term mortality than studies from Europe and the United States (Hamanaka and Mutlu 2018).

A systematic review and meta-analysis by Newell et al. (2017), which included 85 studies from 12 low- and middle-income countries showed an increase of 0.47% (95% CI, 0.34–0.61) for cardiovascular mortality and of 0.57% (95% CI, 0.28–0.86) for respiratory mortality for every 10-μg/m^3 increase in $PM_{2.5}$ levels. On the other hand, an increase of 10 μg/m^3 in PM_{10} levels was associated with a rise in cardiovascular mortality by 0.27% (95% CI, 0.11–0.44) and in respiratory mortality by 0.56% (95% CI, 0.24–0.87) (Newell et al. 2017). The results of this meta-analysis seem surprising, especially considering that the average $PM_{2.5}$ concentrations in low- and middle-income countries were significantly higher (56–179 μg/m^3) than those reported for Europe and the United States (Newell et al. 2017; Hamanaka and Mutlu 2018). There is increasing evidence that the relationship between $PM_{2.5}$ concentrations and mortality is biphasic (Hamanaka and Mutlu 2018). At lower $PM_{2.5}$ levels, a steep concentration-response function is observed, while the curve flattens at higher concentrations (Hamanaka and Mutlu 2018). These conclusions are consistent with the results of a previous study from Beijing, China (Xie et al. 2015), which showed that the relationships between $PM_{2.5}$ and morbidity and mortality from ischemic heart disease are nonlinear, with a steeper dose-response function at lower concentrations and a shallower response at higher concentrations (Xie et al. 2015). In addition, there was no saturation effect for an increased risk of coronary artery disease mortality, even at particulate matter levels as high as 500 μg/m^3 (Xie et al. 2015).

A meta-analysis of 110 time-series studies showed that every 10-μg/m^3 increase in $PM_{2.5}$ was related to an increase of 1.04% (95% CI, 0.52–1.56) in all-cause mortality and of 0.84% (95% CI, 0.41–1.28) in cardiovascular mortality (Atkinson et al. 2014). Interestingly, the rise in all-cause mortality revealed significant regional variations (95% CI, 0.25–2.08). In this meta-analysis, a 10-μg/m^3 increment in $PM_{2.5}$ levels was associated with a 3.36% (95% CI, 0.68–6.10) in the risk of death from coronary artery disease; 1.85% (95% CI, 0.74–2.97), from stroke; and 2.86% (95% CI, −0.12 to 5.93), from chronic obstructive lung disease (Atkinson et al. 2014).

Based on the available studies, Pope and Dockery (2006) concluded that the risk of cardiovascular mortality increases by approximately 1% for every 10-μg/m^3 short-term daily increase in $PM_{2.5}$ levels.

A recent large study conducted by Liu et al. (2019), which included data from 652 urban areas in 24 countries or regions, showed that an increase of 10 μg/m^3 in the 2-day moving average of PM_{10} concentrations was associated with an increase of 0.44% (95% CI, 0.39–0.50) in daily all-cause mortality, 0.36% (95% CI, 0.30–0.43) in daily cardiovascular mortality, and 0.47% (95% CI, 0.35–0.58) in daily respiratory mortality. On the other hand, a 10-μg/m^3 increment in the 2-day moving average $PM_{2.5}$ level was related to an increase of 0.68% (95% CI, 0.59–0.77), 0.55% (95% CI, 0.45–0.66), and 0.74% (95% CI, 0.53–0.95) increase in daily all-cause, cardiovascular and respiratory mortality, respectively (Liu et al. 2019). These relationships remained significant after adjustment for gaseous pollutants (Liu et al. 2019).

Less is known about the short-term effects of gaseous pollutants on all-cause, cardiovascular, and respiratory mortality. In the review of 204 time-series studies of NO_2 and daily mortality and hospital admissions, Mills et al. (2015) found that a 10-μg/m^3 daily increase in nitrogen dioxide concentrations was associated with an increase in cardiovascular mortality by 0.40% to 0.88%. In another systematic review and meta-analysis of 1553 studies from China, Newell et al. (2018) revealed that a 10-μg/m^3 increase in same-day nitrogen oxide and sulfur dioxide levels was related to an increase of 0.92% (95% CI, 0.44–1.39) and 0.50% (95% CI, 0.01–1), respectively, in cardiovascular mortality, whereas authors showed no significant associations for same-day carbon monoxide or ozone and cardiovascular mortality. Also the increase in concentrations of carbon monoxide and nitrogen oxide turned out to be significantly associated with the rise in respiratory mortality by 3.08% (95% CI, 0.76–5.40) and 0.70% (95% CI, 0.01–1.40), respectively (Newell et al. 2018). These conclusions are in contrast to the results of a large, nationwide, time-series study by Liu et al. (2018a) based on data from 272 cities in China between 2013 and 2015. The study showed that an increase in the 2-day average concentration of carbon monoxide by 1 mg/m^3 significantly increases cardiovascular mortality by 1.12% (95% CI, 0.42–1.83), mortality from coronary artery disease by 1.75% (95% CI, 0.85–2.66), and mortality from stroke by 0.88% (95% CI, 0.07–1.69). These associations were significant even after adjustment for region, age, gender, and education level. Interestingly, there was no threshold below which exposure to carbon monoxide had no effect on cardiovascular mortality.

There is no convincing evidence for the relevant effects of short-term variations in ozone concentrations on cardiovascular mortality (Yan et al. 2013; Bourdrel et al. 2017).

5.4.2 *Long-Term Effects on Mortality*

As demonstrated in numerous epidemiological studies, chronic exposure to air pollutants increases the risk of all-cause and cardiovascular mortality. There is strong evidence that the long-term effects of air pollution on mortality are mainly associated with exposure to $PM_{2.5}$ (Newby et al. 2015). Literature data indicate a greater increase in all-cause mortality due to a long-term exposure to $PM_{2.5}$ as compared with short-term exposure (Brook et al. 2010; Newby et al. 2015).

The first large-scale prospective cohort study that showed adverse health effects of long-term exposure to air pollution was the Harvard Six Cities study (Dockery et al. 1993; Brook 2008). The extended 28-year follow-up of over 8000 people living in 6 United States cities showed that the relative risks (RRs) for all-cause and cardiovascular mortality were significantly increased by 1.16 and 1.28, respectively, per a 10-μg/m^3 increase in long-term $PM_{2.5}$ levels (Dockery et al. 1993; Brook 2008). A decrease of $PM_{2.5}$ concentrations during the study resulted in a significant reduction in cardiovascular mortality (RR, 0.69) (Dockery et al. 1993; Brook 2008). It was also noted that the levels of air pollution within 1 year prior to death predicted mortality equally well as those for the entire duration of the study, suggesting a relatively rapid impact of air pollution on health (Dockery et al. 1993; Brook 2008). There is increasing evidence that the annual level of air pollutants is sufficient to reliably predict air pollution related mortality (Dockery et al. 1993; Brook 2008).

The extended 16-year follow-up of over 550,000 adults included in the American Cancer Society II study showed a significant association of both $PM_{2.5}$ and sulfur dioxide levels with all-cause and cardiopulmonary mortality (Pope 3rd et al. 2004). It was demonstrated that every 10-μg/m^3 increase in long-term $PM_{2.5}$ exposure was related to a 4% and 6% increased risk of all-cause and cardiopulmonary mortality, respectively (Pope 3rd et al. 2004). Ischemic heart disease was the most common single cause of death related to air pollution (18% increase in mortality risk) (Pope 3rd et al. 2004; Brook 2008). However, no significant relationship between mortality and the concentrations of PM_{10} and gaseous pollutants other than sulfur dioxide was noted (Pope 3rd et al. 2004).

In the Woman's Health Initiative study including 65,893 healthy postmenopausal women from 36 cities in the United States, a 10-μg/m^3 increase in long-term exposure to $PM_{2.5}$ was associated with a 24% (hazard ratio HR, 1.24; 95% CI, 1.09–1.41) and 76% (HR, 1.76; 95% CI, 1.25–2.47) increase in the risk of a cardiovascular event and cardiovascular mortality, respectively (Miller et al. 2007).

Another large Canadian national-level cohort study showed that chronic exposure to $PM_{2.5}$ was related to cardiovascular mortality, particularly from ischemic heart disease (Crouse et al. 2012). The calculated hazard ratio for mortality due to ischemic heart disease was 1.31 (95% CI, 1.27–1.35) for each increase in $PM_{2.5}$ by 10 μg/m^3. These associations were found with exposure to $PM_{2.5}$ at concentrations that were predominantly lower (mean, 8.7 μg/m^3) than those reported previously (Crouse et al. 2012). Similar conclusions were reported in the 2001 Canadian Census Health and Environment Cohort study (Pinault et al. 2017). The authors demonstrated that the hazard ratio for mortality due to cardiovascular disease was 1.25 (95% CI, 1.19–1.31); ischemic heart disease, 1.36 (95% CI, 1.28–1.44); cardiometabolic disease, 1.27 (95% CI, 1.21–1.33); and chronic obstructive pulmonary disease, 1.24 (95% CI, 1.11–1.39). For ischemic heart disease, cardiometabolic, and respiratory mortality, the shape of the concentration-response curve was supralinear, with larger differences in the relative risk for lower concentrations.

In the NIH-AARP Diet and Health Study (n = 517,043), a long-term exposure to $PM_{2.5}$ was related to higher cardiovascular mortality by 10% (per 100-μg/m^3 increase), despite low $PM_{2.5}$ levels (10–13 μg/m^3) (Thurston et al. 2016). This study showed that chronic $PM_{2.5}$ exposure was significantly associated with all-cause (HR, 1.03; 95% CI, 1.00–1.05) and cardiovascular (HR, 1.10; 95% CI, 1.05–1.15) mortality.

The multicenter ESCAPE study (European Study of Cohorts for Air Pollution Effects), which involved 367,251 participants, showed an increased hazard ratio for all-cause mortality of 1.07 (95% CI, 1.02–1.13) per 5-μg/m^3 increase in $PM_{2.5}$ (Beelen et al. 2014). In this study, the effect of $PM_{2.5}$ on all-cause mortality was about twice that of previous estimates. The relationship between $PM_{2.5}$ concentrations and mortality remained significant even when $PM_{2.5}$ levels were lower than the European annual mean limit value of 25 μg/m^3 (HR, 1.06; 95% CI, 1.00–1.12) and even when they were below 20 μg/m^3 (HR, 1.07; 95% CI, 1.01–1.13). Interestingly, no significant correlations with cardiovascular mortality were found (Beelen et al. 2014).

The Aphekom project (Improving Knowledge and Communication for Decision Making on Air Pollution and Health in Europe) confirmed that the largest health burden in 25 European cities was attributable to the effects of chronic exposure to $PM_{2.5}$ (Pascal et al. 2013). It has been estimated that the benefit of compliance with the World Health Organization guidelines for $PM_{2.5}$ (annual mean <10 μg/m^3) would result in an increase in life expectancy at the age of 30 years from 0 (Stockholm) to 22 months (Bucharest) (median, 5.8 months), depending on the city, which translates into approximately 19,000 postponed deaths annually, including more than 15,000 cardiovascular deaths (Pascal et al. 2013). The results of other multicenter studies, including the ESCAPE study, indicate that this benefit may probably be even greater (Beelen et al. 2014).

A meta-analysis by Hoek et al. (2013) showed that an increase in the annual concentration of $PM_{2.5}$ by 10 μg/m^3 results in an average increase in all-cause mortality by 6% (95% CI, 4–8) and in cardiovascular mortality by 11% (95% CI, 5–16). In another meta-analysis, the strongest relations were found for mortality associated with ischemic heart disease (Bourdrel et al. 2017). Interestingly, some studies have shown that the adverse effects of $PM_{2.5}$ were higher in people with the lowest education status, probably due to an improper diet with a low content of plant-derived antioxidants (Hoek et al. 2013; Bourdrel et al. 2017).

Temperature plays an important role in the health effects of $PM_{2.5}$ in humans (Kioumourtzoglou et al. 2016). A recent study conducted in the United States study on a large population (>35 million Medicare enrollees from 207 U.S. cities) showed a higher association between long-term $PM_{2.5}$ exposure and mortality in warmer cities (Kioumourtzoglou et al. 2016). Although previous studies indicated the effect of temperature on the relationship between short-term air pollution and mortality, this study was the first to prove this phenomenon for chronic exposure to $PM_{2.5}$. Of note, the temperature-related effect was particularly marked in people with low social status. It is not easy to explain this phenomenon.

Kioumourtzoglou et al. (2016) suggested that annual average temperatures could be considered as indicators of other potentially modifying factors, for example, the socioeconomic status. In fact, in this study, the temperature was significantly positively correlated with a high level of poverty as well as a high proportion of black residents and residents without a high school degree. Moreover, the authors reported that temperature can indicate a varying particle composition, as it affects the particulate matter size and composition in the urban air.

Apart from $PM_{2.5}$, also long-term exposure to gaseous co-pollutants has been documented to be associated with all-cause and cardiovascular mortality. Hoek et al. (2013) showed that all-cause mortality was significantly associated with nitrogen dioxide (pooled estimate per 10 $\mu g/m^3$, 5%; 95% CI, 3–8). A meta-analysis by Faustini et al. (2014) showed that also long-term exposure to nitrogen dioxide significantly affects all-cause mortality, regardless of $PM_{2.5}$ levels. In this study, a 13% increase in cardiovascular mortality was found after a 10-$\mu g/m^3$ increase in annual nitrogen dioxide levels (Faustini et al. 2014). The adverse effect of long-term exposure to ozone seems to be less clear (Bourdrel et al. 2017). Some studies have shown a slight increase in cardiopulmonary mortality in response to long-term exposure to ozone, but this was only seen in the warm season and not in the annual analysis (Bourdrel et al. 2017; Atkinson et al. 2016). This can be partly explained by the fact that ozone pollution usually occurs on warm and sunny days, because its formation requires photochemical reactions (Bourdrel et al. 2017). The HRs for cardiovascular and respiratory causes of death were 1.01 (95% CI, 1.00–1.02) and 1.03 (95% CI, 1.01–1.05), respectively, per 10-ppb increase in long-term warm-season exposure to ozone (Atkinson et al. 2016). Finally, chronic exposure to sulfur dioxide was strongly associated with respiratory mortality but its effect on cardiovascular mortality was ambiguous (Bourdrel et al. 2017).

5.5 Air Pollution and Coronary Artery Disease

There is increasing evidence that air pollution contributes to the pathogenesis of cardiovascular diseases. Regarding the pathophysiology of cardiovascular diseases, $PM_{2.5}$ and nitrogen oxides currently attract the greatest attention (Brook et al. 2010; Newby et al. 2015; Robertson and Miller 2018; Integrated Science Assessment for Particulate Matter 2019. United States Environment Protection Agency website; Health Canada. Human health risk assessment for coarse particulate matter 2016a)

5.5.1 Short-Term Exposure Studies

Numerous time-series and case-crossover studies have shown a significant association between short-term changes in air pollution and the incidence of acute coronary syndromes, including acute myocardial infarction, particularly in individuals with preexisting coronary artery disease (Ruidavets et al. 2005; Pope et al. 2015; Rajagopalan et al. 2018).

In a large meta-analysis of case-crossover studies, time-series, and one case-control study, Nawrot et al. (2011), reviewing triggers for nonfatal myocardial infarction, exposure to air pollution and road traffic were the most important factors at the population level. Due to both the magnitude of the risk and the prevalence in the population, it has similar significance as other well-known triggers such as coffee, alcohol, and physical exertion (Nawrot et al. 2011).

Interestingly, Gardner et al. (2014), using data from patients with acute coronary syndrome and case-crossover methods, showed that increased $PM_{2.5}$ concentrations in the hour prior to the onset of acute coronary syndrome triggered ST-segment elevation myocardial infarction (STEMI) but not non-STEMI. In addition, individuals with preexisting hypertension or other cardiovascular diseases appeared to be particularly susceptible. However, not all investigators have reported a significant relationship between short-term $PM_{2.5}$ exposure and the risk of myocardial infarction. In a case-crossover study of 5793 confirmed cases of acute myocardial infarction, Sullivan et al. (2005) did not observe such a link even in patients with preexisting cardiac disease.

In a time-stratified case-crossover study by Pope et al. (2015), which included 16,314 acute coronary syndrome events treated at Intermountain Healthcare hospitals in the urban areas of Utah's Wasatch Front, short-term exposure to elevated $PM_{2.5}$ concentrations contributed to triggering acute coronary events, especially ST-segment elevation myocardial infarction (STEMI), but only in patients with preexisting significant coronary artery stenoses. In individuals with angiographically proven coronary artery disease, the odds ratio (ORs) for a 10-$\mu g/m^3$ increase in concurrent-day $PM_{2.5}$ levels higher than 25 $\mu g/m^3$ were 1.06 (95% CI, 1.02–1.11) for all acute coronary syndrome events, 1.15 (95% CI, 1.03–1.29) for STEMI, and 1.02 (95% CI, 0.97–1.08) for non-STEMI (Pope et al. 2015).

On the other hand, Huang et al. (2017) observed a more complex relationship between air pollution and meteorological factors in terms of health effects. Based on the data from 3164 patients with acute coronary syndromes hospitalized in 39 hospitals, who were included in the Taiwan ACS Full Spectrum Registry, a strong relationship was found between particulate matter (both $PM_{2.5}$ and PM_{10}), carbon monoxide, and climatic changes and the risk of hospitalization for myocardial infarction in Taiwan. High temperatures (>26°C) and low barometric pressure (<1009 hPa) on the day before hospitalization were associated with an increase in the incidence of myocardial infarction. Interestingly, this study also showed that the combinations of high particulate matter or carbon monoxide concentrations with low temperature (<21°C) and of high humidity levels with low temperature were also associated with an increased incidence of acute coronary syndromes. This suggests that air pollutants and meteorological factors have a synergistic effect on the occurrence of acute coronary syndromes (Huang et al. 2017).

Also recent cohort study (n = 3545), with very well characterized population, including the inhabitants of Kraków, Poland (one of the polluted cities in Europe) showed that the short-term elevation in $PM_{2.5}$ concentrations was related to an increased number of daily hospital admissions for myocardial infarction in all age groups (OR, 1.32; 95% CI, 1.01–1.40), regardless of concomitant diseases (Konduracka et al. 2019). In addition, we demonstrated that the adverse effect of PM_{10} was found only with a simultaneous decrease of 1 °C in the mean daily temperature (OR, 1.08; 95% CI, 1.01–1.17). The effect of nitrogen dioxide was relevant only in individuals aged 70 years or older (OR, 1.13; 95% CI, 1.01–1.23) and those with previously diagnosed lung disease (OR, 1.12; 95% CI, 1.01–1.31). Importantly, the average annual concentrations of pollutants in Kraków were 54.2 μg/m^{3}for PM_{10}, 37.6μg/m^3 for $PM_{2.5}$, and 42.3 μg/m^{3}for nitrogen dioxide.

An important aspect of studies on the effect of air pollution on the incidence of myocardial infarction is to determine the time between short-term exposure to air pollutants and development of myocardial infarction (Konduracka et al. 2019). The strongest associations were found with the mean air pollution exposure on the same day or on the last 2 days. Our studies showed that in chronically exposed inhabitants of Kraków (average annual $PM_{2.5}$ concentrations of about 38 μg/m^2; the limit recommended by the World Health Organization <10 μg/m^3), the effect of short-term fluctuations in $PM_{2.5}$ levels on the risk of hospitalizations for myocardial infarction was delayed, with significant effects observed at lags 4 and 6 (Konduracka et al. 2019). It is postulated that in the case of exposure to $PM_{2.5}$, the time required for the destabilization of atherosclerotic plaque and its rupture due to oxidative stress, inflammation, and endothelial dysfunction is longer than for exposure to gaseous pollutants, and may be up to several days since acute exposition (Konduracka et al. 2019). In contrast to our results, other studies have shown adverse effects of $PM_{2.5}$ on the cardiovascular system between 0 and 2 days after exposure to elevated levels of air pollutants (Bhaskaran et al. 2009; Bhaskaran et al. 2013).

While short-term exposure to $PM_{2.5}$ appears to have a clearly negative effect on health, including an increase in the risk of myocardial infarction, the harmful impact of PM_{10} remains inconclusive. The effect of PM_{10} exposure on the risk of myocardial infarction has been assessed by several studies, some of which reported a positive association but most showing a minor or no effect (Peters et al. 2001; Donaldson and Tran 2002; Chen and Nadziejko 2005; Bhaskaran et al. 2009, 2013). However, a systematic review and meta-analysis by Mustafic et al. (2012), which included 17 time-series studies and 17 case-crossover studies, showed an increase in near-term risk of myocardial infarction related to short-term exposure to both $PM_{2.5}$ and PM_{10}. It has been demonstrated that the RR of myocardial infarction associated with a short-term increase in particulate matter concentrations (calculated for each increment of 10 μg/m^3) was 1.025 (95% CI, 1.015–1.036) for $PM_{2.5}$ and 1.006 (95% CI, 1.002–1.009) for PM_{10}.

Similarly, another systematic review and meta-analysis by Luo et al. (2015), which included 31 epidemiological studies published in the electronic databases, showed that the risk of myocardial infarction was significantly associated with short-term 10-μg/m^3 increments in $PM_{2.5}$ (OR, 1.022; 95% CI, 1.015–1.030) and PM_{10} (OR, 1.005; 95% CI, 1.001–1.008). The results of these meta-analyses were confirmed by a recently published, large, time-stratified case-crossover study by Liu et al. (2017), which assessed the relationship between ambient air pollution and daily myocardial infarction in 14 Chinese cities. This multicity study showed that PM_{10} was significantly but weakly associated with increased risk of myocardial infarction on lag2 and lag3 days, and this effect was independent of age ($\geq$65 years and<65 years) and gender (Liu et al. 2017). However, the association of PM_{10} with myocardial infarction was found to be insignificant after adjustment for the other air pollutants, in particular NO_2 and SO_2 (Liu et al. 2017).

Although PM_{10} also contains smaller particles (fine and ultrafine particles), the relationship between exposure to PM_{10} and adverse cardiovascular outcomes is weaker than that observed for $PM_{2.5}$, which is explained by significant differences in the proportion of $PM_{2.5}$ found in PM_{10} depending on the source of these air pollutants. Unlike $PM_{2.5}$ may penetrate the lung alveoli directly into the bloodstream and thus result in more serious adverse cardiovascular effects because of a higher potential to generate oxidative stress and systemic inflammation.

Apart from particulate matter, also gaseous pollutants (particularly nitrogen oxides) play an important role in the pathophysiology of air pollution-related acute coronary syndromes. The large meta-analysis by Mustafic et al. (2012) showed that the relative risk of myocardial infarction related to gaseous co-pollutants was 1.048 (95% CI, 1.026–1.070) for carbon monoxide, 1.011 (95% CI, 1.006–1.016) for nitrogen dioxide, and 1.010 (95% CI, 1.003–1.017) for sulfur dioxide (calculated for each increment of 10 μg/m3 in pollutant concentration). In turn, Liu et al. (2017) found a significant relationship of SO_2, NO_2, and CO concentrations on lag 2 day with a 2.0% (95%CI 1.2%, 2.9%), 2.2% (95% CI 1.4%, 3.1%), and 1.1% (95%CI 0.4%, 1.8%) increases in hospital admissions for myocardial infarction, respectively. However, the effect of CO was weak after adjustment for the NO_2 and SO_2 (Liu et al. 2017). In our study, a short-term increment in nitrogen dioxide concentrations resulted in an increased risk of hospitalization only in patients aged 70 years or older and in those with preexisting respiratory disease (Konduracka et al. 2019). These results are consistent with other previous reports (Luo et al. 2015. On the other hand, Argacha et al. (2016) reported that the risk associated with particulate matter was greater among older people, while younger patients were more prone to harmful effects of exposure to nitrogen dioxide. Each 10-μg/m^3 increment in $PM_{2.5}$, PM_{10}, and nitrogen dioxide levels was associated with an increased OR of STEMI: 1.028 (95% CI, 1.003–1.054), 1.026 (95% CI, 1.005–1.048), and 1.051 (95% CI, 1.018–1.084), respectively. Moreover, an increased risk of STEMI was associated with exposure to PM_{10} in individuals aged 75 years or older (OR, 1.046; 95% CI, 1.002–1.092) as well as with exposure to nitrogen dioxide in those aged 54 years or younger (OR, 1.071; 95% CI, 1.010–1.136) (Argacha et al. 2016). The effect of short-term ozone exposure on the risk of myocardial infarction remains unclear (Mustafic et al. 2012).

5.5.2 Long-Term Exposure Studies

Numerous epidemiological studies have shown that long-term exposure to air pollution, particularly $PM_{2.5}$, promotes premature atherosclerosis and is associated with atherothrombotic events (Cesaroni et al. 2014; Kaufman et al. 2016; Bourdrel et al. 2017; Ljungman et al. 2019).

In their study, Hoffmann et al. (2007) demonstrated that long-term exposure to heavy road traffic is associated with an elevated risk of a high coronary artery calcium score of 63% and 34% for individuals living within 50 m and 50–100 m from a major road, respectively, as compared with those living 200 m away. The reduction of the distance between the place of residence and the main road by half was associated with a 7.0% (95% CI, 0.1–14.4) increase in coronary artery calcification. Also in a cross-sectional analysis of the Heinz Nixdorf Recall Study demonstrated that long-term exposure to fine particulate matter was associated with a 18.1% (95% CI, 6.6–30.9) increase in the burden of thoracic aorta calcification per 2.4 μg/m^3 (Kälsch et al. 2014).

In a prospective 10-year cohort study by Kaufman et al. (2016), which involved 6795 participants of the MESA Air (Multi-Ethnic Study of Atherosclerosis and Air Pollution) in 6 metropolitan areas in the United States, exposure to $PM_{2.5}$ and traffic-related air pollutants was significantly associated with an increased coronary artery calcium score. It was shown that for each 5-μg/m^3 increase in $PM_{2.5}$ levels, coronary calcium progressed by 4.1 Agatston units per year (95% CI, 1.4–6.8) and that for each 40-ppb increase in nitrogen oxide levels, coronary calcium progressed by 4.8 Agatston units per year (95% CI, 0.9–8.7). Of note, exposure to air pollutants was not related to changes in the intima-media thickness.

In a cross-sectional analysis by Künzli et al. (2005), including 798 participants of 2 clinical trials, revealed a relationship between long-term exposure to ambient $PM_{2.5}$ levels (range, 5.2–26.9 μg/m^3) and increased carotid intima-media thickness (CIMT). Each 1-μg/m^3 increase in $PM_{2.5}$ levels was associated with a 5.9% (95% CI, 1–11) increase in CIMT. Another cross-sectional study, which enrolled individuals aged 45–84 years from 6 United States cities, showed that a 21-μg/m^3 increase in

PM_{10} (RR, 1.02; 95% CI, 1.00–1.03) and a 12.5 μg/m^3 increase in $PM_{2.5}$ (RR, 1.01; 95% CI, 1.00–1.02 per) were associated with an increased CIMT (Diez Roux et al. 2008).

A cross-sectional analysis of 3380 participants (age, 45–75 years) of the Heinz Nixdorf Recall Study demonstrated that each 4.2-μg/m^3 increase in $PM_{2.5}$ levels was associated with a 4.3% (95% CI, 1.9%-6.7%) increase in CIMT (Bauer et al. 2010). However, no significant relationship was found between PM_{10} and CIMT in this study. In contrast, Tonne et al. (2012) showed that exposure to PM_{10} was associated with 5.0% (95% CI, 1.9%-8.3%) increase in CIMT even after adjustment for covariates. However, it is important to note than none of those studies reported a relationship between long-term exposure to air pollution (including $PM_{2.5}$, PM_{10}, and gaseous co-pollutants) and subclinical atherosclerosis (Kim et al. 2014; Perez et al. 2015; Dorans et al. 2016).

There is convincing evidence to support a significant relationship between long-term exposure to air pollution and the risk of acute myocardial infarction (Bourdrel et al. 2017; Rajagopalan et al. 2018). Numerous studies, although not all, have demonstrated the effect of long-term exposure to individual air pollutants on an increased incidence of acute coronary syndromes, including myocardial infarction.

A longitudinal cohort study of all manual workers in the Swedish National Census (n = 1,725,671) showed that occupational exposure both to small (adjusted HR, 1.12; 95% CI, 1.09–1.15) and to large (adjusted HR, 1.14; 95% CI, 1.10–1.18) particles was associated with incident myocardial infarction (Wiebert et al. 2012). The risk related to exposure to small particles (<1 μm) was higher among women (HR, 1.30; 95% CI, 1.12–1.51) than among men (HR 1.10; 95% CI, 1.07–1.14).

A recent longitudinal cohort study of Seoul residents (n = 136,094), which used health insurance claims data, revealed that long-term exposure to $PM_{2.5}$ (adjusted HR, 1.36; 95% CI, 1.19–1.56 per 1 μg/m^3), $PM_{2.5-10}$ (adjusted HR, 2.14; 95% CI, 2.02–2.27 per 1 μg/m^3), nitrogen dioxide (adjusted HR, 1.81; 95% CI, 1.46–2.25 per 18.4 ppb), sulfur dioxide (adjusted HR, 1.82; 95% CI, 1.52–2.19 per 2.54 ppb), and carbon monoxide (adjusted HR, 2.12; 95% CI, 1.72–2.61 per 0.25 ppm), but not ozone, was associated with incident myocardial infarction (Kim et al. 2017).

A prospective cohort study and meta-analysis of 11 European cohorts from the ESCAPE study revealed that an annual increase of 5 μg/m^3 in $PM_{2.5}$ levels and of 10 μg/m^3 in PM_{10} levels was related with an increase in the risk of myocardial infarction of 13% (HR, 1.13, 95% CI, 0.98–1.30) and 12% (HR, 1.12, 95% CI, 1.01–1.25), respectively (Cesaroni et al. 2014). Importantly, in this study, positive associations were found below the current annual European threshold of 25 μg/m^3 for $PM_{2.5}$ (HR, 1.18; 95% CI, 1.01–1.39 for a 5-μg/m^3 increment in $PM_{2.5}$) and below 40 μg/m^3 for PM_{10} (HR, 1.12, 95% CI, 1.00–1.27 for a 10-μg/m^3 increment in PM_{10}).

It was also reported that long-term prognosis after acute myocardial infarction is affected by long-term exposure to air pollution, particularly $PM_{2.5}$ (Tonne and Wilkinson 2013; Chen et al. 2016; Konduracka et al. 2019). However, further large epidemiological studies investigating this association are warranted.

5.6 Effect of Air Pollution on Blood Pressure and Arterial Hypertension

Although in most patients hypertension occurs as a result of a genetic predisposition, its progression is strongly influenced by lifestyle and various environmental factors. In recent years, several studies have shown that higher levels of $PM_{2.5}$ and some gaseous air pollutants (nitrogen oxides, ozone, sulfur dioxide) may also lead to increased blood pressure and the development of hypertension. The health effects of both short- and long-term exposure have been described in numerous studies.

5.6.1 Short-Term Exposure to Particulate Matter

Controlled human exposure studies in healthy volunteers provide a special opportunity to explore single pollutants or their mixture with partial control of numerous confounding variables. Urch et al. (2005) compared the effect of known mean levels of $PM_{2.5}$ (147 ± 27 μg/m^3) and ozone (121 ± 3 *vs* 8 ± 5 ppb) on blood pressure. They showed that 2-hour controlled exposure to environmentally relevant concentrations of $PM_{2.5}$ and ozone may rapidly increase diastolic blood pressure. Another study investigated acute exposure to diesel exhaust, which is usually representative of the exhaust levels reached in some cities (Cosselman et al. 2012). This double-blind crossover exposure study demonstrated that diesel exhaust inhalation, maintained at 200 μg/m^3 of fine particulate matter, resulted in a rapid increase in systolic blood pressure of 4.4 mmHg (95% CI, 1.1–7.7 mmHg; $p = 0.0009$) in young nonsmokers. In fact, diesel exhaust is a mixture of particulate matter, nitrogen oxides, and other chemicals, which together may contribute to the observed effect on blood pressure rather than particulate matter alone.

The first large meta-analysis to explore the association between exposure to $PM_{2.5}$ and blood pressure was published by Liang et al. (2014). The authors included the results of 22 panel and cross-sectional studies conducted in Asia, North America, and Europe, which demonstrated a positive relationship between exposure to $PM_{2.5}$ and blood pressure. Most studies ($n = 19$) investigated short-term exposure. Each 10-mg/m^3 increase of $PM_{2.5}$ resulted in an increment of 1.393 mmHg (95% CI, 0.87–1.91) and 0.895 mmHg (95% CI, 0.49–1.29) in systolic and diastolic blood pressure, respectively. The strongest associations for systolic (2.67 mmHg; 95% CI, 0.58–4.76) and diastolic components (1.35 mmHg; 95% CI, 0.126–2.58) were reported at lag 0–5 day. The effect of particulate matter was significant in all geographical regions except Europe.

Two years later, Cai et al. (2016) performed a meta-analysis of 17 studies conducted in Asia, North and South America, and Europe investigating an association of both short- and long-term exposure to different air pollutants with the risk of hypertension. The results of meta-analyses including case-crossover and time-series studies on short-term exposure to $PM_{2.5}$ demonstrated that each 10-μg/m^3 increment of $PM_{2.5}$ increased the risk of hypertension (OR, 1.06; 95% CI, 1.00–1.14). The lag days ranged from 1 to 7. However, no significant results were reported for PM_{10}. Also, in a recent systematic review and meta-analysis of up to 100 studies from 16 countries, Yang et al. (2017) evaluated data on the association of short- and long-term exposure to air pollutants with blood pressure and hypertension. They demonstrated significant results for short-term exposure. Short-term exposure to four (PM_{10}, $PM_{2.5}$, sulfur dioxide, nitrogen dioxide), two ($PM_{2.5}$ and sulfur dioxide), and four air pollutants (PM_{10}, $PM_{2.5}$, nitrogen dioxide, sulfur dioxide) was significantly associated with hypertension (ORs, 1.05–1.10), systolic blood pressure (β values, 0.53–0.75 mmHg), and diastolic blood pressure (β values, 0.15–0.64 mmHg), respectively. The results were stronger for the studies of men, North American and Asian race, and countries with higher air pollution.

5.6.2 Long-Term Exposure to Particulate Matter

A number of studies reported a positive association of long-term exposure to $PM_{2.5}$ with either systolic or diastolic blood pressure or both. In a cohort study (n = 1023), Chuang et al. (2011) revealed that among subjects from Taiwan, aged 54–90 years, an increase in 1-year average concentrations of $PM_{2.5}$ by 20 μg/m^3 resulted in the elevation of systolic blood pressure by 32 mmHg (95% CI, 21.57–42.58) and of diastolic blood pressure by 31 mmHg (95% CI, 25.43–37.14). Also in another large cohort (n = 24845), Dong et al. (2013) showed that long-term exposure to PM_{10}, sulfur dioxide, and ozone in 11 districts of 3 north-eastern cities in China resulted in a slight increase in the risk of arterial blood pressure and hypertension.

Subsequently, Fuks et al. (2014) performed a meta-analysis of 113,926 participants from 15 international cohorts included in the ESCAPE study (European Study of Cohorts for Air Pollution Effects), who were exposed to traffic pollution. No significant associations between $PM_{2.5}$ concentrations and blood pressure were found. Meanwhile, in a meta-analysis, Cai et al. (2016) demonstrated that each 10-μg/m^3 increase in long-term exposure to PM_{10} was associated with a significant increase in the risk of hypertension (OR, 1.034; 95% CI: 1.005–1.063), but the results for $PM_{2.5}$ were non-significant.

In another meta-analysis, including 7 cohorts from the ESCAPE study, Fuks et al. (2017) demonstrated a positive relationship between the incidence of self-reported hypertension and higher $PM_{2.5}$ concentrations (RR, 1.22; 95% CI, 1.08–1.37 per each 5-μg/m^3 increase). Also other studies demonstrated a positive relationship only with one of the blood pressure components. In a cross-sectional study (n = 13,975), Liu et al. (2017) reported that systolic blood pressure increased by 0.60 mmHg (95% CI, 0.05–1.15) per a 41.7 μg/m^3 increase in $PM_{2.5}$. The annual mean exposure to $PM_{2.5}$ for all participants was 72.8 μg/m^3. On the other hand, in a cross-sectional study (n = 2552), Pitchika et al. (2017) showed an increase in diastolic blood pressure by 0.7% (95% CI, 0.2–1.2) per a 1-μg/m^3 increase in $PM_{2.5}$. The association between road traffic noise and hypertension was also included in those studies as a confounding factor. The authors also reported a positive association between $PM_{2.5}$ concentrations and the risk of hypertension. In a study by Liu et al. (2017), the OR of hypertension per a 41.7-μg/m^3 increase in long-term average $PM_{2.5}$ concentrations was 1.11 (95% CI, 1.05–1.17). Author also demonstrated that an interquartile increase in the annual mean $PM_{2.5}$ levels (1 μg/m^3) was significantly associated with a 15% higher prevalence of hypertension (95% CI, 2.5–28.0).

A present large (n = 119,388) prospective cohort study (China-PAR project) provided the first evidence that long-term exposure to $PM_{2.5}$ was independently associated with incident hypertension at relatively high (mean annual $PM_{2.5}$ concentrations, 58.4 μg/m^3) ambient concentrations (Huang et al. 2019). Participants in the highest quartile of the $PM_{2.5}$ concentration had an increased risk of incident hypertension with a HR of 1.77 (95% CI, 1.56–2.00), as compared with those in the lowest quartile. Each 10-μg/m^3 increment of $PM_{2.5}$ concentrations increased the risk of hypertension by 11% (HR, 1.11; 95% CI, 1.05–1.17).

5.6.3 Short-Term Exposure to Ozone

The toxic effect of ozone on the cardiovascular system is much less known than its respiratory and allergenic effects. In some experimental animal studies, short-term exposure to ozone was shown to increase vasoconstriction and reduce vasodilation in response to acetylcholine (Robertson et al. 2013; Snow et al. 2018). In contrast, controlled human exposure studies found no effect of short-term exposure on increased vasoconstriction (Brook et al. 2009), but other investigators reported a significant elevation in systolic blood pressure without changes in diastolic blood pressure after ozone exposure (0.1 and 0.2 ppm) in healthy subjects (Frampton et al. 2015) The effect of acute exposure to ozone was also investigated in a few cohort studies (Chuang et al. 2010; Chen et al. 2012; Weichenthal et al. 2014). In a large Taiwanese cohort study (n = 7578) including a population at a mean age of 43 years, an increase in 3-day ozone levels by 12.15 particles per billion (ppb) was associated with an increase in diastolic blood pressure by 0.37 mmHg (95% CI, 0.04–0.69). On the other hand, no significant changes in systolic blood pressure were reported. The mean daily concentration of ozone was 26.83 ± 9.70 ppb, whereas the mean temperature was 25.49°C ± 2.48°C (Chuang et al. 2010).

Chen et al. (2012) investigated a cohort of 9238 inhabitants of Taiwan (mean age, 58 years) and showed that diastolic blood pressure increased with exposure to ozone at various lag times, whereas systolic blood pressure decreased. Exposure to ozone was measured as a maximum 1-hour concentration and reached about 65 ppb. The mean temperature was similar to that reported by Chuang et al. (2010). Finally, Weichenthal et al. (2014) reported that a 3-hour exposure to ozone was positively correlated with elevated blood pressure in a healthy Canadian woman aged 25 years. Each 24-ppb increase in ozone concentrations resulted in an increase of 2.49% in systolic blood pressure (95% CI, 0.141–4.84) and an increase of 3.26% in diastolic blood pressure (95% CI, 0.012–6.51). Changes in temperature and humidity were not considered as confounding factors. A recent large Korean study (n = 98,577) demonstrated an hourly relationship between ambient air pollution and blood pressure in participants at a mean age of 55 years (Choi et al. 2019). Exposure to the mean ozone concentrations of 17 ppb was associated with a significant increase in systolic blood pressure at 3–5 lag hours after exposure. An unadjusted interquartile range (IQR) increase of ozone (26 ppb) was associated with an increase in systolic blood pressure by 0.55 ± 0.14 mmHg (p = 0.001). Importantly, the effect was more pronounced in patients with hypertension.

5.6.4 Long-term Exposure to Ozone

Only a few studies have demonstrated the relationship between long-term exposure to ozone and blood pressure or prevalence of hypertension. Chuang et al. (2011) reported that a 8.95-ppb increase in ozone levels resulted in an estimated increase in systolic blood pressure by 21.51 mmHg (95% CI, 16.90–26.13) and in diastolic blood pressure by 20.56 mmHg (95% CI, 18.14–22.97). In a study by Dong et al. (2013), including a large cohort of patients (n = 24,845) at a mean age of 45.59 years, systolic blood pressure increased by 0.73 mmHg (95% CI, 0.35–1.11) and diastolic blood pressure increased by 0.37 mmHg (95% CI, 0.14–0.61) per a 22-μg/m^3 increase in ozone levels. A positive relationship between ozone concentrations and hypertension was reported by Coogan et al. (2017). The estimated hazard ratio of hypertension per a 6.7-ppb increase in ozone levels was 1.09 (95% CI, 1.00–1.18). The magnitude of the relationship between ozone and blood pressure remains unclear, but the available evidence indicates that clinical implications may be larger than the effect estimated in clinical studies (Integrated Science Assessment for Ozone and Related Photochemical Oxidants (2020); https://www.epa.gov/isa/integrated-science-assessment-isa-ozone-and-related-photochemical-oxida).

5.6.5 Short-Term Exposure to Nitrogen Dioxide

The main source of nitrogen oxides is combustion (mainly from vehicles). Most emissions of nitrogen oxides are in the form of nitric oxide, which is rapidly converted to nitrogen dioxide. (Health Canada. Human health risk assessment for ambient nitrogen dioxide 2016b).

When evaluating the health impact of nitrogen dioxide, epidemiological studies are more valuable than experimental ones, because they provide the most direct approach to health assessment. Moreover, humans are more heterogeneous than laboratory animals and have a wide range of susceptibilities and coexistent disorders. A few studies reported a positive relationship between exposure to nitrogen dioxides and blood pressure changes, but the results were inconsistent.

In a small cohort study, Dales et al. (2007) investigated 39 healthy volunteers aged 18–50 years and revealed that during exposure to nitrogen dioxide at bus stations in Canada, an increase in nitrogen dioxide by 1 ppb resulted in an increase in diastolic blood pressure by 0.44 mmHg ($p = 0.004$). No effect on systolic blood pressure was noted. The association between nitrogen dioxide levels and blood pressure was adjusted for atmospheric conditions and the time of day. Another two studies reported a positive correlation between exposure to nitrogen dioxide and elevation of blood pressure. Guo et al. (2010) revealed that exposure to nitrogen dioxide was associated with an increased rate of emergency hospital visits for hypertension. For an increase in nitrogen dioxide by 10 $\mu g/m^3$, the odd ratio was 1.10 (95% CI, 1.03–1.16) at lag 3 days. In this study, the mean annual nitrogen dioxide concentration was about 67 $\mu g/m^3$. Szyszkowicz and Rowe (2012) also concluded that even relatively low levels of air pollution may increase the risk of emergency hospital visits for hypertension. For an increase in nitrogen dioxide by 12.8 ppb, the estimated odd ratio was 1.06 (95% CI, 1.00–1.12). During the study, the mean nitrogen dioxide concentration was 22 ppb. Finally, Kubesch et al. (2015) reported that an increase in nitric oxide concentrations was positively associated with elevation in systolic blood pressure.

5.6.6 *Long-Term Exposure to Nitrogen Dioxide*

The effects of long-term exposure to nitrogen dioxide on blood pressure were evaluated by Chuang et al. (2011). A 12.83-ppb increase in nitrogen dioxide levels was associated with an increase in systolic blood pressure by 14.40 mmHg (95% CI, 10.98–17.82) and in diastolic blood pressure by 12.43 mmHg (95% CI, 10.63–14.23).

In a Chinese cross-sectional study on a cohort of 27,752 participants aged above 65 years of age, 1-year exposure to nitric oxide and nitrogen dioxide resulted in an elevation only of diastolic blood pressure. Diastolic blood pressure increased by 0.34 mmHg (95% CI, 0.19–0.50) per a 20-$\mu g/m^3$ increase in nitric oxide levels and by 0.65 mmHg (95% CI, 0.44–0.85) per a 10-$\mu g/m^3$ increase in nitrogen dioxide levels. The 1-year average concentrations of nitric oxide and nitrogen dioxide were 38.4 ± 15.3 $\mu g/m^3$ and 23.7 ± 5.8 $\mu g/m^3$, respectively (Chen et al. 2015). However, no relationship between nitrogen dioxide concentrations and blood pressure was reported in other studies (Dong et al. 2013; Fuks et al. 2014; Pitchika et al. 2017).

In contrast, a negative association between nitric oxide levels and hypertension was observed in a Danish cohort ($n = 57{,}053$) aged 50–64 years (Sorensen et al. 2012). The authors reported a slight decrease of systolic blood pressure (−0.50 mmHg; 95% CI, 0.84–0.16) and of diastolic blood pressure (−0.24 mmHg; 95% CI, 0.42–0.07). Of note, the mean annual concentration of nitric oxide was lower than that reported by Chen et al. (2015). No significant associations between nitrogen dioxide concentrations and elevated blood pressure or hypertension were reported in other individual studies (Chuang et al. 2011; Dong et al. 2013; Foraster et al. 2014; Chan et al. 2015) or in large multi-cohort ESCAPE meta-analyses (Fuks et al. 2014, 2017).

Given the limited supporting data and discrepancies in reported results, the overall evidence is suggestive of but not sufficient to confirm a causal relationship between short-term exposure to ambient nitrogen dioxide and hypertension.

5.6.7 *Short-Term Exposure to Sulfur Dioxide*

The effects of short-term exposure to sulfur dioxide on blood pressure were investigated in a number of studies. In a large (n = 10 459) cross-sectional study from Korea, Choi et al. (2007) revealed that systolic blood pressure was significantly associated with sulfur dioxide concentrations in cold-weather season when the mean 24 hour average sulfur dioxide concentration was 6.9 ppb. A cohort study by Chen et al. (2012), which included 9238 persons over 30 years of age from 6 cities in Taiwan, revealed that a 2-ppb increase in exposure to sulfur dioxide at lag 2 days was associated with an increase in systolic blood pressure by 0.32 mmHg (95% CI, 0.06–0.59) and in diastolic blood pressure by 0.79 mmHg (95% CI, 0.61–0.97).

5.6.8 *Long-Term Exposure to Sulfur Dioxide*

There is a limited number of studies that investigated the impact of sulfur dioxide on blood pressure or hypertension (Chuang et al. 2011; Dong et al. 2013). Chuang et al. (2011) reported a positive association between an increase in sulfur dioxide by 20 μg/m^3 and elevation in systolic blood pressure by 0.80 mmHg (95% CI, 0.46–1.14), whereas diastolic blood pressure increased by 0.31 mmHg (95% CI, 0.10–0.51). The OR for hypertension was 1.11 (95% CI, 1.04–1.18). However, no association between elevated mean sulfur dioxide concentrations (54.42 ± 14.33 mg/m^3) and increased blood pressure was reported by Dong et al. (2013). No correlation was also observed in a study from Serbia (Stanković et al. 2018).

Although the majority of cited studies demonstrated a positive association between air pollutants (especially $PM_{2.5}$) and blood pressure or hypertension, it is important to note that they were not without limitations. First, in most studies, there was an inadequate adjustment for confounding or effect modification by other factors (such as treatment of hypertension and other coexisting disorders, diet, mental stress, obesity, traffic noise, temperature, and the synergistic effect of a mixture of pollutants). In addition, some of the confounders are difficult to evaluate. The studies used different lag patterns of exposure to evaluate immediate, delayed, or late effects of exposure. They also differed in terms of statistical analysis and methods of blood pressure measurement (Wang et al. 2017). Moreover, most studies did not assess individual exposure to pollutants. When estimating the health effects of long-term exposure, details on past exposure, the level of exposure throughout the lifespan, the age at which exposure occurred, and the health status prior to exposure are needed, whereas most studies did not include such information. Finally, although all these studies indicated associations between short- and long-term exposure to air pollutants and changes in blood pressure in different subpopulations, their results are still inconsistent and the evidence remains unclear.

5.7 Effect of Air Pollution on Heart Failure

Heart failure is a consequence of numerous advanced cardiovascular diseases, including end-stage coronary artery disease, hypertension, congenital heart valve defects, or myocarditis. Some studies have indicated a positive association of short- and long-term exposure to particulate matter and selected gaseous pollutants with the risk of hospitalization or death from heart failure. The mechanism of heart failure worsening has not been fully elucidated. It is possible that chemical pollutants cause irritation of the lower respiratory tract and, as a result, increased pulmonary vasoconstriction and elevation of right ventricular diastolic filling pressures. This, in turn, leads to an increase in heart overload, which can be important for patients with heart failure. In addition, when some pollutants enter the bloodstream, it can increase peripheral vascular resistance as well as blood pressure, thus leading to elevation of cardiac afterload, hypercoagulable state, and secondary hypoxia. All these mechanisms may result in exacerbation of heart failure.

5.7.1 Short-Term Exposure and Heart Failure

An important meta-analysis of 35 case-crossover and time-series studies, including the majority of studies investigating the relationship between short-term exposure to air pollutants and heart failure, was published by Shah et al. (2013). The authors reported that acute exposure to air pollutants resulted in an increased risk of hospitalization and death due to heart failure. This association was observed for carbon monoxide (3.52% per 1-ppm increase; 95% CI, 2.52–4.54), sulfur dioxide (2.36% per 10-ppb increase; 1.35–3.38), and nitrogen dioxide (1.70% per 10-ppb increase; 95% CI, 1.25–2.16). However, no effects were noted for ozone. The authors also reported a positive relationship between an increase in $PM_{2.5}$ levels and a rise in the rates of heart failure–related hospitalization or mortality (2.12% per 10 μg/m^3; 95% CI, 1.42–2.82). Importantly, these negative health effects were observed at the following median pollutant concentrations derived from the average daily pollutant concentrations: carbon monoxide, 1.1 (interquartile range IQR, 0.9–1.6) ppm; nitrogen dioxide, 26.4 (IQR, 22.5–30.1) ppb; sulfur dioxide, 6.3 (IQR, 4.7–11.9) ppb; ozone, 23.5 (IQR, 17.6–32.0) ppb; and $PM_{2.5}$, 15.0 (IQR, 10.8–17.6) μg/m^3.

A few years later, Liu et al. (2018b), in a large (n = 105,501) time-stratified case-crossover study, reported a positive relationship between short-term exposure to air pollution and hospital admission for heart failure in 26 large cities in China. An interquartile range increase in PM_{10}, sulfur dioxide, nitrogen dioxide, carbon monoxide, and ozone concentrations on the current day corresponded to an increase of 1.2% (95% CI, 0.5%-1.8%), 1.3% (95% CI, 0.5%-2.0%), 1.0% (95% CI, 0.2%-1.7%), 1.6% (95% CI, 0.6%-2.5%), 1.2% (95% CI, 0.5%-1.9%), and 0.4% (95% CI, 0.9%-1.7%) in heart failure-related hospitalization, respectively.

5.7.2 *Long-Term Exposure and Heart Failure*

The association between long-term exposure to air pollution and heart failure has been described in a few papers based mainly on cohort studies. Atkinson et al. (2013), in a large cohort of patients (n = 836,557) aged 40–89 years and registered in general practices in England, reported a positive association between exposure to air pollutants and incident cases of heart failure. After adjustment for confounders, an increase in PM_{10} by 3 μg/m^3 and in nitrogen dioxide by 10.7 μg/m^3 resulted in an hazard ratio of 1.06 (95% CI, 1.01–1.11). The follow-up was 5 years and the mean 1-year average concentration of PM_{10} was 19.7 μg/m^3, and of nitrogen dioxide, 22.5 μg/m^3.

The relationship between long-term exposure to $PM_{0.1}$ and nitrogen dioxide and increased incidence of heart failure has been also demonstrated in another cohort study by Bai et al. (2019a). Each IQR increase in exposure to $PM_{0.1}$ and nitrogen dioxide was associated with an increased incidence of heart failure (HR, 1.03; 95% CI, 1.02–1.05 and HR, 1.04; 95% CI, 1.03–1.06, respectively). In contrast, Danesh Yazdi et al. (2019), in another large cohort of individuals from the United States (n = 11,084,660) aged 65 years or older, reported that long-term exposure to $PM_{2.5}$ and ozone resulted in an increased frequency of hospitalizations due to exacerbation of existing heart failure. Each 1-μg/m^3 increase in $PM_{2.5}$ led to a 5.3% increase in first hospital admissions due to heart failure. The hazard ratio of hospital admission with heart failure was 1.053 (1.052–1.054) for exposure to average annual $PM_{2.5}$ and 1.023 (1.022–1.024) for ozone. Significant effects were also noted for average annual $PM_{2.5}$ levels below 12 μg/m^3 and ozone levels below 70 ppm. However, the study was limited by the method of estimating the time of exposure to pollutants, because the exposure was assigned based on the participant's mailing zip code, which may have resulted in a measurement error depending on how much time the participant spent in a given area. Moreover, the study did not include other confounders (risk factors of heart failure exacerbation). Bai et al. (2019b), in a large cohort study (n = 5,000,000) from Canada, also demonstrated an association between long-term exposure to $PM_{2.5}$ nitrogen dioxide, and ozone and the incidence of heart failure and myocardial infarction. Each interquartile range increase in exposure to $PM_{2.5}$, nitrogen dioxide, and ozone was associated with an increased incidence of heart failure (HR, 1.05; 95% CI, 1.04–1.05 for $PM_{2.5}$; HR, 1.02; 95% CI, 1.01–04 for nitrogen dioxide; and HR, 1.03; 95% CI, 1.02–1.03 for ozone).

Most studies describing a relationship between air pollution and risk of hospitalization and death due to heart failure have numerous limitations. Data on heart failure diagnosis are obtained from death certificates, autopsy documentation, or hospital discharge records, in which heart failure is very often listed as a secondary and direct cause of death, regardless of the primary cause. However, even in patients with a prior diagnosis of chronic heart failure, it is often difficult to identify the cause of heart failure exacerbation and death, for example, concomitant infection, progression of ischemia, exertion, or excessive salt and fluid intake. None of these confounders were considered in the above studies. Moreover, the studies did not reflect individual exposure to pollutants. Thus, all these uncertainties and limitations resulting from the misclassification of exposure as well as the presence of confounding factors hinder any straightforward interpretation of the available evidence.

5.8 Air Pollution and Stroke

Data from the World Health Organization indicated that about 20% of deaths due to air pollution are related to ischemic stroke (www.who.int/health-topics/air-pollution). Another report has demonstrated a declining trend in the rate of deaths from ischemic stroke attributable to air pollution exposure; however, data varied among the world regions (Babatola 2018). Several studies have reported a positive relationship between short- and long-term exposure to air pollution and the incidence of stroke and stroke-related mortality.

5.8.1 Short-Term Exposure and Stroke

In a systematic review and meta-analysis, Shah et al. (2015) investigated a short-term relationship (up to 7-day lag) between exposure to air pollution and hospitalizations for stroke (mainly ischemic) as well as stroke-related mortality. This meta-analysis included a total of 6.2 million events across 28 countries mainly from East Asia, Latin America, South East Asia, and South Africa, which were reported both by case-crossover and time-series studies. An increase in the concentration of carbon monoxide (RR, 1.015 per 1 ppm; 95% CI, 1.004–1.02), nitrogen dioxide (RR, 1.01 per 10 ppb; 95% CI, 1.009–1.019); sulfur dioxide (RR, 1.019 per 10 ppb, 95% CI, 1.011–1.027), and $PM_{2.5}$ (RR, 1.011 per 10 μg/m^3; 95% CI, 1.011–1.012) was associated with an increase in the number of hospitalizations for stroke and in the rate of stroke-related mortality. The results were nonsignificant for ozone. On the other hand, more recent, but smaller studies reported inconsistent results concerning the associations between the risk of stroke and different air pollutants. In a time-stratified case-crossover study (n = 2742; mean age, 75 years) from Barcelona (Spain), Vivanco-Hidalgo et al. (2018) showed that after adjustment for temperature short-term exposure to traffic pollutant ($PM_{2.5}$) was not associated with overall stroke risk. The median levels of $PM_{2.5}$ during the study were 17.5 μg/m^3 (IQR, 12–25.2).

In another prospective cohort study, conducted in post-menopausal women from the Women's Health Initiative Study (North America), the authors investigated the association between daily variation in air pollution and the ischemic and hemorrhagic subtypes of stroke (Sun et al. 2019). No relationship was observed between day-to-day variation in any pollutant and the risk of overall stroke or ischemic stroke. Only a positive association between the risk of hemorrhagic stroke and nitrogen dioxide and nitrogen oxides in the 3 days prior to stroke with an OR of 1.24 (95% CI, 1.01–1.52) and 1.18 (95% CI, 1.03–1.34) per an interquartile range increase was observed, respectively. The mean concentration of pollutants was 12.4 ± 7.1 μg/m^3 for $PM_{2.5}$, 15.4 ± 8.0 ppb for nitrogen dioxide, 27.8 ± 24.6 ppb for nitrogen oxides, 4.0 ± 2.72 ppb for sulfur dioxide, and 37.9 ± 15 ppb for ozone.

In contrast, in a multi-city case-crossover study conducted in China, Chen et al. (2020) revealed that exposures to $PM_{0.1}$, $PM_{2.5}$, and PM_{10} were significantly associated with hospital admission for ischemic stroke. A total of 68,122 hospital admissions were included in this study. The relative risks correlated with per −10 mg/m^3 increase in each pollutant and the 95% CIs for the cumulative effects of $PM_{0.1}$, $PM_{2.5}$, and PM_{10} on ischemic stroke during lag 0–1 days were 1.014 (1.005–1.0023), 1.007 (1.000–1.014), and 1.005 (1.001–1.009), respectively. In the whole population, exposure to $PM_{0.1}$, $PM_{2.5}$, and PM_{10} concentrations was associated with a higher risk of hospital admissions for ischemic stroke by 3.5%, 3.6%, and 4.1%, respectively. The results were not significantly affected by adjustment for meteorological variables.

5.8.2 *Long-Term Exposure and Stroke*

A relationship between long-term exposure to air pollution and stroke incidence or stroke mortality was evaluated mainly in cohort studies. Most of them reported a positive association, although the results were inconsistent.

Yuan et al. (2019) performed a meta-analysis of 16 cohort studies from Europe, North America, and Asia, including more than 2.2 million participants and more than 49,149 endpoint events, to assess an association between long-term exposure to $PM_{2.5}$ and incident stroke and death from stroke. The pooled hazard ratio for each 5-μg/m^3 increment in $PM_{2.5}$ concentration was 1.11 (95% CI, 1.05–1.17) for incident stroke and 1.11 (95% CI, 1.05–1.17) for mortality from stroke. A significant association between $PM_{2.5}$ and incident stroke was found in North America (HR, 1.09; 95% CI, 1.05–1.14) and Europe (HR, 1.07; 95%CI, 1.05–1.10), while the pooled result for Asia was not significant (HR, 2.31, 95% CI, 0.49–10.95). Interestingly, the results among current smokers were not significant despite the fact that smoking is an independent risk factor for all cardiovascular disorders.

In a more recent prospective cohort (n = 4105 participants; age, 45–76 at baseline, 52.5% of women) from the Heinz Nixdorf Recall study conducted in Germany, Rodins et al. (2020) showed that during 14-year follow-up exposure to traffic-specific particulate matter had a stronger effect on stroke than industry-specific $PM_{2.5}$ (HR, 2.53; 95% CI, 1.07–5.97 and HR, 1.27; 95% CI, 1.03–1.56, respectively, per 1-$\mu g/m^3$ increase). This study included many confounding variables, which increase its quality.

The evaluation of the above meta-analyses and cohort studies suggests a stronger relationship between air pollution and the risk of stroke in relation to particulate matter. This especially concerns individuals over 60 years of age. However, there are discrepancies between the studies in terms of the geographical region studied. Moreover, most investigators focused on ischemic stroke, did not provide a clear differentiation between the types of stroke (hemorrhagic vs ischemic), or provided an incorrect definition of stroke. The studies underestimated the most important stroke confounders, and the authors themselves reported the misclassification of exposure and significant heterogeneity across most of the pollutants. Therefore, the emerging evidence on the association between air pollution and stroke requires further research.

5.9 Air Pollution and the Coagulation System

The impact of air pollutants on coagulation disorders is still being investigated. A great majority of studies focused on the direct or indirect ability of particulate matter to activate the coagulation system and impair fibrinolytic activity.

In vivo animal studies, which have a higher flexibility in terms of examining the mechanisms of air pollution exposure than human studies, demonstrated that inhalation of high $PM_{2.5}$ concentrations was associated with elevation in the levels of interleukin 6, a cytokine with prothrombotic effects such as reduced clotting time, enhanced thrombin generation, and increased levels of coagulation activation markers (thrombin-antithrombin) (Mutlu et al. 2007; Budinger et al. 2011). Moreover, Smyth et al. (2017) showed that inhalation of diesel exhaust particles and nitrogen oxides in a mice model induced platelet aggregation independently of lung inflammation. It seems that exposure to $PM_{2.5}$ can indirectly induce platelet aggregation via inflammation or directly without the involvement of significant inflammatory response (Rudez et al. 2009).

Another interesting mechanism underlying the hematological effects of air pollution is the ability of particulate matter to bind coagulation factor XII, thus activating complement components (mainly C3 and C5), the kallikrein–kinin system, as well as the coagulation system (prothrombin). This concept was also supported by the results of experimental studies (Schmaier 2016; Jin et al. 2019). Other studies suggested that exposure to particulate matter was associated not only with the activation of coagulation but also impairment of fibrinolysis (Budinger et al. 2011; Kodavanti et al. 2011). These experimental findings were also confirmed by controlled human exposure studies, which revealed impaired fibrinolysis by measuring the levels of tissue plasminogen activator after exposure to diesel exhaust as a mixture of particulate matter and nitrogen oxides (Mills et al. 2005, 2007; Barath et al. 2010). Controlled human exposure studies have also demonstrated that the acute exposure of healthy volunteers to dilute diesel exhaust was linked to an increase in the expression of platelet-leukocyte aggregates and increased circulating levels of soluble CD40 ligand (Lucking et al. 2008, 2011). The results suggest that the potentiated thrombus formation in the human body was mediated via platelet activation.

Some epidemiological studies have also demonstrated an association between air pollution exposure and coagulopathy, but the results are highly inconsistent. For example, Hildebrandt et al. (2009), in a very small study on 38 individuals (mean age, 54 years) with chronic obstructive pulmonary disease, reported an increase in the levels of fibrinogen, an important coagulation protein, after short-term exposure to particulate matter and some gaseous air pollutants. However, other larger studies on short- and long-term exposure did not identify a significant association between air pollution and fibrinogen concentrations (Wu et al. 2012; Dadvand et al. 2014; Gong et al. 2014, Viehmann et al. 2015). Importantly, the above studies provided also conflicting results regarding changes in the concentrations of plasma coagulation factors other than fibrinogen.

There are several possible reasons for the discrepancies between study results. First, there was no precise assessment of exposure of participants, as there were no data on individual exposure or spatial differences. Also, exposure to various levels of pollutants, regional and seasonal differences in the chemical composition of pollutants, lack of adjustment for numerous important confounding variables, the use of various methods for the laboratory assessment of coagulation parameters, or limitations due to a small sample size may result in inconsistencies between studies.

Venous thromboembolism includes deep vein thrombosis and pulmonary embolism. The etiology of venous thromboembolism is multifactorial. Apart from the numerous acquired risk factors such as trauma, surgery, immobility, dehydration, malignancy and certain chemotherapy drugs, central venous catheters, oral contraceptive therapy, smoking, obesity, and pregnancy, there are also some hereditary factors that are involved in the development of venous thromboembolism (Rosendaal and Reitsma 2009).

The potential mechanisms whereby air pollution is associated with venous thromboembolism are hypercoagulability and enhanced thrombosis. Some experimental studies suggested an association between hypercoagulability and exposure to $PM_{2.5}$ and an immediate effect of soluble metal compounds (e.g., sulfate and nitrates), which may activate a number of human prothrombotic factors (Sangani et al. 2010; Gray et al. 2015; Pun et al. 2015; Tinkov et al. 2018; Signorelli et al. 2019). The authors revealed that all metal sulfates except nickel reduced the whole-blood coagulation time. Another potential mechanism of hypercoagulability is the release of inflammatory mediators and secondary thrombosis. In an experimental study, Tarantini et al. (2013) investigated 63 steel workers exposed to particulate matter concentrations, as a work-related condition with well-characterized prothrombotic exposure, and revealed that particulate matter induced gene hypomethylation, leading to inflammation and secondary venous thrombosis. Although the relationship between exposure to air pollution and occurrence of venous thromboembolism has been demonstrated in several smaller studies, the results were inconsistent (Spiezia et al. 2016).

Finally, a meta-analysis by Tang et al. (2016) did not support a positive relationship between traffic-related air pollutants (particulate matter, carbon monoxide, sulfur dioxide, ozone, nitrogen dioxide) and the risk of incident venous embolism. However, the meta-analysis included only a few cohort, case-crossover, case-control, and time-series studies conducted in Western countries, where the median $PM_{2.5}$ concentration is lower than 20 μg/m^3, while in Central and Eastern Europe as well as Asia, the median $PM_{2.5}$ concentrations are much higher.

Considering the above results and gaps in current knowledge, the impact of air pollution on venous thromboembolism requires further research. It is not clear whether gaseous pollutants other than $PM_{2.5}$, such as carbon monoxide, sulfur dioxide, and ozone, increase the risk of venous thrombosis. Most studies investigated short-term exposure and more data are needed on the effects of long-term exposure. The available studies differ in terms of pollutant levels, type of pollutants, exposure duration, or particulate matter composition. The number of relevant human studies from other countries is still too low, and they do not reflect individual exposure and vulnerability. Therefore, a definitive link between air pollution and venous thromboembolism cannot be determined as yet.

5.10 Air Pollution and Cardiac Arrhythmias

The relationship between air pollution and cardiac arrhythmias still remains the subject of research. The available papers report discrepant results depending on the study design, the strength of statistical associations between the type of pollutants, levels of chemicals at which a negative health effect is observed, and definition of outcomes. Moreover, some studies reported the results on this relationship only after a short exposure (hours), whereas other papers reported effects after a longer time (days, months).

5.10.1 Air Pollution and Ventricular Arrhythmias

Several studies have reported the positive association between elevated levels of air pollutants (PM_{10}, $PM_{2.5}$, nitrogen dioxide, sulfur dioxide, carbon oxide, and ozone) and the occurrence of ventricular arrhythmias that required the use of temporary implantable cardioverter defibrillators (Rich et al. 2004).

In ARIA, a multicenter prospective study performed in Italy and including 9 cardiological departments, Folino et al. (2017) assessed the relationship between air pollutants and the occurrence of ventricular arrhythmias during long-term follow-up in patients after implantation of pacemakers and cardioverter defibrillators. The authors revealed that $PM_{2.5}$ (but not gaseous pollutants) was associated with the risk of ventricular tachycardia, which increased nonlinearly for the levels higher than 25 μg/m^3 (for 75 μg/m^3 *vs* 25 μg/m^3: OR, 1.59; 95% CI, 1.12–2.25; p = 0.0002). Patients after myocardial infraction were shown to be more susceptible than others.

In their meta-analysis, Yang et al. (2017) included a few original human case-crossover and time-series studies. The authors estimated the relationship between ventricular arrhythmias and pollutant concentrations at lag 0–7 days. They showed a positive but non-significant correlation between the concentrations of carbon oxide, $PM_{2.5}$, sulfur dioxide, and nitrogen dioxide and the risk of ventricular arrhythmias. Additionally, the risk of ventricular arrhythmias was not associated with ozone levels.

5.10.2 Air Pollution and Atrial Fibrillation

Several investigators studied the association between supraventricular arrhythmias and air pollution. They have demonstrated that both short- and long-term exposure to air pollution is related to the occurrence of atrial fibrillation and atrial fibrillation-related hospitalization (Rich et al. 2006; Kwon et al. 2019; Saifipour et al. 2019; Kim et al. 2019).

In a case-crossover study, Saifipour et al. (2019) showed a significant correlation between short-term exposure to nitrogen dioxide (mean daily concentration, 40.53 ± 23.11 ppb) and hospitalization due to atrial fibrillation (OR, 1.26; 95% CI, 1.0–2.1; p = 0.03). The results for other pollutants ($PM_{2.5}$, nitrogen dioxide, sulfur dioxide, carbon monoxide, and ozone) were non-significant. In contrast, Kwon et al. (2019) demonstrated an association between short-term exposure to $PM_{2.5}$ and hospitalization. A 10-μg/m^3 increase in $PM_{2.5}$ was correlated with a significant increase in the rate of hospital admissions (by 4.5% at lag day 3, p = 0.03). The mean daily concentration of $PM_{2.5}$ was 25 μg/m^3. However, the results were not significant for gaseous pollutants. In the second part of this study, which investigated the effect of long-term exposure, the authors did not find significant evidence linking cardiac arrhythmias with air pollution. In another study, Rich et al. (2006) revealed a significant positive association between increased ozone concentrations (22 ppb) in the hour preceding atrial fibrillation and the occurrence of the arrhythmia. However, no significant results were noted for particulate matter and nitrogen dioxide.

A study on a large Asian cohort (n = 432,587) showed a positive association between the occurrence of atrial fibrillation and long-term average levels of $PM_{2.5}$ (HR, 1.179; 95% CI, 1.176–1.183 for 10-μg/m^3 increments, p <0.001). The results were particularly notable in men, elderly individuals, as well as participants with obesity, hypertension, and a history of myocardial infarction (Kim et al. 2019).

Although most studies included some confounders that might have influenced the results, such as weather conditions or the presence of coronary artery disease or hypertension, they have numerous limitations. First, not all substrates for arrhythmia, including its severity or concomitant treatment, were considered. Moreover, the studies failed to evaluate triggers of arrhythmia such as recurrent ischemia, physical exertion, emotional stress, elevation in blood pressure, electrolyte imbalance, tobacco smoking, or alcohol consumption. Although the evaluation of triggers is difficult in practice, especially in retrospective studies, both substrates and triggers are more important factors contributing to the incidence of arrhythmias than air pollution. Some patients with atrial fibrillation and non-sustained ventricular tachycardia may be asymptomatic, and some of the incidents may have gone unnoticed if patients had no implantable arrhythmia recorders. Some studies did not require independent confirmation by the cardiologist that cardioverter defibrillators were activated by ventricular tachycardia. Finally, all these studies are inherently limited by an individual exposure estimation error. In conclusion, the current evidence is insufficient to confirm a strong effect of air pollution on the incidence of cardiac arrhythmias.

5.11 Conclusions and Unresolved Issues

The evaluation of data from published studies show that among different air pollutants, exposure to $PM_{2.5}$ is the main contributing factor to cardiovascular mortality and morbidity. However, studies differ in terms of data reliability and reveal some discrepancies. The strongest evidence exists for the effect of air pollution on all-cause and cardiovascular mortality. The discrepancies between studies result not only from a diversity of chemical composition of particulate matter (size, mass, reactivity), source of emission, and coexistent gaseous pollutants, but also from the fact that studies differ in statistical methods and populations, exposure assessment or classification, as well as adjustment for confounding or effect modification by other factors. It is also difficult to evaluate general individual exposure and susceptibility, while they strongly determine the response to environmental pollution. Moreover, because there is a broad spectrum of individual susceptibility, there is no evidence for numerous pollutants on the safe level below which no adverse health effects occur. Other challenges include the reliability of methods used for the analysis of links between air pollutants and negative health effects; the need to determine causal pathways that may help establish preventive strategies; the necessity to assess pollutant properties that change in time and space and to establish what implications these variations have for the population; identification of people who are particularly vulnerable to pollution; determination of the extent to which confounding factors mask the real health impacts of air pollution; and assessment whether the actions taken to improve air quality result in beneficial health effects.

Health professionals should play an important role in educating patients and political decision makers about the harms of air pollution. Also, it is important to combine the existing knowledge with practical ability to control emission sources. Only interdisciplinary cooperation may help protect the environment and the human right to health preservation, which is guaranteed by the constitution but is not fully respected in terms of environmental pollution. Further research should focus on developing the optimal methods for air pollution reduction and assessing the effects of such interventions on mortality rates and incidence of cardiovascular disorders.

References

Argacha JF, Collart P, Wauters A, Kayaert P, Lochy S, Schoors D, Sonck J, de Vos T, Forton M, Brasseur O, Beauloye C, Gevaert S, Evrard P, Coppieters Y, Sinnaeve P, Claeys MJ (2016) Air pollution and ST-elevation myocardial infarction: a case-crossover study of the Belgian STEMI registry 2009–2013. Int J Cardiol 223:300–305. https://doi.org/10.1016/j.ijcard.2016.07.191

Analitis A, Katsouyanni K, Dimakopoulou K, Samoli E, Nikoloulopoulos AK, Petasakis Y, Touloumi G, Schwartz J, Anderson HR, Cambra K, Forastiere F, Zmirou D, Vonk JM, Clancy L, Kriz B, Bobvos J, Pekkanen J (2006) Short-term effects of ambient particles on cardiovascular and respiratory mortality. Epidemiology 17:230–233. https://doi.org/10.1097/01.ede.0000199439.57655.6b

Araujo JA, Barajas B, Kleinman M, Wang X, Bennett BJ, Gong KW, Navab M, Harkema J, Sioutas C, Lusis AJ, Nel AE (2008) Ambient particulate pollutants in the ultrafine range promote early atherosclerosis and systemic oxidative stress. Circ Res 102:589–596. https://doi.org/10.1161/CIRCRESAHA.107.164970

Atkinson RW, Carey IM, Kent AJ, van Staa TP, Anderson HR, Cook DG (2013) Long-term exposure to outdoor air pollution and incidence of cardiovascular diseases. Epidemiology 24:44–53. https://doi.org/10.1097/EDE.0b013e318276ccb8

Atkinson RW, Kang S, Anderson HR, Mills IC, Walton HA (2014) Epidemiological time series studies of PM2.5 and daily mortality and hospital admissions: a systematic review and meta-analysis. Thorax 69:660–665. https://doi.org/10.1136/thoraxjnl-2013-204492

Atkinson RW, Butland BK, Dimitroulopoulou C, Heal MR, Stedman JR, Carslaw N, Jarvis D, Heaviside C, Vardoulakis S, Walton H, Anderson HR (2016) Long-term exposure to ambient ozone and mortality: a quantitative systematic review and meta-analysis of evidence from cohort studies. BMJ Open 6:e009493. https://doi.org/10.1136/bmjopen-2015-009493

Ayres JG (2006) Research and analysis. COMEAP: cardiovascular disease and air pollution. A report by the committee on the medical effects of air pollutants. Gov.uk website. https://www.gov.uk/government/publications/comeap-cardiovascular-disease-and-air-pollution

Babatola SS (2018) Global burden of diseases attributable to air pollution. J Public Health Afr 9:813. https://doi.org/10.4081/jphia.2018.813

Bai N, Kido T, Suzuki H, Yang G, Kavanagh TJ, Kaufman JD, Rosenfeld ME, van Breemen C, Eeden SF (2011) Changes in atherosclerotic plaques induced by inhalation of diesel exhaust. Atherosclerosis 216:299–306. https://doi.org/10.1016/j.atherosclerosis.2011.02.019

Bai L, Weichenthal S, Kwong JC, Burnett RT, Hatzopoulou M, Jerrett M, van Donkelaar A, Martin RV, Van Ryswyk K, Lu H, Kopp A, Chen H (2019a) Associations of long-term exposure to ultrafine particles and nitrogen dioxide with increased incidence of congestive heart failure and acute myocardial infarction. Am J Epidemiol 188:151–159. https://doi.org/10.1093/aje/kwy194

Bai L, Shin S, Burnett RT, Kwong JC, Hystad P, van Donkelaar A, Goldberg MS, Lavigne E, Copes R, Martin RV, Kopp A, Chen H (2019b) Exposure to ambient air pollution and the incidence of congestive heart failure and acute myocardial infarction: a population-based study of 5.1 million Canadian adults living in Ontario. Environ Int 132:105004. https://doi.org/10.1016/j.envint.2019.105004

Barath S, Mills NL, Lundback M, Tornqvist H, Lucking AJ, Langrish JP, Soderberg S, Boman C, Westerholm R, Londahl J, Donaldson K, Mudway IS, Sandström T, Newby DE, Blomberg A

(2010) Impaired vascular function after exposure to diesel exhaust generated at urban transient running conditions. Part Fibre Toxicol 7:19. https://doi.org/10.1186/1743-8977-7-19

Barbosa CM, Terra-Filho M, de Albuquerque AL, Di Giorgi D, Grupi C, Negrão CE, Rondon MU, Martinez DG, Marcourakis T, dos Santos FA, Braga AL, Zanetta DM, Santos Ude P (2012) Burnt sugarcane harvesting – cardiovascular effects on a group of healthy workers, Brazil. PLoS One 7:e46142. https://doi.org/10.1371/journal.pone.0046142

Bauer M, Moebus S, Möhlenkamp S, Dragano N, Nonnemacher M, Fuchsluger M, Kessler C, Jakobs H, Memmesheimer M, Erbel R, Jöckel KH, Hoffmann B (2010) Urban particulate matter air pollution is associated with subclinical atherosclerosis: results from the HNR (Heinz Nixdorf recall) study. J Am Coll Cardiol 56:1803–1808. https://doi.org/10.1016/j.jacc.2010.04.065

Beelen R, Raaschou-Nielsen O, Stafoggia M, Andersen ZJ, Weinmayr G, Hoffmann B, Wolf K, Samoli E, Fischer P, Nieuwenhuijsen M, Vineis P, Xun WW, Katsouyanni K, Dimakopoulou K, Oudin A, Forsberg B, Modig L, Havulinna AS, Lanki T, Turunen A, Oftedal B, Nystad W, Nafstad P, De Faire U, Pedersen NL, Östenson CG, Fratiglioni L, Penell J, Korek M, Pershagen G, Eriksen KT, Overvad K, Ellermann T, Eeftens M, Peeters PH, Meliefste K, Wang M, Bueno-de-Mesquita B, Sugiri D, Krämer U, Heinrich J, de Hoogh K, Key T, Peters A, Hampel R, Concin H, Nagel G, Ineichen A, Schaffner E, Probst-Hensch N, Künzli N, Schindler C, Schikowski T, Adam M, Phuleria H, Vilier A, Clavel-Chapelon F, Declercq C, Grioni S, Krogh V, Tsai MY, Ricceri F, Sacerdote C, Galassi C, Migliore E, Ranzi A, Cesaroni G, Badaloni C, Forastiere F, Tamayo I, Amiano P, Dorronsoro M, Katsoulis M, Trichopoulou A, Brunekreef B, Hoek G (2014) Effects of long-term exposure to air pollution on natural-cause mortality: an analysis of 22 European cohorts within the multicentre ESCAPE project. Lancet 383:785–795. https://doi.org/10.1016/S0140-6736(13)62158-3

Bhaskaran K, Hajat S, Haines A, Herrett E, Wilkinson P, Smeeth L (2009) Effects of air pollution on the incidence of myocardial infarction. Heart 95:1746–1759. https://doi.org/10.1136/hrt.2009.175018

Bhaskaran K, Gasparrini A, Hajat S, Smeeth L, Armstrong B (2013) Time series regression studies in environmental epidemiology. Int J Epidemiol 42:1187–1195. https://doi.org/10.1093/ije/dyt092

Bourdrel T, Bind MA, Béjot Y, Morel O, Argacha JF (2017) Cardiovascular effects of air pollution. Arch Cardiovasc Dis 110:634–642. https://doi.org/10.1016/j.acvd.2017.05.003

Briggs D (2003) Environmental pollution and the global burden of disease. Br Med Bull 68:1–24. https://doi.org/10.1093/bmb/ldg019

Brook RD (2008) Cardiovascular effects of air pollution. Clin Sci 115:175–187. https://doi.org/10.1042/CS20070444

Brook RD, Urch B, Dvonch JT, Bard RL, Speck M, Keeler G, Morishita M, Marsik FJ, Kamal AS, Kaciroti N, Harkema J, Corey P, Silverman F, Gold DR, Wellenius G, Mittleman MA, Rajagopalan S, Brook JR (2009) Insights into the mechanisms and mediators of the effects of air pollution exposure on blood pressure and vascular function in healthy humans. Hypertension 54:659–667. https://doi.org/10.1161/HYPERTENSIONAHA.109.130237

Brook RD, Rajagopalan S, Pope CA III, Brook JR, Bhatnagar A, Diez-Roux AV, Holguin F, Hong Y, Luepker RV, Mittleman MA, Peters A, Siscovick D, Smith SC Jr, Whitsel L, Kaufman JD (2010) Particulate matter air pollution and cardiovascular disease: an update to the scientific statement from the American Heart Association. Circulation 121:2331–2378. https://doi.org/10.1161/CIR.0b013e3181dbece1

Brook RD, Newby DE, Rajagopalan S (2017) Air Pollution and Cardiometabolic Disease: An Update and Call for Clinical Trials. Am J Hypertens 31:1–10. https://doi.org/10.1093/ajh/hpx109

Budinger GR, McKell JL, Urich D, Foiles N, Weiss I, Chiarella SE, Gonzalez A, Soberanes S, Ghio AJ, Nigdelioglu R, Mutlu EA, Radigan KA, Green D, Kwaan HC, Mutlu GM (2011) Particulate matter-induced lung inflammation increases systemic levels of PAI-1 and activates coagulation through distinct mechanisms. PLoS One 6:e18525. https://doi.org/10.1371/journal.pone.0018525

Buteau S, Goldberg MS (2016) A structured review of panel studies used to investigate associations between ambient air pollution and heart rate variability. Environ Res 148:207–247. https://doi.org/10.1016/j.envres.2016.03.013

Cai Y, Zhang B, Ke W, Feng B, Lin H, Xiao J, Zeng W, Li X, Tao J, Yang Z, Ma W, Liu T (2016) Associations of short-term and long-term exposure to ambient air pollutants with hypertension. Hypertension 68:62–70. https://doi.org/10.1161/HYPERTENSIONAHA.116.07218

Cesaroni G, Forastiere F, Stafoggia M, Andersen ZJ, Badaloni C, Beelen R, Caracciolo B, de Faire U, Erbel R, Eriksen KT, Fratiglioni L, Galassi C, Hampel R, Heier M, Hennig F, Hilding A, Hoffmann B, Houthuijs D, Jöckel KH, Korek M, Lanki T, Leander K, Magnusson PK, Migliore E, Ostenson CG, Overvad K, Pedersen NL, Juha Pekkanen J, Penell J, Pershagen G, Pyko A, Raaschou-Nielsen O, Ranzi A, Ricceri F, Sacerdote C, Salomaa V, Swart W, Turunen AW, Vineis P, Weinmayr G, Wolf K, de Hoogh K, Hoek G, Brunekreef B, Peters A (2014) Long term exposure to ambient air pollution and incidence of acute coronary events: prospective cohort study and meta-analysis in 11 European cohorts from the ESCAPE Project. BMJ 348:f7412. https://doi.org/10.1136/bmj.f7412

Chan SH, Van Hee VC, Bergen S, Szpiro AA, DeRoo LA, London SJ, Marshall JD, Kaufman JD, Sandler DP (2015) Long-term air pollution exposure and blood pressure in the Sister Study. Environ Health Perspect 123:951–958. https://doi.org/10.1289/ehp.1408125

Chen LC, Nadziejko C (2005) Effects of subchronic exposures to concentrated ambient particles (CAPs) in mice. V. CAPs exacerbate aortic plaque development in hyperlipidemic mice. Inhal Toxicol 17:217–224. https://doi.org/10.1080/08958370590912815

Chen SY, Su TC, Lin YL, Chan CC (2012) Short-term effects of air pollution on pulse pressure among non-smoking adults. Epidemiology 23:341–348. https://doi.org/10.1097/EDE.0b013e3182452f1d

Chen SY, Wu CF, Lee JH, Hoffmann B, Peters A, Brunekreef B, Chu DC, Chan CC (2015) Associations between long-term air pollutant exposures and blood pressure in elderly residents of taipei city: a cross-sectional study. Environ Health Perspect 123:779–784. https://doi.org/10.1289/ehp.1408771

Chen H, Burnett RT, Copes R, Kwong JC, Villeneuve PJ, Goldberg MS, Brook RD, van Donkelaar A, Jerrett M, Martin RV, Brook JR, Kopp A, Tu JV (2016) Ambient fine particulate matter and mortality among survivors of myocardial infarction: population-based cohort study. Environ Health Perspect 124:1421–1428. https://doi.org/10.1289/EHP185

Chen L, Zhang Y, Zhang W, Chen G, Lu P, Guo Y, Li S (2020) Short-term effect of PM_1 on hospital admission for ischemic stroke: a multi-city case-crossover study in China. Environ Pollut 260–263:113776. https://doi.org/10.1016/j.envpol.2019.113776

Chin MT (2015) Basic mechanisms for adverse cardiovascular events associated with air pollution. Heart 101:253–256. https://doi.org/10.1136/heartjnl-2014-306379

Choi JH, Xu QS, Park SY, Kim JH, Hwang SS, Lee KH, Lee HJ, Hong YC (2007) Seasonal variation of effect of air pollution on blood pressure. J Epidemiol Commun Health 6:314–318. https://doi.org/10.1136/jech.2006.049205

Choi YJ, Kim SH, Kang SH, Kim SY, Kim OJ, Yoon CH, Lee HY, Youn TJ, Chae IH, Kim CH (2019) Short-term effects of air pollution on blood pressure. Sci Rep 9:20298–20299. https://doi.org/10.1038/s41598-019-56413-y

Chuang KJ, Yan YH, Cheng TJ (2010) Effect of air pollution on blood pressure, blood lipids, and blood sugar: a population-based approach. J Occup Environ Med 52:258–262. https://doi.org/10.1097/JOM.0b013e3181ceff7a

Chuang KJ, Yan YH, Chiu SY, Cheng TJ (2011) Long-term air pollution exposure and risk factors for cardiovascular diseases among the elderly in Taiwan. Occup Environ Med 68:64–68. https://doi.org/10.1136/oem.2009.052704

Coogan PF, White LF, Yu J, Brook RD, Burnett RT, Marshall JD, Bethea TN, Rosenberg L, Jerrett M (2017) Long-term exposure to NO_2 and ozone and hypertension incidence in the black women's health study. Am J Hypertens 30:367–372. https://doi.org/10.1093/ajh/hpw168
Cosselman KE, Krishnan RM, Oron AP, Jansen K, Peretz A, Sullivan JH, Larson TV, Kaufman JD (2012) Blood pressure response to controlled diesel exhaust exposure in human subjects. Hypertension 59:943–948. https://doi.org/10.1161/HYPERTENSIONAHA.111.186593
Crouse DL, Peters PA, van Donkelaar A, Goldberg MS, Villeneuve PJ, Brion O, Khan S, Atari DO, Jerrett M, Pope CA, Brauer M, Brook JR, Martin RV, Stieb D, Burnett RT (2012) Risk of nonaccidental and cardiovascular mortality in relation to long-term exposure to low concentrations of fine particulate matter: a Canadian national-level cohort study. Environ Health Perspect 120:708–714. https://doi.org/10.1289/ehp.1104049
Dadvand P, Nieuwenhuijsen MJ, Agustí À, de Batlle J, Benet M, Beelen R, Cirach M, Martinez D, Hoek G, Basagana X, Ferrer A, Ferrer J, Rodriguez-Roisin R, Sauleda J, Guerra S, Antó JM, Garcia-Aymerich J (2014) Air pollution and biomarkers of systemic inflammation and tissue repair in COPD patients. Eur Respir J 44:603–613. https://doi.org/10.1183/09031936.00168813
Dales R, Liu L, Szyszkowicz M, Dalipaj M, Willey J, Kulka R, Ruddy TD (2007) Particulate air pollution and vascular reactivity: the bus stop study. Int Arch Occup Environ Health 81:159–164
Danesh Yazdi M, Wang Y, Di Q, Zanobetti A, Schwartz J (2019) Long-term exposure to PM2.5 and ozone and hospital admissions of Medicare participants in the Southeast USA. Environ Int 130:104879. https://doi.org/10.1016/j.envint.2019.05.073.
De Brouwere K, Buekers J, Cornelis C, Schlekat CE, Oller AR (2012) Assessment of indirect human exposure to environmental sources of nickel: oral exposure and risk characterization for systemic effects. Sci Total Environ 419:25–36. https://doi.org/10.1016/j.scitotenv.2011.12.049
Diez Roux AV, Auchincloss AH, Franklin TG, Raghunathan T, Barr RG, Kaufman J, Astor B, Keeler J (2008) Long-term exposure to ambient particulate matter and prevalence of subclinical atherosclerosis in the multi-ethnic study of atherosclerosis. Am J Epidemiol 167:667–675. https://doi.org/10.1093/aje/kwm359
Dockery DW, Pope CA 3rd, Xu X, Spengler JD, Ware JH, Fay ME, Ferris BG Jr, Speizer FE (1993) An association between air pollution and mortality in six U.S. cities. N Engl J Med 329:753–759. https://doi.org/0.1056/NEJM199312093292401
Dominici F, McDermott A, Daniels M, Zeger SL, Samet JM (2005) Revised analyses of the national morbidity, mMortality, and air pollution study: mortality among residents of 90 cities. J Toxicol Environ Health A 68:1071–1092. https://doi.org/10.1080/15287390590935932
Donaldson K, Tran CL (2002) Inflammation caused by particles and fibers. Inhal Toxicol 14:5–27. https://doi.org/10.1080/089583701753338613
Dong GH, Qian ZM, Xaverius PK, Trevathan E, Maalouf S, Parker J, Yang L, Liu MM, Wang D, Ren WH, Ma W, Wang J, Zelicoff A, Fu Q, Simckes M (2013) Association between long-term air pollution and increased blood pressure and hypertension in China. Hypertension 61:578–584. https://doi.org/10.1161/HYPERTENSIONAHA.111.00003
Dorans KS, Wilker EH, Li W, Rice MB, Ljungman PL, Schwartz J, Coull BA, Kloog I, Koutrakis P, D'Agostino RB Sr, Massaro JM, Hoffmann U, O'Donnell CJ, Mittleman MA (2016) Residential proximity to major roads, exposure to fine particulate matter, and coronary artery calcium: the framingham heart study. Arterioscler Thromb Vasc Biol 36:1679–1685. https://doi.org/10.1161/ATVBAHA.116.307141
Faustini A, Rapp R, Forastiere F (2014) Nitrogen dioxide and mortality: review and meta-analysis of long-term studies. Eur Respir J 44:744–753. https://doi.org/10.1183/09031936.00114713
Folino F, Buja G, Zanotto G, Marras E, Allocca G, Vaccari D, Gasparini G, Bertaglia E, Zoppo F, Calzolari V, Suh RN, Ignatiuk B, Lanera C, Benassi A, Gregori D, Iliceto S (2017) Association between air pollution and ventricular arrhythmias in high-risk patients (ARIA study): a multicentre longitudinal study. Lancet Planet Health 1:e58–e64. https://doi.org/10.1016/S2542-5196(17)30020-7
Foraster M, Basagana X, Aguilera I, Rivera M, Agis D, Bouso L, Deltell A, Marrugat J, Ramos R, Sunyer J, Vila J, Elosua R, Künzli N (2014) Association of long-term exposure to traffic-related

air pollution with blood pressure and hypertension in an adult population-based cohort in Spain (the REGICOR study) Environ. Health Perspect 122:404–411. https://doi.org/10.1289/ehp.1306497

Frampton MW, Pietropaoli A, Dentler M, Chalupa D, Little EL, Stewart J, Frasier L, Oakes D, Wiltshire J, Vora R, Utell MJ (2015) Cardiovascular effects of ozone in healthy subjects with and without deletion of glutathione-S-transferase M1. Inhal Toxicol 27:113–119. https://doi.org/10.3109/08958378.2014.996272

Fuks KB, Weinmayr G, Foraster M, Dratva J, Hampel R, Houthuijs D, Oftedal B, Oudin A, Panasevich S, Penell J, Sommar JN, Sorensen M, Tiittanen P, Wolf K, Xun WW, Aguilera I, Basagana X, Beelen R, Bots ML, Brunekreef B, Bueno-de-Mesquita HB, Caracciolo B, Cirach M, de Faire U, de Nazelle A, Eeftens M, Elosua R, Erbel R, Forsberg B, Fratiglioni L, Gaspoz JM, Hilding A, Jula A, Korek M, Krämer U, Künzli N, Lanki T, Leander K, Magnusson PK, Marrugat J, Nieuwenhuijsen MJ, Ostenson CG, Pedersen NL, Pershagen G, Phuleria HC, Probst-Hensch NM, Raaschou-Nielsen O, Schaffner E, Schikowski T, Schindler C, Schwarze PE, Sogaard AJ, Sugiri D, Swart WJ, Tsai MY, Turunen AW, Vineis P, Peters A, Hoffmann B (2014) Arterial blood pressure and long-term exposure to traffic-related air pollution: an analysis in the European study of cohorts for air pollution effects (ESCAPE). Environ Health Perspect 122:896–905. https://doi.org/10.1289/ehp.1307725

Fuks KB, Weinmayr G, Basagana X, Gruzieva O, Hampel R, Oftedal B, Sørensen M, Wolf K, Aamodt G, Aasvang GM, Aguilera I, Becker T, Beelen R, Brunekreef B, Caracciolo B, Cyrys J, Elosua R, Eriksen KT, Foraster M, Fratiglioni L, Hilding A, Houthuijs D, Korek M, Künzli N, Marrugat J, Nieuwenhuijsen M, Östenson CG, Penell J, Pershagen G, Raaschou-Nielsen O, WJR S, Peters A, Hoffmann B (2017) Long-term exposure to ambient air pollution and traffic noise and incident hypertension in seven cohorts of the European study of cohorts for air pollution effects (ESCAPE). Eur Heart J 38:983–990. https://doi.org/10.1093/eurheartj/ehw413

Gardner B, Ling F, Hopke PK, Frampton MW, Utell MJ, Zareba W, Cameron SJ, Chalupa D, Kane C, Kulandhaisamy S, Topf MC, Rich DQ (2014) Ambient fine particulate air pollution triggers ST-elevation myocardial infarction, but not non-ST elevation myocardial infarction: a case-crossover study. Part Fibre Toxicol 11:1. https://doi.org/10.1186/1743-8977-11-1

Ghio AJ, Kim C, Devlin RB (2000) Concentrated ambient air particles induce mild pulmonary inflammation in healthy human volunteers. Am J Respir Crit Care Med 162:918–981. https://doi.org/10.1164/ajrccm.162.3.9911115

Gong J, Zhu T, Kipen H, Wang G, Hu M, Guo Q, Ohman-Strickland P, Lu SE, Wang Y, Zhu P, Rich DQ, Huang W, Zhang J (2014) Comparisons of ultrafine and fine particles in their associations with biomarkers reflecting physiological pathways. Environ Sci Technol 48:5264–5273. https://doi.org/10.1021/es5006016

Gray DL, Wallace LA, Brinkman MC, Buehler SS, La Londe C (2015) Respiratory and cardiovascular effects of metals in ambient particulate matter: a critical review. Rev Environ Contam Toxicol 234:135–203. https://doi.org/10.1007/978-3-319-10638-0-3

Guo Y, Tong S, Zhang Y, Barnett AG, Jia Y, Pan X (2010) The relationship between particulate air pollution and emergency hospital visits for hypertension in Beijing, China. Sci Total Environ 408:4446–4450. https://doi.org/10.1016/j.scitotenv.2010.06.042

Ha MH, Lee DH, Jacobs DR (2007) Association between serum concentrations of persistent organic pollutants and self-reported cardiovascular disease prevalence: results from the national health and nutrition examination survey, 1999–2002. Environ Health Perspect 115:1204–1209. https://doi.org/10.1289/ehp.10184

Hamanaka RB, Mutlu GM (2018) Particulate matter air pollution: effects on the cardiovascular system. Front Endocrinol 9:680. https://doi.org/10.3389/fendo.2018.00680

Health Canada. Human health risk assessment for coarse particulate matter (2016a). https://www.canada.ca/en/health-canada/services/publications/healthy-living/human-health-risk-assessment-coarse-particulate-matter-executive-summary.html

Health Canada. Human health risk assessment for ambient nitrogen dioxide (2016b). https://www.canada.ca/en/health-canada/services/publications/healthy-living/human-health-risk-assessment-ambient-nitrogen-dioxide.html

Hildebrandt K, Rückerl R, Koenig W, Schneider A, Pitz M, Heinrich J, Marder V, Frampton M, Oberdörster G, Wichmann HE, Peters A (2009) Short-term effects of air pollution: a panel study of blood markers in patients with chronic pulmonary disease. Part Fibre Toxicol 6:25. https://doi.org/10.1186/1743-8977-6-25

Hoek G, Krishnan RM, Beelen R, Peters A, Ostro B, Brunekreef B, Kaufman JD (2013) Long-term air pollution exposure and cardio-respiratory mortality: a review. Environ Health 12:43. https://doi.org/10.1186/1476-069X-12-43

Hoffmann B, Moebus S, Möhlenkamp S, Stang A, Lehmann N, Dragano N, Schmermund A, Memmesheimer M, Mann K, Erbel R, Jöckel KH, Heinz Nixdorf Recall Study Investigative Group (2007) Residential exposure to traffic is associated with coronary atherosclerosis. Circulation 116:489–496. https://doi.org/10.1161/CIRCULATIONAHA.107.693622

Huang CH, Lin HC, Tsai CD, Huang HK, Lian IB, Chang CC (2017) The interaction effects of meteorological factors and air pollution on the development of acute coronary syndrome. Sci Rep 7:44004. https://doi.org/10.1038/srep44004

Huang K, Yang X, Liang F, Liu F, Li J, Xiao Q, Chen J, Liu X, Cao J, Shen C, Yu L, Lu F, Wu X, Zhao L, Wu X, Li Y, Hu D, Huang J, Liu Y, Lu X, Gu D (2019) Long-term exposure to fine particulate matter and hypertension incidence in China. Hypertension 73:1195–1201. https://doi.org/10.1161/HYPERTENSIONAHA.119.12666

Integrated Science Assessment for Ozone and Related Photochemical Oxidants (2020) United States Environment Protection Agency website. https://www.epa.gov/isa/integrated-science-assessment-isa-ozone-and-related-photochemical-oxida

Integrated Science Assessment for Particulate Matter (2019) United States Environment Protection Agency website. https://www.epa.gov/isa/integrated-science-assessment-isa-particulate-matter

Jiang XQ, Mei XD, Feng D (2016) Air pollution and chronic airway diseases: what should people know and do? J Thorac Dis 8:E31–E40. https://doi.org/10.3978/j.issn.2072-1439.2015.11.50

Jin X, Ma Q, Sun Z, Yang X, Zhou Q, Qu G, Liu Q, Liao C, Li Z, Jiang G (2019) Airborne fine particles induce hematological effects through regulating the crosstalk of the kallikrein-kinin, complement, and coagulation systems. Environ Sci Technol 53:2840–2851. https://doi.org/10.1021/acs.est.8b05817

Katsouyanni K, Samet JM, Anderson HR, Atkinson R, Le Tertre A, Medina S, Samoli E, Touloumi G, Burnett RT, Krewski D, Ramsay T, Dominici F, Peng RD, Schwartz J, Zanobetti A (2009) HEI Health Review Committee. Air pollution and health: a European and North American approach (APHENA). Res Rep Health Eff Inst 142:5–90

Kaufman JD, Adar SD, Barr RG, Budoff M, Burke GL, Curl CL, Daviglus ML, Diez Roux AV, Gassett AJ, Jacobs DR Jr, Kronmal R, Larson TV, Navas-Acien A, Olives C, Sampson PD, Sheppard L, Siscovick DS, Stein JH, Szpiro AA, Watson KE (2016) Association between air pollution and coronary artery calcification within six metropolitan areas in the USA (the Multi-Ethnic Study of Atherosclerosis and Air Pollution): a longitudinal cohort study. Lancet 388:696–704. https://doi.org/10.1016/S0140-6736(16)00378-0

Kälsch H, Hennig F, Moebus S, Möhlenkamp S, Dragano N, Jakobs H, Memmesheimer M, Erbel R, Jöckel KH, Hoffmann B (2014) Are air pollution and traffic noise independently associated with atherosclerosis: the Heinz Nixdorf recall study. Eur Heart J 35:853–860. https://doi.org/10.1093/eurheartj/eht426

Kim SY, Sheppard L, Kaufman JD, Bergen S, Szpiro AA, Larson TV, Adar SD, Diez Roux AV, Polak JF, Vedal S (2014) Individual-level concentrations of fine particulate matter chemical components and subclinical atherosclerosis: a cross-sectional analysis based on 2 advanced exposure prediction models in the multi-ethnic study of atherosclerosis. Am J Epidemiol 180:718–728. https://doi.org/10.1093/aje/kwu186

Kim H, Kim J, Kim S, Kang SH, Kim HJ, Kim H, Heo J, Yi SM, Kim K, Youn TJ, Chae IH (2017) Cardiovascular effects of long-term exposure to air pollution: a population-based

study with 900 845 person-years of follow-up. J Am Heart Assoc 6. https://doi.org/10.1161/JAHA.117.007170
Kim IS, Yang PS, Lee J, Yu HT, Kim TH, Uhm JS, Pak HN, Lee MH, Joung B (2019) Long-term exposure of fine particulate matter air pollution and incident atrial fibrillation in the general population: a nationwide cohort study. Int J Cardiol 283:178–183. https://doi.org/10.1016/j.ijcard.2018.12.048
Kioumourtzoglou MA, Schwartz J, James P, Dominici F, Zanobetti A (2016) PM2.5 and mortality in 207 US Cities: modification by temperature and city characteristics. Epidemiology 27:221–227. https://doi.org/10.1097/EDE.0000000000000422
Kodavanti UP, Thomas R, Ledbetter AD, Schladweiler MC, Shannahan JH, Wallenborn JG, Lund AK, Campen MJ, Butler EO, Gottipolu RR, Nyska A, Richards JE, Andrews D, Jaskot RH, McKee J, Kotha SR, Patel RB, Parinandi NL (2011) Vascular and cardiac impairments in rats inhaling ozone and diesel exhaust particles. Environ Health Perspect 119:312–318. https://doi.org/10.1289/ehp.1002386
Konduracka E (2019) A link between environmental pollution and civilization disorders: a mini review. Rev Environ Health 34:227–233. https://doi.org/10.1515/reveh-2018-0083
Konduracka E, Niewiara Ł, Guzik B, Kotynia M, Szolc P, Gajos G, Nessler J, Podolec P, Żmudka K (2019) Effect of short-term fluctuations in outdoor air pollution on the number of hospital admissions due to acute myocardial infarction among inhabitants of Kraków, Poland. Pol Arch Intern Med 129:88–96. https://doi.org/10.20452/pamw.4424
Kubesch N, De Nazelle A, Guerra S, Westerdahl D, Martinez D, Bouso L, Carrasco-Turigas G, Hoffmann B, Nieuwenhuijsen MJ (2015) Arterial blood pressure responses to short-term exposure to low and high traffic-related air pollution with and without moderate physical activity. Eur J Prev Cardiol 22:548–557. https://doi.org/10.1177/2047487314555602
Künzli N, Jerrett M, Mack WJ, Beckerman B, LaBree L, Gilliland F, Thomas D, Peters J, Hodis HN (2005) Ambient air pollution and atherosclerosis in Los Angeles. Environ Health Perspect 113:201–206. https://doi.org/10.1289/ehp.7523
Kwon OK, Kim SH, Kang SH, Cho Y, Oh IY, Yoon CH, Kim SY, Kim OJ, Choi EK, Youn TJ, Chae IH (2019) Association of short- and long-term exposure to air pollution with atrial fibrillation. Eur J Prev Cardiol 26:1208–1216. https://doi.org/10.1177/2047487319835984
Lelieveld J, Klingmüller K, Pozzer A, Pöschl U, Fnais M, Daiber A, Münzel T (2019) Cardiovascular disease burden from ambient air pollution in Europe reassessed using novel hazard ratio functions. Eur Heart J 40:1590–1596. https://doi.org/10.1093/eurheartj/ehz135
Li H, Cai J, Chen R, Zhao Z, Ying Z, Wang L, Chen J, Hao K, Kinney PL, Chen H, Kan H (2017) Particulate matter exposure and stress hormone levels: a randomized, double-blind, crossover trial of air purification. Circulation 136:618–627. https://doi.org/10.1161/CIRCULATIONAHA.116.026796
Li J, Zhou C, Xu H, Brook RD, Liu S, Yi T, Wang Y, Feng B, Zhao M, Wang X, Zhao Q, Chen J, Song X, Wang T, Liu S, Zhang Y, Wu R, Gao J, Pan B, Pennathur S, Rajagopalan S, Huo Y, Zheng L, Huang W (2019) Ambient air pollution is associated with HDL (high-density lipoprotein) dysfunction in healthy adults. Arterioscler Thromb Vasc Biol 39:513–522. https://doi.org/10.1161/ATVBAHA.118.311749
Liang R, Zhang B, Zhao X, Ruan Y, Lian H, Fan Z (2014) Effect of exposure to PM2.5 on blood pressure: a systematic review and meta-analysis. J Hypertens 32:2130–2141. https://doi.org/10.1097/HJH.0000000000000342
Lind L, Lind PM (2012) Can persistent organic pollutants and plastic-associated chemicals cause cardiovascular disease? J Intern Med 271:537–553. https://doi.org/10.1111/j.1365-2796.2012.02536
Liu C, Chen R, Zhao Y, Ma Z, Bi J, Liu Y, Meng X, Wang Y, Chen X, Li W, Kan H (2017) Associations between ambient fine particulate air pollution and hypertension: a nationwide cross-sectional study in China. Sci Total Environ 584–585:869–874. https://doi.org/10.1016/j.scitotenv.2017.01.133

Liu C, Yin P, Chen R, Meng X, Wang L, Niu Y, Lin Z, Liu Y, Liu J, Qi J, You J, Kan H, Zhou M (2018a) Ambient carbon monoxide and cardiovascular mortality: a nationwide time-series analysis in 272 cities in China. Lancet Planet Health 2:e12–e18. https://doi.org/10.1016/S2542-5196(17)30181-X

Liu H, Tian Y, Song J, Cao Y, Xiang X, Huang C, Li M, Hu Y (2018b) Effect of ambient air pollution on hospitalization for heart failure in 26 of China's largest cities. Am J Cardiol 121:628–633. https://doi.org/10.1016/j.amjcard.2017.11.039

Liu C, Chen R, Sera F, Vicedo-Cabrera AM, Guo Y, Tong S, Coelho MSZS, Saldiva PHN, Lavigne E, Matus P, Valdes Ortega N, Osorio Garcia S, Pascal M, Stafoggia M, Scortichini M, Hashizume M, Honda Y, Hurtado-Díaz M, Cruz J, Nunes B, Teixeira JP, Kim H, Tobias A, Íñiguez C, Forsberg B, Astrom C, Ragettli MS, Guo YL, Chen BY, Bell ML, Wright CY, Scovronick N, Garland RM, Milojevic A, Kyselý J, Urban A, Orru H, Indermitte E, Jaakkola JJK, Ryti NRI, Katsouyanni K, Analitis A, Zanobetti A, Schwartz J, Chen J, Wu T, Cohen A, Gasparrini A, Kan H (2019) Ambient particulate air pollution and daily mortality in 652 Cities. N Engl J Med 381:705–715. https://doi.org/10.1056/NEJMoa1817364

Ljungman PLS, Andersson N, Stockfelt L, Andersson EM, Nilsson Sommar J, Eneroth K, Gidhagen L, Johansson C, Lager A, Leander K, Molnar P, Pedersen NL, Rizzuto D, Rosengren A, Segersson D, Wennberg P, Barregard L, Forsberg B, Sallsten G, Bellander T, Pershagen G (2019) Long-term exposure to particulate air pollution, black carbon, and their source components in relation to ischemic heart disease and stroke. Environ Health Perspect 127:107012. https://doi.org/10.1289/EHP4757

Lopez-Campos JL, Marquez-Martin E, Soriano JB (2016) The role of air pollution in COPD and implications for therapy. Expert Rev Respir Med 10:849–859. https://doi.org/10.1080/17476348.2016.1191356

Lucking AJ, Lundback M, Mills NL, Faratian D, Barath SL, Pourazar J, Cassee FR, Donaldson K, Boon NA, Badimon JJ, Sandstrom T, Blomberg A, Newby DE (2008) Diesel exhaust inhalation increases thrombus formation in man. Eur Heart J 29:3043–3051. https://doi.org/10.1093/eurheartj/ehn464

Lucking AJ, Lundback M, Barath SL, Mills NL, Sidhu MK, Langrish JP, Boon NA, Pourazar J, Badimon JJ, Gerlofs-Nijland ME, Cassee FR, Boman C, Donaldson K, Sandstrom T, Newby DE, Blomberg A (2011) Particle traps prevent adverse vascular and prothrombotic effects of diesel engine exhaust inhalation in men. Circulation 123:1721–1728. https://doi.org/10.1161/CIRCULATIONAHA.110.987263

Luo C, Zhu X, Yao C, Hou L, Zhang J, Cao J, Wang A (2015) Short-term exposure to particulate air pollution and risk of myocardial infarction: a systematic review and meta-analysis. Environ Sci Pollut Res Int 22:14651–11462. https://doi.org/10.1007/s11356-015-5188-x

Mazurek H, Badyda A (2018) Smog-konsekwencje zdrowotne zanieczyszczeń powietrza [Smog – the health effects of air pollution]. PZWL Wydawnictwo Lekarskie, Stryków, Poland, pp 89–143

Miller KA, Siscovick DS, Sheppard L, Shepherd K, Sullivan JH, Anderson GL, Kaufman JD (2007) Long-term exposure to air pollution and incidence of cardiovascular events in women. N Engl J Med 356:447–458. https://doi.org/10.1056/NEJMoa054409

Miller MR, Newby DE (2020) Air pollution and cardiovascular disease: car sick. Cardiovasc Res 116:279–294. https://doi.org/10.1093/cvr/cvz228

Mills NL, Tornqvist H, Robinson SD, Gonzalez M, Darnley K, MacNee W, Boon NA, Donaldson K, Blomberg A, Sandstrom T, Newby DE (2005) Diesel exhaust inhalation causes vascular dysfunction and impaired endogenous fibrinolysis. Circulation 112:3930–3936. https://doi.org/10.1161/CIRCULATIONAHA.105.588962

Mills NL, Tornqvist H, Gonzalez MC, Vink E, Robinson SD, Soderberg S, Boon NA, Donaldson K, Sandstrom T, Blomberg A, Newby DE (2007) Ischemic and thrombotic effects of dilute diesel-exhaust inhalation in men with coronary heart disease. N Engl J Med 357:1075–1082. https://doi.org/10.1056/NEJMoa066314

Mills IC, Atkinson RW, Kang S, Walton H, Anderson HR (2015) Quantitative systematic review of the associations between short-term exposure to nitrogen dioxide and mortality and hospital admissions. BMJ Open 5:e006946. https://doi.org/10.1136/bmjopen-2014-006946

Min JY, Cho JS, Lee KJ, Park JB, Park SG, Kim JY, Min KB (2011) Potential role for organochlorine pesticides in the prevalence of peripheral arterial diseases in obese persons: results from the national health and nutrition examination survey 1999-2004. Atherosclerosis 218:200–206. https://doi.org/10.1016/j.atherosclerosis.2011.04.044

Mustafic H, Jabre P, Caussin C, Murad MH, Escolano S, Tafflet M, Périer MC, Marijon E, Vernerey D, Empana JP, Jouven X (2012) Main air pollutants and myocardial infarction: a systematic review and meta-analysis. JAMA 307:713–721. https://doi.org/10.1001/jama.2012.126

Mutlu GM, Green D, Bellmeyer A, Baker CM, Burgess Z, Rajamannan N, Christman JW, Foiles N, Kamp DW, Ghio AJ, Chandel NS, Dean DA, Sznajder JI, Budinger GR (2007) Ambient particulate matter accelerates coagulation via an IL-6-dependent pathway. J Clin Invest 117:2952–2961. https://doi.org/10.1172/JCI30639

Nawrot TS, Perez L, Künzli N, Munters E, Nemery B (2011) Public health importance of triggers of myocardial infarction: a comparative risk assessment. Lancet 377:732–740. https://doi.org/10.1016/S0140-6736(10)62296-9

Newby DE, Mannucci PM, Tell GS, Baccarelli AA, Brook RD, Donaldson K, Forastiere F, Franchini M, Franco OH, Graham I, Hoek G, Hoffmann B, Hoylaerts MF, Künzli N, Mills N, Pekkanen J, Peters A, Piepoli MF, Rajagopalan S, Storey RF (2015) Expert position paper on air pollution and cardiovascular disease. Eur Heart J 36:83–93. https://doi.org/10.1093/eurheartj/ehu458

Newell K, Kartsonaki C, Lam KBH, Kurmi OP (2017) Cardiorespiratory health effects of particulate ambient air pollution exposure in low-income and middle-income countries: a systematic review and meta-analysis. Lancet Planet Health 1:e368–e380. https://doi.org/10.1016/S2542-5196(17)30166-3

Newell K, Kartsonaki C, Lam KBH, Kurmi O (2018) Cardiorespiratory health effects of gaseous ambient air pollution exposure in low and middle income countries: a systematic review and meta-analysis. Environ Health 17:41–47. https://doi.org/10.1186/s12940-018-0380-3

Novotna B, Topinka J, Solansky I, Chvatalova I, Lnenickova Z, Sram RJ (2007) Impact of air pollution and genotype variability on DNA damage in Prague policemen. Toxicol Lett 172:37–47. https://doi.org/10.1016/j.toxlet.2007.05.013

Ohlwein S, Kappeler R, Kutlar Joss M, Kunzli N, Hoffmann B (2019) Health effects of ultrafine particles: a systematic literature review update of epidemiological evidence. Int J Public Health 64:547–559. https://doi.org/10.1007/s00038-019-01202-7

Pascal M, Corso M, Chanel O, Declercq C, Badaloni C, Cesaroni G, Henschel S, Meister K, Haluza D, Martin-Olmedo P, Medina S, Aphekom group (2013) Assessing the public health impacts of urban air pollution in 25 European cities: results of the Aphekom project. Sci Total Environ 449:390–400. https://doi.org/10.1016/j.scitotenv.2013.01.077

Perez L, Wolf K, Hennig F, Penell J, Basagaña X, Foraster M, Aguilera I, Agis D, Beelen R, Brunekreef B, Cyrys J, Fuks KB, Adam M, Baldassarre D, Cirach M, Elosua R, Dratva J, Hampel R, Koenig W, Marrugat J, de Faire U, Pershagen G, Probst-Hensch NM, de Nazelle A, Nieuwenhuijsen MJ, Rathmann W, Rivera M, Seissler J, Schindler C, Thiery J, Hoffmann B, Peters A, Künzli N (2015) Air pollution and atherosclerosis: a cross-sectional analysis of four European cohort studies in the ESCAPE study. Environ Health Perspect 123:597–605. https://doi.org/10.1289/ehp.1307711

Peters A, Dockery DW, Muller JE, Mittleman MA (2001) Increased particulate air pollution and the triggering of myocardial infarction. Circulation 103:2810–2815. https://doi.org/10.1161/01.cir.103.23.2810

Pinault LL, Weichenthal S, Crouse DL, Brauer M, Erickson A, Donkelaar AV, Martin RV, Hystad P, Chen H, Finès P, Brook JR, Tjepkema M, Burnett RT (2017) Associations between fine particulate matter and mortality in the 2001 Canadian census health and environment cohort. Environ Res 159:406–415. https://doi.org/10.1016/j.envres.2017.08.037

Pitchika A, Hampel R, Wolf K, Kraus U, Cyrys J, Babisch W, Peters A, Schneider A (2017) Long-term associations of modeled and self-reported measures of exposure to air pollution and noise at residence on prevalent hypertension and blood pressure. Sci Total Environ 593-594:337–346. https://doi.org/10.1016/j.scitotenv.2017.03.156

Pope CA 3rd, Burnett RT, Thurston GD, Thun MJ, Calle EE, Krewski D, Godleski JJ (2004) Cardiovascular mortality and long-term exposure to particulate air pollution: epidemiological evidence of general pathophysiological pathways of disease. Circulation 109:71–77. https://doi.org/10.1161/01.CIR.0000108927.80044.7F

Pope CA, Dockery DW (2006) Health effects of fine particulate air pollution: lines that connect. J Air Waste Manag Assoc 56:709–742. https://doi.org/10.1080/10473289.2006.10464485

Pope CA, Muhlestein JB, Anderson JL, Cannon JB, Hales NM, Meredith KG, Le V, Horne BD (2015) Short-term exposure to fine particulate matter air pollution is preferentially associated with the risk of ST-segment elevation acute coronary events, J Am Heart Assoc:4. https://doi.org/10.1161/JAHA.115.002506

Pun VC, Kabrhel C, Camargo CA Jr, Baccarelli AA, Laden F (2015) Prospective study of ambient particulate matter exposure and risk of pulmonary embolism in the nurses' health study cohort. Environ Health Perspect 123:1265–1270. https://doi.org/10.1289/ehp.1408927

Rajagopalan S, Al-Kindi SG, Brook RD (2018) Air pollution and cardiovascular disease: JACC state-of-the-art review. J Am Coll Cardiol 72:2054–2070. https://doi.org/10.1016/j.jacc.2018.07.099

Rich KE, Petkau J, Vedal S, Brauer M (2004) A case-crossover analysis of particulate air pollution and cardiac arrhythmia in patients with implantable cardioverter defibrillators. Inhal Toxicol 16:363–372. https://doi.org/10.1080/08958370490439515

Rich DQ, Mittleman MA, Link MS, Schwartz J, Luttmann-Gibson H, Catalano PJ, Speizer FE, Gold DR, Dockery DW (2006) Increased risk of paroxysmal atrial fibrillation episodes associated with acute increases in ambient air pollution. Environ Health Perspect 114:120–123. https://doi.org/10.1289/ehp.8371

Rich DQ, Kipen HM, Huang W, Wang G, Wang Y, Zhu P, Ohman-Strickland P, Hu M, Philipp C, Diehl SR, Lu SE, Tong J, Gong J, Thomas D, Zhu T, Zhang JJ (2012) Association between changes in air pollution levels during the Beijing Olympics and biomarkers of inflammation and thrombosis in healthy young adults. JAMA 307:2068–2078. https://doi.org/10.1001/jama.2012.3488

Robertson S, Colombo ES, Lucas SN, Hall PR, Febbraio M, Paffett ML, Campen MJ (2013) CD36 mediates endothelial dysfunction downstream of circulating factors induced by O3 exposure. Toxicol Sci 134:304–311. https://doi.org/10.1093/toxsci/kft107

Robertson S, Miller MR (2018) Ambient air pollution and thrombosis. Part Fibre Toxicol 15:1. https://doi.org/10.1186/s12989-017-0237-x

Rodins V, Lucht S, Ohlwein S, Hennig F, Soppa V, Erbel R, Jöckel KH, Weimar C, Hermann DM, Schramm S, Moebus S, Slomiany U, Hoffmann B (2020) Long-term exposure to ambient source-specific particulate matter and its components and incidence of cardiovascular events – The Heinz Nixdorf recall study. Environ Int 142:105854. https://doi.org/10.1016/j.envint.2020.105854

Rosendaal FR, Reitsma PH (2009) Genetics of venous thrombosis. J Thromb Haemost 7:301–304. https://doi.org/10.1111/j.1538-7836.2009.03394.x

Rossner P Jr, Svecova V, Milcova A, Lnenickova Z, Solansky I, Santella RM, Sram RJ (2007) Oxidative and nitrosative stress markers in bus drivers. Mutat Res 617:23–32. https://doi.org/10.1016/j.mrfmmm.2006.11.033

Rudez G, Janssen NA, Kilinc E, Leebeek FW, Gerlofs-Nijland ME, Spronk HM, ten Cate H, Cassee FR, de Maat MP (2009) Effects of ambient air pollution on hemostasis and inflammation. Environ Health Perspect 117:995–1001. https://doi.org/10.1289/ehp.0800437

Ruidavets JB, Cournot M, Cassadou S, Giroux M, Meybeck M, Ferrières J (2005) Ozone air pollution is associated with acute myocardial infarction. Circulation 111:563–569. https://doi.org/10.1161/01.CIR.0000154546.32135.6E

Samek L, Stegowski Z, Styszko K, Furman L, Zimnoch M, Skiba A, Kistler M, Kasper-Giebl A, Rożański K, Konduracka E (2020) Seasonal variations of chemical composition of PM2.5 fraction in the urban area of Krakow, Poland: PMF source attribution. Air Quality Atmosphere Health 13:89–96. https://doi.org/10.1007/s11869-019-00773-x

Saifipour A, Azhari A, Pourmoghaddas A, Hosseini SM, Jafari-Koshki T, Rahimi M, Nasri A, Shishehforoush M, Lahijanzadeh A, Sadeghian B, Moazam E, Mohebi MB, Ezatian V, Rabiei K, Sarrafzadegan N (2019) Association between ambient air pollution and hospitalization caused by atrial fibrillation. ARYA Atheroscler 15:106–112. https://doi.org/10.22122/arya.v15i3.1843

Samoli E, Aga E, Touloumi G, Nisiotis K, Forsberg B, Lefranc A, Pekkanen J, Wojtyniak B, Schindler C, Niciu E, Brunstein R, Dodic Fikfak M, Schwartz J, Katsouyanni K (2006) Short-term effects of nitrogen dioxide on mortality: an analysis within the APHEA project. Eur Respir J 27:1129–1138. https://doi.org/10.1183/09031936.06.00143905

Samoli E, Peng R, Ramsay T, Pipikou M, Touloumi G, Dominici F, Burnett R, Cohen A, Krewski D, Samet J, Katsouyanni K (2008) Acute effects of ambient particulate matter on mortality in Europe and North America: results from the APHENA study. Environ Health Perspect 116:1480–1486. https://doi.org/10.1289/ehp.11345

Sangani RG, Soukup JM, Ghio AJ (2010) Metals in air pollution particles decrease whole-blood coagulation time. Inhal Toxicol 22:621–626. https://doi.org/10.3109/08958371003599037

Shah AS, Langrish JP, Nair H, McAllister DA, Hunter AL, Donaldson K, Newby DE, Mills NL (2013) Global association of air pollution and heart failure: a systematic review and meta-analysis. Lancet 382:1039–1048. https://doi.org/10.1016/S0140-6736(13)60898-3

Shah AS, Lee KK, McAllister DA, Hunter A, Nair H, Whiteley W, Langrish JP, Newby DE, Mills NL (2015) Short term exposure to air pollution and stroke: systematic review and meta-analysis. BMJ 350:h1295. https://doi.org/10.1136/bmj.h1295

Schmaier AH (2016) The contact activation and kallikrein/kinin systems: pathophysiologic and physiologic activities. J Thromb Haemost 14:28–39. https://doi.org/10.1111/jth.13194

Signorelli SS, Oliveri Conti G, Zanobetti A, Baccarelli A, Fiore M, Ferrante M (2019) Effect of particulate matter-bound metals exposure on prothrombotic biomarkers: a systematic review. Environ Res 177:108573. https://doi.org/10.1016/j.envres.2019.108573

Smyth E, Solomon A, Birrell MA, Smallwood MJ, Winyard PG, Tetley TD, Emerson M (2017) Influence of inflammation and nitric oxide upon platelet aggregation following deposition of diesel exhaust particles in the airways. Br J Pharmacol 174:2130–2139. https://doi.org/10.1111/bph.13831

Snow SJ, Cheng WY, Henriquez A, Hodge M, Bass V, Nelson GM, Carswell G, Richards JE, Schladweiler MC, Ledbetter AD, Chorley B, Gowdy KM, Tong H, Kodavanti UP (2018) Ozone-induced vascular contractility and pulmonary injury are differentially impacted by diets enriched with coconut oil, fish oil, and olive oil. Toxicol Sci 163:57–69. https://doi.org/10.1093/toxsci/kfy003

Sorensen M, Hoffmann B, Hvidberg M, Ketzel M, Jensen SS, Andersen ZJ, Tjønneland A, Overvad K, Raaschou-Nielsen O (2012) Long-term exposure to traffic-related air pollution associated with blood pressure and self-reported hypertension in a Danish cohort. Health Perspect 120:418–424. https://doi.org/10.1289/ehp.1103631

Spiezia L, Campello E, Maggiolo S, Tormene D, Simioni P (2016) Short-term exposure to high levels of air pollution (nickel) and the risk of unprovoked proximal deep vein thrombosis in the legs. Intern Emerg Med 11:159–162. https://doi.org/10.1007/s11739-015-1363-9

Stanković A, Bogdanović D, Nikolić M, Anđelković Apostolović M (2018) Does short-term air pollution exposure have effects on blood pressure and heart rate in healthy women in the city of Niš, Serbia? Cent Eur J Public Health 26:310–315. https://doi.org/10.21101/cejph.a5104

Sullivan J, Sheppard L, Schreuder A, Ishikawa N, Siscovick D, Kaufman J (2005) Relation between short-term fine-particulate matter exposure and onset of myocardial infarction. Epidemiology 16:41–48. https://doi.org/10.1097/01.ede.0000147116.34813.56

Sun Q, Wang A, Jin X, Natanzon A, Duquaine D, Brook RD, Aguinaldo JG, Fayad ZA, Fuster V, Lippmann M, Chen LC, Rajagopalan S (2005) Long-term air pollution exposure and acceleration of atherosclerosis and vascular inflammation in an animal model. JAMA 294:3003–3010. https://doi.org/10.1001/jama.294.23.3003

Sun S, Stewart JD, Eliot MN, Yanosky JD, Liao D, Tinker LF, Eaton CB, Whitsel EA, Wellenius GA (2019) Short-term exposure to air pollution and incidence of stroke in the xomen's health initiative. Environ Int 132:105065. https://doi.org/10.1016/j.envint.2019.105065

Szyszkowicz M, Rowe BH, Brook RD (2012) Even low levels of ambient air pollutants are associated with increased emergency department visits for hypertension. Can J Cardiol 28:360–366. https://doi.org/10.1016/j.cjca.2011.06.011

Tang L, Wang QY, Cheng ZP, Hu B, Liu JD, Hu Y (2016) Air pollution and venous thrombosis: a meta-analysis. Sci Rep 6:32794. https://doi.org/10.1038/srep32794

Tarantini L, Bonzini M, Tripodi A, Angelici L, Nordio F, Cantone L, Apostoli P, Bertazzi PA, Baccarelli AA (2013) Blood hypomethylation of inflammatory genes mediates the effects of metal-rich airborne pollutants on blood coagulation. Occup Environ Med 70:418–425. https://doi.org/10.1136/oemed-2012-101079

Terzano C, Di Stefano F, Conti V, Graziani E, Petroianni A (2010) Air pollution ultrafine particles: toxicity beyond the lung. Eur Rev Med Pharmacol Sci 14:809–821

Tinkov AA, Filippini T, Ajsuvakova OP, Skalnaya MG, Aaseth J, Bjørklund G, Gatiatulina ER, Popova EV, Nemereshina ON, Huang PT, Vinceti M, Skalny AV (2018) Cadmium and atherosclerosis: a review of toxicological mechanisms and a meta-analysis of epidemiologic studies. Environ Res 162:240–260. https://doi.org/10.1016/j.envres.2018.01.008

Thurston GD, Ahn J, Cromar KR, Shao Y, Reynolds HR, Jerrett M, Lim CC, Shanley R, Park Y, Hayes RB (2016) Ambient particulate matter air pollution exposure and mortality in the NIH-AARP Diet and Health cohort. Environ Health Perspect 124:484–490. https://doi.org/10.1289/ehp.1509676

Tonne C, Yanosky JD, Beevers S, Wilkinson P, Kelly FJ (2012) PM mass concentration and PM oxidative potential in relation to carotid intima-media thickness. Epidemiology 23:486–494. https://doi.org/10.1097/EDE.0b013e31824e613e

Tonne C, Wilkinson P (2013) Long-term exposure to air pollution is associated with survival following acute coronary syndrome. Eur Heart J 34:1306–1311. https://doi.org/10.1093/eurheartj/ehs480

Urch B, Silverman F, Corey P, Brook JR, Lukic KZ, Rajagopalan S, Brook RD (2005) Acute blood pressure responses in healthy adults during controlled air pollution exposures. Environ Health Perspect 113:1052–1055. https://doi.org/10.1289/ehp.7785

Viehmann A, Hertel S, Fuks K, Eisele L, Moebus S, Möhlenkamp S, Nonnemacher M, Jakobs H, Erbel R, Jöckel KH, Hoffmann B (2015) Long-term residential exposure to urban air pollution, and repeated measures of systemic blood markers of inflammation and coagulation. Occup Environ Med 72:656–663. https://doi.org/10.1136/oemed-2014-102800

Vinzents P, Moller P, Sorensen M, Knudsen LE, Hertel O, Jensen FP, Schibye B, Loft S (2005) Personal exposure to ultrafine particles and oxidative DNA damage. Environ Health Perspect 113:1485–1490. https://doi.org/10.1289/ehp.7562

Vivanco-Hidalgo RM, Wellenius GA, Basagaña X, Cirach M, González AG, Ceballos P, Zabalza A, Jiménez-Conde J, Soriano-Tarraga C, Giralt-Steinhauer E, Alastuey A, Querol X, Sunyer J, Roquer J (2018) Short-term exposure to traffic-related air pollution and ischemic stroke onset in Barcelona, Spain. Environ Res 162:160–165. https://doi.org/10.1016/j.envres.2017.12.024

Wang Q, Li C, Guo Y, Barnett AG, Tong S, Phung D, Chu C, Dear K, Wang X, Huang C (2017) Environmental ambient temperature and blood pressure in adults: a systematic review and meta-analysis. Sci Total Environ 575:276–286. https://doi.org/10.1016/j.scitotenv.2016.10.019

Wang W, Zhou J, Chen M, Huang X, Xie X, Li W, Cao Q, Kan H, Xu Y, Ying Z (2018) Exposure to concentrated ambient PM2.5 alters the composition of gut microbiota in a murine model. Part Fibre Toxicol 15:17–20. https://doi.org/10.1186/s12989-018-0252-6

Weichenthal S, Hatzopoulou M, Goldberg MS (2014) Exposure to traffic-related air pollution during physical activity and acute changes in blood pressure, autonomic and micro-vascular function in women: a cross-over study. Part Fibre Toxicol 11:70. https://doi.org/10.1186/s12989-014-0070-4.

Wiebert P, Lönn M, Fremling K, Feychting M, Sjögren B, Nise G, Kauppinen T, Plato N, Gustavsson P (2012) Occupational exposure to particles and incidence of acute myocardial infarction and other ischaemic heart disease. Occup Environ Med 69:651–657. https://doi.org/10.1136/oemed-2011-100285

World Health Organization (2020) Air pollution. World Health Organization. http://who.int/health-topics/air-pollution

Wu S, Deng F, Wei H, Huang J, Wang H, Shima M, Wang X, Qin Y, Zheng C, Hao Y, Guo X (2012) Chemical constituents of ambient particulate air pollution and biomarkers of inflammation, coagulation and homocysteine in healthy adults: a prospective panel study. Part Fibre Toxicol 9:49–60. https://doi.org/10.1186/1743-8977-9-49

Xie W, Li G, Zhao D, Xie X, Wei Z, Wang W, Wang M, Li G, Liu W, Sun J, Jia Z, Zhang Q, Liu J (2015) Relationship between fine particulate air pollution and ischaemic heart disease morbidity and mortality. Heart 101:257–263. https://doi.org/10.1136/heartjnl-2014-306165

Yan M, Liu Z, Liu X, Duan H, Li T (2013) Meta-analysis of the Chinese studies of the association between ambient ozone and mortality. Chemosphere 93:899–905. https://doi.org/10.1016/j.chemosphere.2013.05.040

Yang HJ, Liu X, Qu C, Shi SB, Liang JJ, Yang B (2017) Main air pollutants and ventricular arrhythmias in patients with implantable cardioverter-defibrillators: a systematic review and meta-analysis. Chronic Dis Transl Med 3:242–251. https://doi.org/10.1016/j.cdtm.2017.09.001

Yuan S, Wang J, Jiang Q, He Z, Huang Y, Li Z, Cai L, Cao S (2019) Long-term exposure to $PM_{2.5}$ and stroke: a systematic review and meta-analysis of cohort studies. Environ Res 177:108587. https://doi.org/10.1016/j.envres.2019.108587

Zhang L, Nichols RG, Correll J, Murray IA, Tanaka N, Smith PB, Hubbard TD, Sebastian A, Albert I, Hatzakis E, Gonzalez FJ, Perdew GH, Patterson AD (2015) Persistent organic pollutants modify gut microbiota-host metabolic homeostasis in mice through aryl hydrocarbon receptor activation environ. Health Perspect 123:679–688. https://doi.org/10.1289/ehp.1409055

Index

N. Morin-Crini et al. (eds.), *Emerging Contaminants Vol. 1*, Environmental Chemistry for a Sustainable World 65,
https://doi.org/10.1007/978-3-030-69079-3

The manufacturer's authorised representative in the EU is Springer
Nature Customer Service Centre GmbH, Europaplatz 3, 69115 Heidelberg,
Germany. If you have any concerns regarding our products, please
contact PrcductSafety@springernature.com

Printed and bound by CPI Group (UK) Ltd, Croydon, CR0 4YY
30/09/2026
02225705-0004